OXFORD PAPERBACK REFERENCE

A Dictionary of
Weights, Measures, and Units

Donald Fenna is Professor Emeritus, Faculty of
Medicine, University of Alberta. He is the author of
numerous publications in journals covering
mathematics, computing, and medicine, as well as
Elsevier's Encyclopedic Dictionary of Measures (1998).

Oxford
Paperback
Reference

The most authoritative and up-to-date reference books for both students and the general reader.

Abbreviations
ABC of Music
Accounting
Allusions*
Archaeology
Architecture
Art and Artists
Art Terms
Astronomy
Better Wordpower
Bible
Biology
British History
Buddhism*
Business
Card Games
Catchphrases*
Celtic Mythology
Chemistry
Christian Art
Christian Church
Classical Literature
Computing
Concise Quotations
Dance
Dates
Dynasties of the World
Earth Sciences
Ecology
Economics
Engineering*
English Etymology
English Folklore
English Grammar
English Language
English Literature
English Place-Names
Euphemisms*
Everyday Grammar
Finance and Banking
First Names
Food and Drink
Food and Nutrition
Foreign Words and Phrases
Geography
Handbook of the World
Humorous Quotations
Idioms
Irish Literature
Jewish Religion
Kings and Queens of Britain
Language Toolkit*
Law
Linguistics
Literary Quotations
Literary Terms
Local and Family History

London Place-Names
Mathematics
Medical
Medicines
Modern Design*
Modern Quotations
Modern Slang
Music
Nursing
Opera
Philosophy
Phrase and Fable*
Physics
Plant-Lore
Plant Sciences
Pocket Fowler's Modern
    English Usage
Political Biography
Political Quotations
Politics
Popes
Proverbs
Psychology
Reverse Dictionary
Rhyming*
Rhyming Slang*
Sailing Terms
Saints
Science
Scientists
Shakespeare
Ships and the Sea
Slang
Sociology
Spelling
Statistics
Superstitions
Surnames*
Synonyms and Antonyms
Theatre
Twentieth-Century Art
Twentieth-Century Poetry
Twentieth-Century World
    History
Weather
Weights, Measures, and Units
Who's Who in the Classical
    World
Who's Who in the Twentieth
    Century
World History
World Mythology
World Religions
Writers' Dictionary
Zoology

*forthcoming

A Dictionary of

Weights, Measures, and Units

DONALD FENNA

Oxford   New York

OXFORD UNIVERSITY PRESS

OXFORD
UNIVERSITY PRESS

Great Clarendon Street, Oxford OX2 6DP

Oxford University Press is a department of the University of Oxford.
It furthers the University's objective of excellence in research, scholarship,
and education by publishing worldwide in

Oxford New York
Auckland Bangkok Buenos Aires Cape Town Chennai
Dar es Salaam Delhi Hong Kong Istanbul Karachi Kolkata
Kuala Lumpur Madrid Melbourne Mexico City Mumbai Nairobi
São Paulo Shanghai Singapore Taipei Tokyo Toronto

with an associated company in Berlin

Oxford is a registered trade mark of Oxford University Press
in the UK and in certain other countries

Published in the United States
by Oxford University Press Inc., New York

British Library Cataloguing in Publication Data
Data available

Library of Congress Cataloging in Publication Data
Data available

ISBN 019–860522–6

1 3 5 7 9 10 8 6 4 2

Typeset in Swift and Frutiger by Kolam Information Services Pvt. Ltd,
Pondicherry, India

Printed in Great Britain by Clays Ltd, St Ives plc

This work is dedicated to my wife Sydney.

Firstly for persuading me to convert a hobby item into one fit for publication, then for her tireless direct and indirect support to that endeavour.

Contents

Preface

This dictionary addresses the units of measure used in science and engineering over the last two centuries (plus associated terminology) and includes a full exposition of the Système International (SI) – the contemporary version of the metric system founded in France in the 1790s. Each entry provides the historical information necessary for interpreting the headword throughout the intervening time.
For SI terms, the entry includes the motions passed by the international authorities, reproduced verbatim with kind permission from *La Système International d'Unités* (Sèvres, France; Bureau International de Poids et Mesures; 1985 & 2000) for which appreciation is expressed.

Entries

Each unit of measure is accorded its own entry, though related units are usually considered collectively as regards history and relationships. In addition, selected terms specific to metrology have explanatory entries, as have various terms of mathematics involved with counting and measurement.

The typical entry shows the symbol and other synonyms for the unit, the quantity (e.g. length, magnetic flux) that it measures, and the system within which the unit fits. For metric the system title differentiates SI and the various forms that preceded our contemporary form. The British Imperial and US Customary measures also are recognized as systems, though the former is now essentially obsolete. Each of these national systems is translated into metric, while to aid the American reader metric units are translated into US-C.

Numbers: styling and size.

The punctuation style for numbers is the established compromise between traditional English-speaking and European styles, where the dot on the line has the crucial role of decimal point, while the space is used to demarcate both fractional and integer trios of digits; the comma is abjured. Some numbers are exact but most are rounded; none are truncated. Numbers have a swung-dash appended at line level to indicate incompleteness. Thus the foot, now defined internationally as precisely 304.8 millimetres, appears as 304.8 mm or 0.3048 m, but until 1959 in the USA it was about one part in a million longer, and in the UK a similar amount shorter than the new precise figure – both expressible only as an unending fraction. Expressed to 8 significant figures (5 decimal places for millimetres), for the USA the value was 304.800 61∿ mm, and for the UK 304.799 73∿ mm, the small wavy line after each last digit indicating the incompleteness of the stated value. The effect of rounding can be illustrated by taking each value and progressively reducing the number of decimal places expressed:

Significant figures	USA	UK
7	304.800 6∿ mm	304.799 7∿ mm
6	304.801∿ mm	304.800∿ mm
5	304.80∿ mm	304.80∿ mm
4	304.8∿ mm	304.8∿ mm

The wavy equals sign ≈ is used to indicate looser approximation, i.e. that even allowing for any indicated rounding the later digits shown may be incorrect.

Words

The word 'volume' is used where 'capacity' would often be found, while the word 'speed' is used where some readers would use 'velocity'. The word 'else' is also used in preference to 'or' where there are mutually exclusive choices.

Final words

This work is derived from the author's comprehensive work on the world's weights and measures‡, but includes both updated and additional material of concern to scientists and engineers. For both old and new the author is indebted to many individuals for their advice and information, to all of whom appreciation and thanks are re-expressed. The further contributions of Doug Hube, Robert Morse, and Doug Fenna were particularly appreciated.

No proprietary ownership is claimed over the multitudinous facts embodied in this work, nor is any authority claimed in their correctness. The author has endeavored to cover its subjects both completely and accurately, but is well aware that such an endeavor will not be fully realized. Any criticism, destructive or constructive, is welcome. In particular, any information that would enhance the coverage would be welcome as material for any further editions. If you have any criticism or advice, please

email: don.fenna@ualberta.ca
fax:+1 780 662 4699
telephone:+1 780 662 2280.

‡ *Elsevier's Encyclopedic Dictionary of Measures* [Amsterdam, Elsevier Science NV; 1998]

A, a *Metric* As a lower-case prefix **a-**, *see* atto-.
length Metric Correctly **Å**, *see* angstrom.
area As **a**, *see* are; *see* SI alphabet for prefixes, notably ha = hectare.
time As **a**, *see* annum, i.e. year.
electromagnetics As **A**, *see* ampere; *see also* SI alphabet for prefixes.
astronomy As **A**, *see* astronomical unit (of length).
informatics In †hexadecimal notation as **A** for 10, the first digit after 9.
music See pitch.

Å, Å* *See* angstrom.

a- *Metric* As symbol *see* A, a.

a$_{anom}$ [annum] *time* Anomalistic year; *see* anomalistic and year.

a$_{astr}$ *time* Astronomical †year.

abA *See* abampere.

abampere [absolute ampere] (**biot**) *electric current*. Symbol abA.
Metric-c.g.s.-e.m.u 1 abA = the constant current that produces, when
maintained in two parallel conductors of negligible circular section and
of infinite length placed 1 centimetre apart, a force of 2 dynes per
centimetre between the two conductors. 1 abA = 10 A.

abC *See* abcoulomb.

abcoulomb *electric charge*. Symbol abC. *Metric-c.g.s.* The †absolute
†coulomb of the †e.m.u. system, for a †steady current identically s·abA. 1
abC = 10 C.

abD *See* abdaraf.

abdaraf *electric elastance*. Symbol abD. *Metric-c.g.s.* The †absolute
†daraf of the †e.m.u. system, for a †steady current identically abV/abC,
and identically reciprocal abfarard. 1 abD = 1 abF^{-1}.

abF *See* abfarad.

abfarad *electric capacitance*. Symbol abF. *Metric-c.g.s.* The †absolute †farad of the †e.m.u. system, for a †steady current identically abC/abV, and identically reciprocal abdaraf. 1 abF = 1 GF.

abH *See* abhenry.

abhenry *electromagnetic inductance*. Symbol abH. *Metric-c.g.s.* The †absolute †henry of the †e.m.u. system, for a †steady current identically abV/abA/second. 1 abH = 1 nH.

abmho *electric conductance* *Metric-c.g.s.* The †absolute †mho of the †e.m.u. system, for a †steady current identically abA/abV, and identically reciprocal abΩ. 1 abmho = 1 GS.

abohm *electric resistance*. Symbol abΩ. *Metric-c.g.s.* The †absolute †ohm of the †e.m.u. system, for a †steady current identically abV/abA, and identically reciprocal abmho. 1 abΩ = 1 nΩ.

absolute Within metrology generally, the qualifier 'absolute' indicates using mass rather than weight, i.e. force, as a basic dimension; hence an absolute system contrasts with a †gravitational system. Within metric, it tended to mean the †c.g.s. system.[1]

electromagnetics *Metric* The ampere and other electrical units agreed by the †CIPM in 1946, to take effect at the start of 1948, with the same names that had been in use for many years, are often labelled 'absolute' when discrimination is necessary; they replaced the †international units. With the formal creation of the †SI system in 1960 they became identically the SI ampere, etc. With the established use of the SI, no such prefixing should be practised. If any discrimination is necessary, the other unit rather than the SI unit should be qualified or, if essential, the SI unit should be labelled as such; SI definitions include no term 'absolute'.

It should be noted that the ampere and other such units defined in the †e.m.u. system in the 19th century (the abampere etc.) had that prefix because they were seen as 'absolute'; *see* absolute system. However, the terms 'absolute ampere', etc., today never mean the 'abampere', etc.; if anything they mean the SI ampere, etc. (despite the fact that the ampere itself is a base unit and not absolute in the original sense of being defined in mechanical units).

absolute centesimal thermodynamic scale *temperature* *See* kelvin scale.

absolute magnitude *astronomy.* Symbol M. *See* stellar magnitude.

absolute practical system *electromagnetics Metric See* absolute system and practical units.

absolute system *electromagnetics Metric* The evolution of telecommunications and other applications of electricity in the 19th century led to a plethora of units for associated measurement. Concomitant realization that the electrical and magnetic effects could be represented in mechanical terms led to the expression of electromotive force and current, and thereby resistance, in terms of the metre, gram, and second. Such expression of the electrical in terms purely of the established mechanical units was called 'absolute'.

Gauss is seen as pioneering this absolutism, within the context of his millimetre–milligram–second system,[72] but it became significant only in the context of the †c.g.s. system. The first such unit was the BA unit defined by the British Association in 1861 for resistance, computed but then enshrined as a wire; which to be realistic was set at 10^7 times the fully †coherent theoretical value within the natural †m.g.s. system. Subsequently the Association accepted that the c.g.s. system was preferable to the m.g.s. system, producing units that gave more comparable readings in a given circumstance; it also accepted the name ohm for the previously established unit.

The first International Electrical Conference of 1881 made both of these accepted systems international, and moved to establish a fuller system. This system was to be 'practical', i.e. with units of the size that occurred in normal work. Its basis was the absolute c.g.s. units, which became the abohm, abvolt, etc., of the †e.m.u. system. Its scaling would be of the order of the BA unit, and had to provide units that were mutually †coherent. Termed pedantically the absolute practical system, the units became familiarly known as the †practical units.

absolute temperature *temperature* The reading on a scale with zero at the †thermodynamic null, most particularly the †kelvin scale (attuned to †Celsius, i.e. having identically sized units but different zero points), but also the †Rankine scale (attuned to †Fahrenheit).

absolute zero *temperature* The thermodynamic null.

abstatampere, etc. *See* statampere, etc.

abV *See* abvolt.

abvolt *electric potential, electromotive force*. Symbol abV. *Metric-c.g.s.* The †absolute †volt of the †e.m.u. system, defined as 1 abV = the potential difference that exists between two points when the work done to transfer one abcoulomb of charge between them equals 1 erg. 1 abV = 10 nV.

abΩ *electromagnetics See* abohm.

ac. *See* acre, *also* ac. ohm.

acceleration of free fall, acceleration of gravity *See* standard gravity.

acid number *biochemistry* For measuring the amount of fatty acid in a fat or wax, the number of mg of potassium hydroxide required to neutralize 1 g of substance.
Metric-m.k.s. Pressure in Pa = $N \cdot m^{-2}$, speed in $m \cdot s^{-1}$, result in $N \cdot s \cdot m^{-3}$ = $kg \cdot s^{-1} \cdot m^{-2}$ ($m^{-2} \cdot kg \cdot s^{-1}$).
Metric-c.g.s. Pressure in $dyn \cdot cm^{-2}$, speed in $cm \cdot s^{-1}$, result in $dyn \cdot s \cdot cm^{-3}$ = $g \cdot s^{-1} \cdot cm^{-2}$ ($cm^{-2} \cdot g \cdot s^{-1}$).

ac. ohm *See* acoustic ohm.

acoustic ohm, acoustical ohm *acoustics*. Symbol ac ohm. *Metric* A unit for the impedance to the flow of sound energy (e.g. by a constricting orifice); quantitatively the ratio of the average effective sound pressure (assuming harmonic quantities) to the volume velocity (i.e. area times speed †orthogonally thereto) through it[2] – a parallel of the electric ohm (but *compare* mechanical ohm[3]). E.g.
Metric-m.k.s. $Pa \cdot (m^2 \cdot m \cdot s^{-1})^{-1} = N \cdot m^{-2} \cdot (m^{-1} \cdot s^1) = N \cdot s \cdot m^{-3}$ = $m^{-2} \cdot kg \cdot s^{-1}$,
Metric-c.g.s. $dyn \cdot cm^{-2} \cdot (cm^2 \cdot cm \cdot s^{-1})^{-1} = dyn \cdot s \cdot cm^{-3} = cm^{-2} \cdot g \cdot s^{-1}$.

acre [Lat: 'field'] *area*. Symbol ac. *BI, US-C, Australia, Canada, New Zealand, etc.* For land measure (originally the area that could be worked by an animal team in a day) = $4\,840\,yd^2$ ($0.404\,685\sim$ ha) = $\frac{1}{640}\,mi^2$. The area of a chain by a furlong, and of a square with sides of $69.570\,11\sim$ yd (= $22 \times \sqrt{10}$ yd). See Table 1.

acre·foot (usually, but improperly, **acre-foot**) *volume* For irrigation water, the volume of 1 ft × 1 ac = $43\,560\,ft^3$ ($1\,233.482\sim m^3$, $325\,851.\sim$ US gal).

acre·inch (usually, but improperly, **acre-inch**) *volume* For irrigation

Table 1

BI, US-C, Australia, Canada, New Zealand, etc.					SI	US-C
ft²						
272.25	sq. rod	...	...	...	25.3~ m²	30.25 yd²
10 890	40	rood	...	...	0.101~ ha	1 210 yd²
43 560	160	4	acre	...	0.405~ ha	4 840 yd²
			640	sq. mile	259.~ ha	640 ac

water, the volume of 1 in × 1 ac = 3 630 ft³ (102.790 2~ m³, 27 154.~ US gal).

AD *time* [Anno Domini, i.e. 'the year of the Lord'] The label prefixed to the number of a year, e.g. AD 500, but *see* CE.

admiralty mile *See* nautical mile.

aeolian frequency *rheology See* Strouhal number.

Ag amp, Ag ampere Silver ampere. *See* ampere.

a_gauss *time* Gaussian †year.

age *geology* A smaller, informal unit in the context of the †geochronologic scale, being usually less than an epoch, but used also to span epochs.

Al *rheology See* Alfvén number.

aleph [Anglicised name of ℵ, first letter of Hebrew alphabet] *See* infinity.

Alfvén number [H. O. G. Alfvén; Sweden 1908–95] *rheology*. Symbol *Al*. A dimensionless quantity characterizing the steady flow past a fixed object of a conducting fluid (such as a physical plasma) subject to a magnetic field parallel to the direction of flow. The number is the ratio of the speed of flow to the **Alfvén speed**, which is the ratio of the magnetic flux density to the square root of magnetic permeability times volumic mass.[61]

algebraic *mathematics* The **algebraic value** of a number is the number inclusive of any negative sign; *compare* arithmetic. *See also* negative number.

For **algebraic number** *see* number.

aliquant part, aliquot part The term aliquot indicates a part such that the whole is an exact multiple of it, e.g. three apples from a dozen, or one piece of a pie that has been cut exactly into quarters. Aliquant is its antonym, describing a part of which the whole is not an exact multiple.

alphanumeric character [alphabet + numeric] Strictly any †character that is in the alphabet else is a numeric †digit, but often, by normal extension, another graphic, e.g. a familiar or other punctuation mark. Usually the space character is also included.

alpha particle mass *sub-atomic physics.* Symbol m_α. 6.644 655 98(52) $\times 10^{-27}$ kg = 7 294.299 508(16) m_e with †relative standard uncertainties 7.9×10^{-8} and 2.1×10^{-9}.[4]

AM, a.m. [ante meridian, i.e. before the meridian] *time* Indicative of a time before noon, i.e. before the Sun nominally reaches the meridian, so the time is before the meridian. *See* p.m. for discussion.

amagat density, amagat unit, amagat volume [E. H. Amagat; Netherlands, France 1841–1915] *molecular density, molecular volume* For pressurised gases, the **amagat volume** is that of 1 mole of the gas at †s.t.p., which is 22.414 0~ L (the †standard volume) for ideal gas and close to that figure for real gases; the **amagat unit for volume** for a real gas is the ratio of its figure to that of ideal gas. The reciprocal of the above figure shows the **amagat density** to be 44.614 98~ mol·kL^{-1} for the ideal gas. The **amagat unit for density** is the ratio of the amagat density of real gas to that of the ideal, hence the reciprocal of the amagat unit for volume.

American cut, American run *textiles See* yarn units.

American Wire Gage, American Wire Gauge (AWG) *See* gauge.

amp *electrics* A common but deprecated abbreviation of †ampere.

ampere [A. M. Ampère; France 1775–1836] *electric current strength.* Symbol A. The amperes of †steady current crossing any cross-section of a circuit equals the ratio of the charge in coulombs to the time in seconds, identically the amperes of steady current produced between two points of a conductor equals the ratio of the potential difference in volts across these points to the intervening resistance in ohms (the conductor not being the seat of any electromotive force). However, this unit is defined as a base unit in any m.k.s. A. system (including the SI), and was for the †e.m.u. system.

SI, Metric-m.k.s. A. 1948 the base unit for all electromagnetic units, defined as the constant current which, if maintained in two straight parallel conductors of infinite length and of negligible cross-section, and placed 1 metre apart in vacuum, would produce between these conductors a force equal to 2×10^{-7} newton per metre of length. (This number effectively set the magnetic permeability of vacuum at $4\pi \times 10^{-7}\,H \cdot m^{-1}$; *see* permittivity.) The following are among the coherent derived units:

- $A \cdot m^{-1}$ for magnetic field strength, magnetization;
- $A \cdot m^{-2}$ for current density;
- $A \cdot s$ = coulomb for quantity of electricity;
- $A \cdot s \cdot V^{-1}$ = farad for electric capacitance;
- $A \cdot V^{-1}$ = siemens for electric conductance;
- $A \cdot H$ = weber for magnetic flux;
- $A \cdot turn$ for magnetomotive force.

This ampere is equatable with $6.241\,45 \sim \times 10^{18}$ electronic charges per second.

Metric-c.g.s. See abampere and statampere. *See also* practical unit.

History

The name 'ampère' was agreed, along with related units and the use of the †c.g.s. system, in 1881 at the first International Electrical Conference,[33] as the 'current produced by a volt in an ohm', with the implication that there should be both an absolute form and a corresponding †practical unit. The former, later discriminated as the abampere, falls within the †e.m.u. system, and is fundamentally definable in terms of purely mechanical units. The **practical ampere** $= 10^{-1}$ abampere.

To make it a base unit instead of a derived unit, a specification for a laboratory realization of the ampere was established. This was expressed in terms of the rate of electrolytic deposition of silver, so has often been called the **silver ampere** or **Ag ampere**; the definition was 'the unvarying current which deposits 1.118 mg of silver by electrolysis from a silver nitrate solution in one second'. The specification was subsequently shown to have made the ampere slightly smaller than intended,[5] prompting the adoption by the IEC of 1908 of the distinct name **international ampere**, with no reference to it being either absolute or practical (though it was the latter). Because of experimental vagaries, the value for conversions is normally referred to as the mean international ampere $= 0.999\,85 \sim$ A. There is also the **US international ampere** $= 0.999\,835 \sim$ A.

At the implementation of the Metric-m.k.s. A. system in 1948, with the ampere as the base electrical unit but its definition made compatible with the original absolute units, the modern ampere became essentially the old practical ampere; this became identically the ampere of the SI (again the base electric unit).

The calibration of reference electrical instruments from the fundamental definition presents obvious practical problems with accuracy, as well as the impossibility of literally infinite length. Until the 1980s the method involved weighing on a balance the magnetic force between two coils of carefully measured copper wire; this gave an accuracy of barely 1 in 10^5. Discovery of the Josephson effect, then of the quantum Hall effect, applying at very low temperatures with superconductors, together with subsequent development of the moving-coil balance and related work with the volt, improved accuracies about a thousandfold for the ampere, volt, ohm, etc.[6] For maximum accuracy, the ampere has been realized via the watt, by comparison of electrical power and mechanical power.[7]

Previous to adoption of the ampere, the names weber, oersted, and oerstedt were applied to units of electric current strength.

1893 International Electrical Conference: international ampere The unvarying current which deposits 1.118 00 mg of silver by electrolysis from a silver nitrate solution in 1 second.

1946 CIPM '*Ampere* (unit of electric current) The ampere is that constant current which, if maintained in two parallel conductors of infinite length, of negligible circular cross-section, and placed 1 metre apart in vacuum, would produce between these conductors a force equal to 2×10^{-7} MKS unit of force (i.e. newton) per metre of length.'[8]

ampere · turn (A · turn) *magnetomotive force Metric* The number of ampere · turns of magnetomotive force of a †steady current passing through a coil equals the product of the number of amperes of current and the number of †turns of the circuit.

amu *See* unified atomic mass unit.

angstrom [A. J. Ångström; Sweden 1814–74] *physics*. Symbol Å. The length 10^{-10} m = 0.1 nm, though expressed as $1.000\,000\,2{\sim} \times 10^{-10}$ prior to the minor re-sizing of the metre in 1960.

History

Originally (in 1868 for solar radiations) and still used for expression of wavelengths in the †electromagnetic spectrum. After elaborate measurement of the wavelength of the red line of cadmium at 6438.4696 $\times 10^{10}$ m, it was agreed internationally in 1907 that the angstrom be defined by assigning the value 6438.4696 Å to that wavelength[9] (in dry air containing 0.03% carbon dioxide, at standard atmospheric pressure and a temperature of 15°C). This was the pioneer use of light to define length, presaging the application to the metre in 1960, using krypton. By that time the 1907 definition had been shown to have made the angstrom 1.000 000 2~ $\times 10^{-10}$ m relative to the extant prototype metre; the 1960 re-definition changed the length of the metre to make it precisely 10^{10} Å as defined in 1907.

The 1978 decision of the CIPM considering it acceptable to continue to use the angstrom with the SI still stands.

See also angström star; tenth metre.

angstrom star *physics*. Symbol Å*. 1.000 015 01(90) $\times 10^{-10}$ m = 100.001 501~ pm with †relative standard uncertainty 9.0 $\times 10^{-7}$.[4] Based on the Kα_1 line of tungsten, Å* is sometimes used for expressing the wavelength of x-rays.

annum [Lat: 'year'] Per annum (p.a.) means per year.

annum fictus ['annum' + 'fictitious'] *See* Besselian year.

anomalistic ['uneven' or 'deviant'] *astronomy* The term 'anomaly' refers to the angular distance of a planet from its perihelion (the point of closest approach to the Sun), i.e. the angle at the Sun between the radial lines to the planet's current position and to its position at perihelion. The adjective 'anomalistic' implies being referenced to the moment of perihelion, the anomalistic year being the period between successive perihelions.

antilog, antilogarithm *See* logarithm.

ap. *mass USA* Indicates unit of †apothecaries' scale.

apgar [V. Apgar; USA 1909–74] *medicine* A scoring scheme for new-born infants, being the total of five scores in the range 0 to 2, relating to respiratory effort, heart rate, skin colour (i.e. normal to blue), muscle tone, and reflex reaction to olfactory stimulation of the nose.

API gravity [American Petroleum Institute] (degree API, °API)
petroleum processing A scale, expressed in degrees, for the volumic
mass of petroleum liquids. Used for liquids less dense than water, it
is structured identically with the Baumé scale for such liquids,
except that it has its set temperature at 60°F rather than 10°
Réaumur (12.5°C, 54.5°F); readings are thus about 1% greater than
Baumé.

If S is volumic mass in $kg \cdot m^{-3}$, and D is the API degrees, the two relate
as follows:

$S \times (131.5 + D) = 141.5$,   hence $D = (141.5/S) - 131.5$.

Since the volumic mass of water is virtually $1\,000 kg \cdot m^{-3}$, the same
equations effectively apply if S is interpreted as specific gravity. Practical
values of API gravity range from 10° to around 100°, higher values
meaning lighter oils, e.g.

Specific gravity	1	0.95	0.9	0.8	0.7
API gravity (°)	10	17.45	25.72	45.38	70.64

The volumic mass of natural crude oils varies by more than 10%, so the
mass of a standard barrel of 42 US gal varies likewise. The densities are
expressed in API degrees, but are often called **gravity**. Oil of 40° API has
$6.923\sim$ barrels (bbl) to the short ton ($7.631\sim bbl \cdot t^{-1}$).

apostilb [Gk: 'away' + 'glitter'] (**blondel**) *luminance, irradiance.*
Symbol asb. *Metric-m.k.s.* The name for a point on a diffuser, identically
candela·steradian per square metre, $= cd \cdot sr \cdot m^{-2}$. (There is no
corresponding specially named unit in the SI, but the lux for illuminence
is basically the same as the apostilb.) For a perfect diffuser, 1 asb
corresponds to the emission of $\frac{1}{\pi} cd \cdot m^{-2} = 0.3183\sim cd \cdot m^{-2}$. The
corresponding c.g.s. unit was the lambert, $= 10^4$ asb.

apoth. *mass UK* Indicates unit of †apothecaries' scale.

apothecaries' scale *mass* The traditional, and in the USA
contemporary, scheme used by apothecaries, with units distinguished,
where necessary, by the trailing qualifier **ap**. in the USA, **apoth**. in the
UK, e.g. oz ap.

Unlike the more usual avoirdupois scale with its pound of 16 ounces
and 7 000 grains (making its ounce 437.5 gr), the apothecaries' scale,
like the †troy scale, has a pound of only 12 ounces each of 480 grains,
giving a total of 5 760 grains, the †grain being the one unit common to all
three scales. The distinctive scale for apothecaries' units is shown in
Table 2.

Table 2

Bl-apoth, US-C-ap			Internat values:	SI	US-C-av
grain	...	...	...	64.8~ μg	1 gr
20	scruple	...	...	1.30~ g	20 gr
60	3	drachm, dram	...	3.89~ g	60 gr
480	24	8	ounce	31.1~ g	1.10~ oz
5 760	288	96	12 pound	373.~ g	13.2~ oz

As with the avoirdupois units, the apothecaries' units have for centuries been very close to their current international value, probably the same to at least six significant figures. Current values are based on the international grain, adopted in 1959, of 64.798 91 mg.

This scale was removed entirely from UK measures in 1970.[28]

apparent magnitude *astronomy See* stellar magnitude.

Aquarius *astronomy See* zodiac; right ascension.

Arabic numerals I.e. Indo-Arabic numerals; *see* numerals.

arbitrary Describing a unit defined by a physical †prototype, rather than being a †natural unit theoretically reproducible anywhere.[41]

arcdeg *geometry* Degree of arc, i.e. †degree in the angular sense.

arcmin *geometry* Minute of arc, i.e. †minute in the angular sense.

arcsec *geometry* For plane angle, = second of arc, i.e. †second in the angular sense, but beware of the meaning for reciprocal secant, akin to arcsin for reciprocal sine.

are *area. Metric 1879* 100 m^2 ($= \frac{100}{0.836}$ 127 36 yd^2 = 119.599 00~ yd^2), better known by its 100-multiple, the hectare. The 1978 decision of the †CIPM considering it acceptable to continue to use the are with the SI still stands.

Aries *astronomy See* zodiac; right ascension.

arithmetic *mathematics* The **arithmetic value** of a number is the number stripped of any negative sign. For number x it is expressed as $|x|$, termed the †modulus of x; *compare* algebraic. The **arithmetic mean** of n numbers is their collective algebraic sum (i.e. respectful of signs) divided by n; *compare* geometric.

Applied to a series of numbers, 'arithmetic' indicates that adjacent members differ by a constant additive increment, the 'common difference' (any finite number). The arithmetic series with difference b has the form

$a, \quad a + b, \quad a + 2b, \quad a + 3b, \quad \ldots$

for some value a. *Compare* geometric.

For measurement scales, 'arithmetic' means that a step of any one size in the scale value represents the same amount of additive change in the measured item, regardless of place on the scale. The traditional British and US scales for shoe size, where 1 equates to a third of an inch, is arithmetic.

(It should be noted that a simple arithmetic increase in a diameter has a squared effect on circular cross-section and spherical surface area, a cubed effect on volume and hence mass, etc.)

arithmetic value *See* negative number.

arpent *length Canada See* perche.

ASA [American Standards Association] *photography See* film speed.

asb *See* apostilb.

ascension *astronomy See* right ascension.

A Series, A4, etc. *paper and printing See* paper size.

a$_{sid}$ [annum] *time* Sidereal year; *see* sidereal and year.

assay ton Symbol AT. For assaying of precious ores, originally the quantity of ore, expressed in tons, required to produce one troy ounce of noble metal: now translated to the equivalent ratio of metric units. Since 1 oz troy = 480 gr and 1 lb av = 7 000 gr, one gets:

UK: one long ton of 2 240 lb = 15 680 000 gr = 32 666.67~ oz troy.

USA: one short ton of 2 000 lb = 14 000 000 gr = 29 166.67~ oz troy.

The number of assay tons in a given situation is the multiple of such mass that is required to produce 1 troy ounce of ore. With the tons expressed in troy ounces, these numbers are just the ratios of mass of ore to that of extracted metal, and applicable using any unit. Taking the gram instead of the troy ounce, the number of assay tons can be equally expressed as the multiple of 32 666.67~ g in UK terms, else 29 166.67~ g in US terms, of ore required to produce 1 g of noble metal. Using the convenience of the kilogram to express the amount of ore gives, relative to 1 gram of metal:

UK: assay tons = 0.030 612 245~ times the kg of ore.

USA: assay tons = 0.034 285 714~ times the kg of ore.

(The figures are equally applicable to grams of ore per milligram of metal, etc.)

The reciprocal of assay ton, e.g. ounces troy per ton, is the **assay value**.

astronomical day *astronomy* Traditionally the mean solar day beginning at 12:00 noon within the civil day of the same date. *See* astronomical day system.

astronomical day system *astronomy* A calendrical scheme comprising a simple day count measured from a pre-historic instant, thus providing a reference base independent of perceived eras and years of the various †calendars. Unlike the familiar day, the **astronomical day** begins at noon, its date that of the half-preceding regular day. Applied holistically around Earth, rather than by local time zones, each day begins at 12:00 †Universal Time. Termed the **Julian day number** and identified by a prefixed **JD**, the count has its zero at 01 January 4713 BCE. The day number is augmented by fractions to represent instants within a day where needed. Values for the beginning and end instants of a selection of Gregorian dates are:

JD 2 452 275.5	1 Jan 2002	JD 2 452 276.5
JD 2 452 640.5	1 Jan 2003	JD 2 452 641.5
JD 2 453 005.5	1 Jan 2004	JD 2 453 006.5
JD 2 453 371.5	1 Jan 2005	JD 2 453 372.5

To simplify reckoning, almanacs[10] cite 0-numbered days within calendars, e.g. 0 January 2003, which ends at JD 2 452 640.5.

Since all Julian dates from 1860 CE to 2130 CE begin with the same two digits, a modified scheme omitting those digits has been adopted. Termed the **modified Julian date (MJD)**, it also reverts to having the day begin at midnight, thus $MJD = JD - 2\,400\,000.5$, giving, for example, 52 640 for 1 January 2003.

Astronomical time-keeping uses †Terrestrial Time.

astronomical unit *astronomy*. Symbol AU. Effectively the mean distance of the Sun to Earth, internationally standardized for general use in 1964 at 149.6 Gm (499.0119~ ls) = 10^{-6} siriometre. Conceived as the mean distance from Earth to the Sun, it was previously specified as the radius of a circular orbit about the Sun of a nominal body with an orbital year equal to that of Earth.

For preparing and using the data of the *Astronomical Almanac*[10] it is defined as the length that, together with the astronomical units of mass and time, gives a value of 0.017 202 098 95 for the †Gaussian gravitational constant. It was set at 149.597 870~ $\times 10^6$ km in 1976, but is now 149.597 870 66~ $\times 10^6$ km (499.004 784~ ls, 92.955 807 3~ $\times 10^6$ mi).

astronomical unit of mass Symbol *S*. Mass of the Sun = 1.9891~ $\times 10^{30}$ kg.

astronomical unit of time Symbol *D*. Day of 86 400 s.

astronomical year *time See* year.

atmo-metre [metre-atmosphere] *physics* A measure of areal concentration of gas, being the length in metres of a column of the gas when at †s.t.p. Numerically it is the partial pressure of a gas relative to that exerted by a 1-metre column at s.t.p. Since the pressure exerted by a collection of gas molecules is independent of the particular gas, the number of molecules in the column per square metre of cross-section equals the atmo-metres times †Loschmidt's number.

atmos., atmosphere *pressure* The characteristic level of atmospheric pressure at Earth's surface, most particularly the †standard atmosphere, now taken to be 101.325 kPa. The distinct **technical atmosphere** = 1 kgf·cm^2 = 98.0665 kPa, the figure being from the original †standard gravity.

Atmospheric pressure declines with increasing altitude (and ever more rapidly) because of the thinning of the atmosphere and the declining gravitational acceleration. The decline is significant even over the altitude range of major human habitation, exceeding 10% at 1 000 m (3 200.~ ft) – below the altitude of Denver, Mexico City, Nairobi, and many other cities. Weather maps and reports give values adjusted to implied sea level values (and aneroid barometers must be adjusted accordingly to the site).

While the specific pressure at any altitude depends on current atmospheric conditions, just as it does at ground level, the routine progressive change with altitude allows for measurement of altitude by measurement of pressure. Long advantageous in mountain climbing, this is of great relevance now in flying; the standard settings for aircraft altimeters are shown in Table 3.

See also head of liquid.

Table 3

Altitude		Pressure		
m	ft	kPa	mm of Hg	p.s.i.
−305	−1 000	105.040	787.86	15.235
0	0	101.325	760.00	14.696
305	1 000	97.716	732.93	14.173
610	2 000	94.213	706.66	13.664
914	3 000	90.812	681.15	13.171
1 219	4 000	87.910	659.38	12.750
1 524	5 000	84.307	632.25	12.228
3 048	10 000	69.681	522.65	10.106
6 096	20 000	46.563	349.25	6.753
9 144	30 000	30.089	225.69	4.364
12 192	40 000	18.754	140.67	2.720
15 240	50 000	11.598	86.99	1.682

The negative figure applies to the Red Sea zone, etc.

atomic mass constant, atomic mass unit *See* unified atomic mass unit.

atomic number The number of protons in the nucleus of an atom; more correctly the number of unit charges of such a nucleus measured arithmetically in terms of the charge of an electron but with opposite sign. Such a figure equates with the number of protons under current theory, and equals the number of surrounding electrons under non-ionized conditions. The term was first used in 1865, but only effectively to number the elements sequentially, for placement in the periodic table; it was nearly 50 years later that its charge basis was realized, hence the scientific basis of the periodic table. The values of atomic number range from 1 for hydrogen and 2 for helium to 6 for carbon, 8 for oxygen, 26 for iron, 79 for gold, 92 for uranium, and up to 103 for lawrencium, all being strictly integers, with no gap. (The 'atomic weight' includes the chargeless neutrons plus the relatively minor mass of the electrons; for hydrogen this gives 1.008; generally atomic weight, correctly called †relative atomic mass, is more than twice the atomic number.)

atomic system of units *See* atomic unit.

atomic unit *physics*. Symbol a.u. Any one of the †coherent set of units based on the †Bohr radius plus other †natural sub-atomic units

Table 4

Base units: atomic unit of			
action	$\hbar$	the †Planck constant over 2π	
charge	e	the †elementary charge	
length	a_0	†Bohr radius	
mass	m_e	†electron mass	
Derived units: atomic unit of			
charge density	e/a_0^3	$1.081\,202\,285(43) \times 10^{12}$ C·m^{-3}	4.0
current	$e \cdot E_h/\hbar$	$6.623\,617\,53(26) \times 10^{-3}$ A	3.9
electric dipole moment	$e \cdot a_0$	$8.478\,352\,67(33) \times 10^{-30}$ C·m	3.9
electric field	$E_h/e \cdot a_0$	$5.142\,206\,24(20) \times 10^{11}$ V·m^{-1}	3.9
electric field gradient	$E_h/e \cdot a_0^2$	$9.717\,361\,53(39) \times 10^{21}$ V·m^{-2}	4.0
electric polarizability	$e^2 \cdot a_0^2/E_h$	$1.648\,777\,251(18) \times 10^{-41}$ C^2·m^2·J^{-1}	1.1
electric potential	E_h/e	$2.721\,138\,34(11) \times 10^1$ V	3.9
electr. quadrupole moment	$e \cdot a_0^2$	$4.486\,551\,00(18) \times 10^{-40}$ C·m^2	4.0
energy (Hartree)	$E_h = m_e \cdot e^4/\hbar^2$	$4.359\,743\,81(34) \times 10^{-18}$ J	7.8
force	E_h/a_0	$8.238\,721\,81(64) \times 10^{-8}$ N	7.8
magnetic dipole moment	$e \cdot \hbar/m_e$	$1.854\,801\,799(75) \times 10^{-23}$ J·T^{-1}	4.0
magnetic flux density	$\hbar/e \cdot a_0^2$	$2.350\,517\,349(94) \times 10^5$ T	4.0
magnetizability	$e^2 \cdot a_0^2/m_e$	$7.891\,036\,41(14) \times 10^{-29}$ J·T^{-2}	1.8
momentum	$\hbar/a_0$	$1.992\,851\,51(16) \times 10^{-24}$ kg·m·s^{-1}	7.8
permittivity	$e^2/a_0 \cdot E_h$	$1.112\,650\,056\sim \times 10^{-10}$ F·m^{-1}	(exact)
time	$\hbar/E_h$	$2.418\,884\,326\,500(18) \times 10^{-17}$ s	0.0*
velocity (or speed)	$a_0 \cdot E_h/\hbar$	$2.187\,691\,252\,9(80) \times 10^6$ m·s^{-1}	0.37

The asterisked figure is 0.000 76.

introduced by †Hartree in 1927.[78] (*Compare* natural unit.) Named generically **atomic unit of action**, etc. (and earlier sometimes the **Hartree unit of action**, etc., though that personal name is used now to mean the atomic unit of energy), the members and their current values (with 10^8 times the †relative standard uncertainty appended to each)[4] of the derived units are shown in Table 4.

atomic weight Now correctly called relative atomic mass.

a$_{\text{trop}}$ *time* Tropical year; *see* tropical; year.

att- *SI* Contracted form of †atto-.

atto- [Danish: 'eighteen'] Symbol a-. *Metric* The 10^{-18} multiplier, e.g. 1 attogram = 1 ag = 10^{-18} g; contractable to att- before a vowel.

AU *astronomy See* astronomical unit.

a.u. *sub-atomic physics See* atomic unit.

av., avdp. *See* †avoirdupois, in American and British usage, respectively.

average *statistics* For a set of numbers, a synonym for the arithmetic **mean**, i.e. the sum of those numbers divided by their count. The term referred originally to the damage, partial loss, else taxation of ships' cargo, the sharing of such costs between the shareholders in the cargo presumably prompting its general use in statistics. Other central values sometimes seen as averages, though not properly called such, include the **median** (the value such that there are just as many numbers greater than it as there are less than it, by convention the mean of the nearest values for a set with an even count) and the **midrange** (the mean of the two extreme-valued numbers in the set).

Clearly all three of these definitions can result in values that are not in the set, indeed values that could not be in the set. The average family with 2.2 children is the best-known unreal result.

No average can give more than a cursory picture of the set; it ignores the dispersion of the member numbers, due to erratic distribution and overall spread. The derived parameter †standard deviation provides a measure of the dispersion.

Avogadro constant [L. R. A. C. Avogadro; Italy 1776–1856] *fundamental constant*. Symbol N_A, L. From Avogadro's number = $6.022\ 141\ 99(47) \times 10^{23}$ mol^{-1} with †relative standard uncertainty 7.9×10^{-8} (i.e. $602.\sim$ per zeptomole).[4]

Avogadro's number The number of molecules contained in one mole of any substance, = $6.022\sim \times 10^{23}$, but *see* Avogadro constant. *See also* Loschmidt's number.

avoirdupois ['goods by weight'] (av, avdp) *UK* (avdp), *USA* (av) The traditional weight scale for all goods except precious metals and jewels (which use the †troy scale) and pharmaceuticals (which use the †apothecaries' scale). First recognized in statute in 1532, avoirdupois probably reflected the wide adoption of †weighing instead of volumetric measure. *See* pound.

AWG *USA* For steel wire, American Wire Gage; *see* gauge.

B

B, b *physics* As **b**, *See* Wien displacement law constant.

    informatics As **B**, *See* byte, e.g. kB = kilobytes.

        As **b**, *See* bit, e.g. kb = kilobits.

In †hexadecimal notation as **B** for 11, the second digit after 9.

    music See pitch.

B dose *radiation physics See* pastille dose.

Babylonian numerals *See* sexagesimal.

bag, bale *volume US-C* For cement, $1\,ft^3 \approx 94\,lb\,(42.638\sim kg) = \frac{1}{4}$ barrel.

ball *oceanography* The unit of the †Zhubov scale.

Balling [K. J. F. Balling; Bohemia 1805–64] *liquor and food processing* A scale, expressed in degrees, for the percentage by mass of sucrose in water solution, at 17.5°C; revised as the †Brix scale and degree.

bar [barometer] *pressure Metric* Defined to be the decimal power of standard metric units closest to the typical atmospheric pressure at Earth's surface $= 10^5 Pa = 100\,kPa\,(\frac{1}{1.01325}$ std atmos, 14.503 77~ p.s.i.), originally defined under metric-c.g.s. as $10^6 dyn \cdot cm^{-2}$ (but also interpreted as $1\,dyn \cdot cm^{-2}$ in the USA into the 1920s).[11] Introduced in 1911,[12] the bar competed for decades with the more visible but problematic †mmHg,[13] which is still used for many pressure readings. The 1978 decision of the †CIPM considering it acceptable to continue to use the bar with the SI still stands. However, there is no official acceptance of it being abbreviated to its initial letter or otherwise, hence no official acceptance of mb for millibar or μb for microbar; the correct forms are mbar and μbar. Usage for atmospheric pressure is normally in the form of millibars, even when exceeding 1 000 mbar. Since the millibar equals 100 Pa, i.e. the hectopascal, this, in the symbolic form hPa, is often used in place of the millibar, producing such quaint expressions as 800 hPa for what is 80 kPa.

    The term barye was introduced for this unit prior to bar. Both bar and

barye have also been used, most notably in acoustics, for the fully
†coherent c.g.s. unit of $1 \text{dyn} \cdot \text{cm}^{-2}$, i.e. what is correctly the microbar
(and called also the barad).

A distinct old metric unit for atmospheric pressure was the torr.

barad [from bar] *pressure Metric-c.g.s.* Identically dyne per square
centimetre = $\text{dyn} \cdot \text{cm}^{-2}$, = $0.1 \text{Pa}(1 \mu\text{bar}, 0.000\ 014\ 504\sim$ p.s.i.). Note
that this is precisely the microbar, the confusable bar being related in
size to the normal atmospheric pressure, at $100 \text{dyn} \cdot \text{cm}^{-2}$. Accordingly
barad was not abbreviated, so occurs prefixed as in cbarad =
centibarad.

Despite being the †coherent unit for pressure in c.g.s., barad was
probably much less common than the non-coherent bar.

Sometimes called barye, etc., a name also used for bar.

barie *pressure See* barrie.

barn *sub-atomic physics* 10^{-28}m^2, being the order of magnitude of the
cross-sectional area of the nucleus, hence of relevance in probabilistic
studies for interactions. Now better expressed just as 100fm^2. The 1978
decision of the †CIPM considering it acceptable to continue to use the
barn with the SI still stands, though it is deprecated by other authorities.

baromil [barometer mil] *meteorology USA* The height difference of
mercury in a barometer corresponding to a pressure difference of 1
millibar at 0°C and at sea level at a latitude of 45°. [14] These conditions
are very close to those for the †standard atmosphere of 760 mm of
mercury, which has a pressure of 1 013.25 mbar, giving 0.750 06$\sim$ mm of
mercury per mbar.

barrel (bbl.) A bulk-measure cask, with established volumes and
quantities for various commodities in historic marketplaces, = $\frac{1}{8}$ tun.
   volume US-C liq Generally = 31.5 US gal (119.24$\sim$ L), but **brewers'**
   barrel = 31 US gal (117.35$\sim$ L), and **petroleum barrel** = 42 US gal
   (158.99$\sim$ L); for energy *see* b.o.e.
US-C dry Generally = 7 056 in^3(115.63$\sim$ L, 3.281$\sim$ US bu), but for cement
   = 4 ft^3 = 6 912 in^3 (113.26$\sim$ L) but defined by mass.
   mass As accepted values for a nominal barrel.
US-C For cement, = 4 bags of 94 lb = 376 lb (170.55$\sim$ kg).
Canada For cement, = 350 lb (158.76$\sim$ kg).

barrel bulk For maritime use, typically 5 ft^3(0.1416$\sim$ m^3); *see* shipping
ton.

barrel of oil equivalent *See* b.o.e.

barrie, bary, baryd, barye, baryed (barie) *pressure Metric*
Ambiguously used to mean both the bar (= 100 kPa) and barad, i.e.
$dyn \cdot cm^{-2}$ (= $0.1Pa$ = $1\mu bar$), though barye was a former name for bar.
Spellings involving the letter d seem generally to refer to barad.

base box *engineering* A reference surface area for metallic coating,
typically of 31 360 in^2, comprising 112 two-sided plates of 10 in × 14 in.
The amount of coating is then expressed as pounds per base box, 1 lb per
box representing $0.073\,469\sim oz \cdot ft^{-2}$ ($22.419\sim g \cdot m^{-2}$).

base unit Any comprehensive system of units of measure must be
founded on a set of units equal in number to the dimensions[15]
addressed, and they must be independent to encompass those
dimensions. Length, mass, and time form the primeval set of dimensions
and of base units; they have usually been augmented by volt else ampere
as a base unit to extend coverage to the electromagnetic domain (though
electric charge, represented in the SI by the coulomb, might be seen as
the essential pertinent dimension). However, as with these latter, any of
the initial trio could be replaced by combinations: for instance speed, the
ratio of length to time, could replace either one of those two
components.
  To serve adequately, each base unit must be founded on reference
standards accurately measurable to the precision appropriate to the
conceived purposes of the system. In the †SI, for time the second is
defined as an appropriate multiple of the period of the hyperfine
transitions of the caesium atom giving the second; for length the metre is
defined as the distance travelled by light in a specified number of
seconds; for mass the kilogram is defined by a physical †prototype; and
the ampere is defined as the electric current strength that maintains a
specific force in a specified circumstance. Earlier versions of the †metric
system used other definitions, and some used the volt instead of the
ampere. Yet earlier, the †absolute system, ignoring the phenomenon of
electric charge, managed with only the first three dimensions, while the
metre was defined originally to have a set count along a quadrant of
Earth. The gram was the original base unit for mass, defined as the mass
of 1 cubic centimetre of water, but it soon became one thousandth of a
prototype kilogram.
  Ideally the standards should be indestructible and be invariable over
time and place; to be practical they should also be reproducible and

easily measured, accurately, to the required precision. Ideally they should be defined and measurable independently of each other. While the modern standards defining the second and the metre come close to requirements, those for the kilogram and ampere fall well short.

Most fundamental constants are indestructible, invariant (though see below) and measurable, within sophisticated laboratories, to the level of precision required. Either the †Bohr radius else the †Compton wavelength could provide a standard for length, from which the †speed of light in a vacuum could provide for time, and the †Newtonian constant of gravitation for mass. The †elementary charge of the proton provides directly for electric charge. All of these, like the period of transition of caesium and the wavelength of light that underlie the definitions of the second and the metre, are seen as †natural units.

Given the facility to name derived units as combinations of base units, the crucial question is which are to be the standards, with the secondary question of what multiples of the chosen standards shall be employed as fundamental units (as in the numbers involved in defining the second and the metre, and the practice of dividing the †Planck constant by 2π in some systems).

The SI includes the kelvin for thermodynamic temperature, the candela for luminous intensity, and the †mole for the amount of a substance as base units too, and consideration has been given to including units for plane and solid angles with them. Since 1980 the angles have been accepted as not independent but merely derived units, specifically ratios of like-dimensioned measurements, hence dimensionless. The **Kalantaroff system** avoids mass as a fundamental, augmenting length, time, and electric charge with magnetic flux instead (the metre, second, coulomb, and weber being extant units for these quantities).[16]

Various schemes use selected fundamental physical constants entirely as their standards,[17] sometimes directly as base units, but otherwise such that the selected constants have value 1, referred to as being normalized, in the pertinent system. The †Planck length, mass, and time are base units in our familiar dimensions that are sized to normalize three such constants (though the scheme would need scaling to be practical). The schemes described under †atomic units and †natural units use selected fundamental physical constants entirely as their standards, indeed directly as base units. McWeeny [18] advocated the following as base units:

†electron mass ($\approx 9.11 \times 10^{-31}$ kg),

†elementary charge of the proton ($\approx 1.60 \times 10^{-19}$C),

†Planck constant over 2π ($\approx 1.05 \times 10^{-34}$J·s) and

†permittivity of free space ($\approx 8.86 \times 10^{-12}$m^{-3}·kg^{-1}·s^2·C^2).

The first two give valued units of mass and charge directly; the third adds time to the dimensions, the last length.

Ludovici,[19] following Dirac,[20] saw all atomic units except for elementary charge as not assuredly invariable, especially allowing for the Heisenberg uncertainty principle, so advocated the following as fundamental units:

elementary charge of the proton ($\approx 1.60 \times 10^{-19}$C),

Newtonian constant of gravitation ($\approx 6.67 \times 10^{-11}$m^3·kg^{-1}·s^{-2}),

permittivity of free space ($\approx 8.86 \times 10^{-12}$m^{-3}·kg^{-1}·s^2·C^2) and

†permeability of free space ($\approx 1.26 \times 10^{-6}$m·kg·C^{-2}),

all unvarying. The first gives a valued unit of charge directly. From that, the product of the next two gives a value for mass. From those two values, the fourth choice gives a value for length, then either of the middle two the value for time.

The †f.p.s. system was the main non-metric system. *See* gravitational system for further variations.

basis point A unit at the level to which a particular figure is routinely expressed, regardless of the position of the decimal point. Thus, if interest rates are normally expressed (as a percentage) to two decimal places, then a rise from 5.41 to 5.61 would be termed a rise of 20 basis points (and had it risen by 19, one should say to 5.60 rather than 5.6). That rise of 20 basis points could also be described as a rise of 0.2 †percentage points, this latter expression being independent of the number of decimal places cited. (A rise of 0.2 from 5.41 is a rise of $0.2 \times 100\%/5.41 = 3.7\%$ of the interest rate itself.)

basis weight *paper and printing* USA The 'weight', i.e. mass, in pounds of a defined amount of paper, boxboard, etc., as a measure of its substance. The US Government Printing Office uses 1 000 sheets, but elsewhere it is usually a ream of 500 else 480 sheets, the sheets being of a specified named size, referred to as **basic size**. For many materials the size is 17 in × 22 in, for which the mass in pounds of 500 sheets is called the **substance** or **substance number**. For boxboard the standard is a ream of 500 sheets of 25 in × 40 in, i.e. 500 000 in^2, while for paperboard the value is normally expressed directly as pounds per 1 000 ft^2. All are ultimately expressible in this way, or in grams per square metre, which is the usual practice in Europe.

baud [J. M. E. Baudot; France 1845–1903] *electromagnetics, informatics*
In telegraphy and other digital signalling, the time between
representative samplings of the signal, during which any transition of
signal state occurs, for example a hundredth of a second. A transmission
line with such a rate of transition/reading would invariably be called a
'100-baud' line; in such an expression the baud becomes a rate rather
than the original time element. Since virtually all telegraphy and most
similar transmission until recently used only two levels for transition,
each baud accommodated just 1 bit; hence a 100-baud line was often
referred to as being of 100 bits per second. In reality, most lines of such
magnitude dissipate from 10 to over 30% of their capacity on
asynchronous signal control, so the translation to bits is quite
exaggerated to the user. Modern high-speed circuits dissipate very little,
so the nominal figure is not significantly misleading; however, its
transition/reading scheme may be very different, with perhaps eight
levels communicating three bits per baud on a '4 800-baud' line (i.e.
sampled every $\frac{1}{4800}$ s = 208.33~ μs), giving 14.4 kilobits per second
(numerically far in excess of the 4 000 Hz bandwidth usually provided for
a standard telephone circuit). More levels, and hence greater
multiplication from baud rate to bit rate, are increasingly common as
greater transmission rates are pursued. There is no absolute limit to the
number of discernible transitions (hence bits per baud) or to the number
of bauds or samples per second for any line circuit. The quality of the
circuit and its equipment is the sole arbiter. Higher rates often involve
error detection/correction techniques that reduce the effective transfer
rate through both control and re-transmission, but often use data
compression techniques that enhance the effective rate even more.
   See also kibi-.

Baumé [A. Baumé; France 1728–1804] *liquor and food processing*.
Symbol Bé. A scheme, expressed in degrees, that measures the †specific
gravity of liquids, primarily in saccherimetry. Many such scales have
been used. All employ a hydrometer graduated such that, for some
constant m (the modulus), if d is the reading in a solution of specific
gravity s, then $(d - d_w)s = (s - 1)m$, where d_w is the reading chosen for
pure water, usually zero. What has been labelled the 'old' Baumé scale
uses $m = 144$, apparently based on an interpretation of Baumé's original
scale as having 66° Bé for strong sulphuric acid at 15°C, with $s = 1.8427$,
hence $m = 144.32$. A later interpretation, with 15° Bé for a 15% salt
solution at 17.5°C, gave $m = 146.78$; this became the 'new' or **Gerlach**

scale.[117] The modern standard for liquids denser than water is $m = 145$ with 15° Bé for 15% salt by mass in water but a temperature of 10° Réaumur (12.5°C, 54.5°F); example values are:

Specific gravity	1	1.01	1.03	1.1	1.5	2.0
°Baumé	0.72	1.44	4.22	13.18	48.3	72.5

A scale with 0° Bé for 10% salt and 10° Bé for pure water and $m = -140$ is used for liquids less dense than water (hence having readings greater than 10° Bé); example values are:

Specific gravity	1	0.95	0.9	0.8	0.7	0.6
°Baumé	10	17.37	25.56	45.	70.	103.3

The scheme for †API gravity (for petroleum) is structured identically with the less dense Baumé but has a slightly different reference temperature, making its readings generally about 1% greater.

BA unit [British Association] *electromagnetics See* †metric system.

bbl. *See* barrel.

BC [Before Christ] *time* The label suffixed to the number of a year, e.g. 500 BC, counted backwards from the nominal year for the birth of Christ. But *see* BCE.

BCE [Before the Common Era, i.e. before the Christian era] *time* An equivalent to BC, used, with CE instead of AD, to avoid the unwanted religious connotations of BC and AD while using what is effectively the universal †calendar. Written after the number, e.g. 500BCE or 500 BCE.

b.c.f. [billion cubic feet] *volume North America* For natural gas, 10^6m.c.f. $= 10^9$ft^3 ($28.317\sim \times 10^6$m^3).

bd. ft. *See* board foot.

Beaufort scale [F. Beaufort; UK 1774–1857] *meteorology* A code for wind speed/force, originally defined as twelve progressively rougher sea conditions, then re-expressed by G. C. Simpson as values 1, 2, … , 12 (and 0 for calm), with scale value B related to wind speed S (in knots) approximately by the formula $S^2 = 3.5\ B^3$. An international variant was agreed in 1939, but the older one survives in Britain, the USA, and related countries, extended by the USA in 1955 to value 17 to elaborate the cover of hurricanes. (*See also* Saffir–Simpson Hurricane Scale.) The international scale specifies an altitude of 10 m (originally 6 m) for measurement, the other 36 ft (11 m). The feathered arrows of meteorological maps represent the Beaufort scale, with one full-thickness feather being two

Table 5

International		Min Speed knots	British and American	
calm	0	0	0	light
light air	1	1	1	light
light breeze	2	4	2	light
gentle breeze	3	7	3	gentle
moderate breeze	4	11		
		13	4	moderate
fresh breeze	5	17		
		19	5	fresh
strong breeze	6	22		
		25	6	strong ('force 6 wind')
near gale	7	28		
		32	7	strong ('force 7 wind')
gale	8	34		
		39	8	gale ('force 8 gale')
strong gale	9	41		
		47	9	gale ('force 9 gale')
severe gale	10	48		
		55	10	whole gale ('force 10 gale')
violent storm	11	56		
hurricane	12	64	11	whole gale ('force 11 gale')
		73	12	hurricane
		83	13	hurricane
		93	14	hurricane
		104	15	hurricane
		115	16	hurricane
		126	17	hurricane

Beaufort points. The respective scales, showing the minimal wind speed in knots for each value and corresponding names, are shown in Table 5.

becquerel [A. H. Becquerel; France 1852–1908] *radiation physics*. Symbol Bq. *SI* For the activity of a radionuclide, the rate of decay or other nuclear transformations per second ($= s^{-1}$ in base terms).

Note that this is the number of disintegrations or other nuclear transformations, not the number of particles produced or other measure of the product of transformation, which can be any combination of (doubly charged) ^{4}He nuclei from alpha decay, electrons from beta decay, photons from gamma decay, or any other form of electromagnetic radiation, with each at any energy level. The unit of absorbed dose (i.e. the amount of energy imparted to matter by such radiation) is the gray; the unit of ionizing radiation relating to gamma rays and x-rays is the roentgen.

The becquerel was defined only in 1975, by the 15th †CGPM, to succeed the curie[21] and the uncommon rutherford as the measure of radioactivity.

$$1Bq = 10^{-6} \text{ rutherford,} = (3.7 \times 10^{10})^{-1} \text{ curie} = 2.702\ 703\sim\times\ 10^{-11} \text{ curie.}$$

1975　15th CGPM: re ionizing radiations '*adopts* the following special name for the SI unit of activity: *becquerel*, symbol Bq, equal to one reciprocal second.'[8]

bel [A. G. Bell; UK, Canada, USA 1847–1922] *See* decibel.

Besselian year [F. W. Bessel; Germany 1784–1846] *astronomy* Now the †Julian year, i.e. 365.25 days, $= \frac{1}{100}$ Julian century of 36 525 days.

Originally the †year of a †calendar scheme established by Bessel, the year beginning when the mean Sun reaches 18 h 40 min of †right ascension (which currently occurs on 1 January). Variablity of the length of the year (which is essentially that of the tropical year but more varied because of being demarcated close to a solstice), and having its calendar offset from the noon moment that demarcates the Julian day number, led to the practical demise of this pioneer scheme for identifying astronomical †epochs. Being based on a fictitious Sun, the Besselian year was also called the **annus fictus** or **fictitious year**[22] and, ambiguously, because of other meanings for the name, the **astronomical year**.

BeV [billion electron volts] *sub-atomic physics* USA GeV, i.e. giga-electron-volts, which is preferred.

BG *UK* For steel sheet, Birmingham Gauge; *see* gauge.

BHP, b.h.p. Brake horse power; *see* horse power.

Bi *electromagnetics* Metric-c.g.s. Biot. See abampere.

BI, BI-apoth, BI-avdp, BI-f.p.s., BI-troy *See* British Imperial system.

bicron [bi micron] (also **stigma**) *length* 10^{-12} m $=$ 1pm, the numeric factor being the square of that of the micron (an SI-deprecated name), along with stigma, now correctly the picometre.

b.i.d. [Lat: bis in die] *medicine* Twice per day. *See also* o.d.

BI-f.p.s. BI foot-pound-second; *see* f.p.s. system.

billion [Lat: 'two'] Symbol bn. Often used to mean a huge number, but specifically defined traditionally as:

$North\ America = 1\ 000\ million = (1\ 000)^2\ thousand, = 10^{(3\times2)+3} = 10^9.$

$UK = 1\ 000\ 000\ million = (million)^2, = 10^{6\times2} = 10^{12}.$

The index value 2 following the bracket in each is the respective etymological factor. The billion is clearly ambiguous and confusing in the intercontinental context, so should be avoided. However, the escalation of so many countable entities, from money to humans, makes such large terms of growing appropriateness. The increasingly pervasive use of the American form (despite the 9th †CGPM in 1948 recommending the extant British/European form[162]) has displaced the traditional British form even in the UK. With this trend, the term has surviving value; however, such terms are yielding in either sense to expressions formed from the prefixes of the †SI system (e.g. 'megabucks'), that for 10^9 being giga (G-). Use of billion in a scientific context is undesirable. *See* thousand for general discussion.

binary [Lat: 'two'] Relating to anything having just two alternatives, or based on such (see below), in contrast with the steps of 10 and its powers for †decimal and its compounds, of 12 for †duodecimal, 16 for †hexadecimal, 20 for †vigesimal, 60 for †sexagesimal, etc. In both the material and the electrical worlds, from the presence or absence of an item (e.g. a hole in a card, the on/off of electricity) to the north/south polarity of magnetism, the binary state is of widespread natural significance. Hence, although efforts to mimic decimal notation have been made, the binary state epitomizes computers, and hence the electromagnetic equivalents of †binary notation are their normal internal form for representing numbers. The immediate compounds of binary are quaternary for steps of 4, †octal for steps of 8, and †hexadecimal for steps of 16.

binary notation, binary number A style of expressing numeric values similar to †decimal notation except for being based on 2 rather than 10, hence using only the digits 0 and 1, but requiring over three times as many digit positions for a given number, e.g. the number 301 in the decimal system

$$= 1 \times 256 + 0 \times 128 + 0 \times 64 + 1 \times 32 + 0 \times 16 + 1 \times 8 \ + 1 \times 4 \ + 0 \times 2 \ + 1 \times 1$$
$$= 1 \times 2^8 \ \ + 0 \times 2^7 \ \ + 0 \times 2^6 + 1 \times 2^5 + 0 \times 2^4 + 1 \times 2^3 + 1 \times 2^2 + 0 \times 2^1 + 1 \times 2^0$$

is written in binary as 100101101. Just as with the decimal version, the individual positions are called digits, so the binary representation would be said to be of nine digits, against the three digits of the decimal representation 301. To distinguish, one can say binary digits (contracted to 'bits') for the one, decimal digits for the other.

Since $8 = 2^3$ and $16 = 2^4$, both the 8-based †octal notation and the 16-based †hexadecimal notation effectively group into single graphic characters successive strings of three and four binary digits respectively. Octal notation was once the dominant alternative to the cumbersome pure binary number, but, since the introduction of the †byte, hexadecimal has come to predominate.

biot [J. B. Biot; France 1774–1862] *electromagnetics*. Symbol *Bi*. *See* abampere.

    rheology For **biot number** *see* Nusselt number *Nu* for heat transport.

BIPM [Bureau International des Poids et Mesures] The world bureau for 'weights and measures', established by the International Convention of the Metre in 1875. Situated in the Paris suburb of Sèvre, supervised by the †CIPM, and under the policy and dictates of the †CGPM, it administers the various elements of the †metric system, now solely in the form of the †SI.

Birmingham Gauge, etc. *See* gauge.

bit [binary digit] The single two-valued digit in the †binary number system.

    informatics From the above, the fundamental measure of memory and transmission capacities, corresponding to a single two-state element, effectively on/off, $+/-$, etc. Externally in punched cards and paper tape this became hole/no-hole. In information theory it becomes the fundamental measure of information content, with the hartley being the amount required to hold one decimal digit, i.e. $= \log_2 10 = 3.321\ 928\sim$ bits. The typical names for groupings of multiple bits in computers are now as shown in Table 6. *See* word for a discussion of the variable sizing of this unit.

    For **bits per second** *see* baud and kibi-.

Table 6

bit					
4	nibble				
8	2	byte			
16	4	2	half word		
32	8	4	2	full word	
64	16	8	4	2	double word

blink *time* Equated by some to 10^{-5} day $= 0.864$ s.

blondel [A. E. Blondell; France 1863–1938] *photics See* apostilb.[23]

bn. *See* billion.

board foot (**bd ft, FBM, foot board measure, superfoot**) (pertaining to board as a piece of lumber, a plank) *volume* For sawn lumber, 1 ft × 1 ft × 1 in or equivalent $= \frac{1}{12}$ft^3(2 359.7∼ cm^3, 144 in^3). Usually applied to undressed size for dressed material. Identical with the †shipping cubic inch. *See also* FBM.

Board of Trade unit [UK governmental department of trade] (sometimes **BTU**, an abbreviation used much more commonly for †British thermal unit) *electromagnetics UK* An early name for a kilowatt · hour.[11]

body mass index *health* The ratio of a person's mass in kilograms to the square of their height in metres. Providing a rough correction for mass relative to height, it is approximately 20 for a 'trim' person, 25 for overweight, 30 for obese.

BOE, b.o.e. *engineering* Barrel of oil equivalent. For fuels other than oil it is the amount having the energy content equal to that of a standard †barrel of oil, which is accepted as 6.119 GJ, hence about 164.∼ m^3 (5.8 m.c.f., 5 800ft^3) of gas, though often taken as 10 m.c.f.

Each form of fossil fuel varies markedly in energy content per unit volume or mass. (Even when refined to remove the gases other than methane, natural gas as piped to consumers varies; it is normally priced on an energy basis, applied as a multiplier of volume in any billing.) However, for planning and comparison several standard values are used. Table 7 gives the conventional gigajoule values for key units, plus their relative sizes (with e standing for equivalent).

Table 7

		GJ	b.o.e.	m.t.o.e.	m³gas	m.c.f.	m.t.c.e.
b.o.e.	barrel of oil	6.119	1	0.1367	164.2	5.800	0.2089
m.t.o.e.	metric ton of oil	44.76	7.315	1	1201	42.43	1.528
m³gas	1m³ of gas	0.037 26	0.006 089	0.000 832 3	1	0.035 32	0.001 272
m.c.f.	10³ft³ of gas	1.055	0.172 4	0.023 57	28.35	1	0.036 02
m.t.c.e.	metric ton of coal	29.29	4.787	0.654 4	786.1	27.76	1

The generic unit quad (10^{15} B.t.u., 1.055 4~ $\times 10^{18}$ J) equates almost exactly to the content of a trillion m.c.f. of natural gas.

bohr [N. H. D. Bohr; Denmark 1885–1962] *physics See* Bohr radius.

Bohr magneton Symbol μ_B. *See* magneton.

Bohr radius *fundamental constant*. Symbol bohr. Actually the first Bohr radius, the minimal radius of an electron orbit in Bohr's model of the atom, $= 0.529\ 177\ 208\ 3(19) \times 10^{-10}$m with †relative standard uncertainty 3.7×10^{-9}.[4]

Boltzmann constant, Boltzmann universal conversion factor [L. E. Boltzmann; Austria 1844–1906] *thermodynamics*. Symbol k, k_B. The ratio of the universal gas constant to †Avogadro's number (so effectively gas constant per molecule) $= 1.380\ 650\ 3(24) \times 10^{-23}$J·K^{-1} with †relative standard uncertainty 1.7×10^{-6}.[4]

BOT, B.o.T. *See* Board of Trade unit.

bottle *volume* An obviously general and hence variable unit, but often fixed for a specific context. The most notable case is the wine bottle, now widely standardized at 750 mL (26.398~ BI fl oz, 25.362~ US liq oz) and applying to liquor generally. This is 99% of the long-used UK **reputed quart** of $\frac{1}{6}$ BI gal and of the US **fifth** of $\frac{1}{5}$ US gal.

bougie decimale *photic luminosity France* For flame luminosity, $= \frac{1}{10}$ carcel ≈ 1 cd.

boulder *geology* A †particle size, typically exceeding 256 mm or 10 in.

box *mass UK 1975* For herrings, $= 25$ kg $= \frac{1}{4}$ herring unit.

Boyle, boyle [R. Boyle; Ireland. UK 1627–91] *pressure See* deciboyle.

BP [Before the Present] *time* An acultural measure into the past, expressed in years. Clearly it has a moving reference point, but its scale of use makes this of little relevance, and indeed rapidly indistinguishable from BC and [†]BCE (e.g. in the [†]geochronologic scale).

B.p.s. Bytes per second. *Compare* b.p.s.

b.p.s. Bits per second; *see also* baud. *Compare* B.p.s.

Bq *radiation physics* SI *See* becquerel; *see also* SI alphabet for prefixes.

brewster [D. Brewster; UK 1781–1868] *optics*. Symbol B. Originally coined for the unit '10^{-10} cm^2 per gramme weight' [24] in 1910, for the expression of stress-optical coefficients, the brewster is now regarded as 10^{-12}m$^2 \cdot$N^{-1}. The unit applies to the differential effect of [†]orthogonal stress on the components of a linearly polarized light beam as it passes through an elastic medium. The coefficient is the ratio of the relative retardation in angstroms to the product of distance travelled through the glass and the applied stress. The brewster requires distance in millimetres and stress in units of 10^5 Pa, and gives
$$Å \cdot (10^{-3}m^1 \cdot 10^5 N \cdot m^{-2})^{-1} = 10^{-10} \ m \cdot 10^3 m^{-1} \cdot 10^{-5} N^{-1} \cdot m^2$$
$$= 10^{-12} m^2 \cdot N^{-1}.$$
If distance is measured in inches and stress in pounds per square inch (but retardation still in angstroms), the result must be divided by 1.7513~ to give brewsters.

The **fringe value** is the ratio of the wavelength in angstroms to the coefficient in pound and inch terms, i.e. the stress in pounds per square inch required to cause a retardation of one wavelength per inch.

brig, brigg [H. Briggs; UK 1561–1630] The ratio of two quantities expressed [†]logarithmically to the base 10, e.g. for quantities Q_1 and Q_2, the number of brigs = $\log_{10} Q_1/Q_2$. Hence 1 brig represents a ten-fold ratio. A generalization of the bel.[25]

briggsian logarithm [from brig] *See* common logarithm.

Brinell hardness number [J. A. Brinell; Sweden 1849–1925] *See* hardness numbers.

British Apothecaries' system (BI-apoth) *See* grain for table of units.

British Imperial system (**BI, Imperial system**) *UK* The system established by 'An act for ascertaining and establishing Uniformity of Weights and Measures' assented to by George IV on 17 June 1824, which became effective on the first day of 1825 and was subsequently common throughout the British realm. It was founded on a brass [†]prototype yard constructed in 1760 (the 'Standard Yard'), the last in a long tradition of metal prototypes, and a brass prototype pound of the troy scale, constructed in 1758 (the 'Imperial Standard Troy Pound'). The grain of $\frac{1}{5760}$ lb troy was defined as the elemental unit of 'weight', i.e. mass, and 7 000 such grains the 'Pound Avoirdupois'; *see* grain regarding the three distinct (though closely interrelated) schemes for weight: †avoirdupois (**BI-avdp**), †troy (**BI-troy**), and †apothecaries' (**BI-apoth**). The standard measure for capacity (volume) 'as well for Liquids as for dry Goods not measured by Heaped Measure shall be the gallon of Ten Pounds Avoirdupois Weight of Distilled Water'. *See also* f.p.s. system.

British thermal unit *mechanics See* B.t.u.

British Wire Gauge *See* gauge.

Brix [A. F. W. Brix; Germany] *liquor and food processing* A revised †Balling scale, expressed in degrees, that measures the percentage by mass of sucrose in water solution, at 17.5°C; the degree of such a scale is identically the percentage point. *See also* relative volumic mass. Some illustrative values for volumic mass ('specific gravity') as Brix degrees are:

Brix °	25.0	45.0	63.0	79.0	93.0
Volumic mass	1.103~	1.202~	1.303~	1.404~	1.500~

B Series, B4, etc., *paper and printing See* paper size.

BSI [British Standards Institute] *photography See* film speed.

B. Th.U. *See* B.t.u.

BTU, B.t.u, Btu *electromagnetics UK See* Board of Trade unit.
   energy BI (**British thermal unit**, also sometimes **B.Th.U.** in UK to differentiate it from †Board of Trade unit) The energy required to raise the temperature of 1 pound of water by 1 degree Fahrenheit. Since the precise amount is dependent on the starting temperature, any use of the term requires an acknowledgement of this. The figure is denoted by a suffix, e.g. **B.t.u.$_{59}$** or **B.t.u.$_{59/60}$** denotes the rise from 59 to 60°F; **B.t.u.$_{mean}$** denotes the mean over the range 32 to 212°F, i.e. $\frac{1}{180}$ of the energy required to raise 1 pound of liquid water from freezing point to boiling point. Another form is based on the international steam tables, and is termed the **international B.t.u., B.t.u.$_{IT}$**.
   1 B.t.u.$_{mean}$ = 1055.87~ J,

1 B.t.u. (thermochemical) = 1054.350∼ J,

1 B.t.u.$_{IT}$ = 1055.055 26∼ J (0.367 717 ft^3·atmos).

History
The full name 'British thermal unit' appears to have been adopted in the 1870s to differentiate it from other thermal units. Previously and continuingly it was also called the **British unit of heat** and plainly **heat unit**, the latter accorded the symbol **u**. The modern name prevailed by 1895, but was synonymously the **pound degree fahrenheit**.[26] The joule was adopted in 1947 by the International Union of Physicists as the standard heat unit.[27]

bu., bushel *volume British/American* Measure for grains, etc.
BI 8 BI gal (36.368 72∼ L). The Act of 1824 that introduced BI specified striken measurement for liquids and 'for dry Goods not measured by heaped Measure', going on to cite coal, lime, fish, potatoes, and fruits as specific exceptions. For these, the bushel measure had to have a flat base and a round top of external diameter 19½ in; heaping must be in the form of a cone, of at least 6 in height. The bushel was removed from official UK measures in 1970.[28]

US-C dry The volume of a cylindrical vessel of 18½ in diameter and 8 in deep (identically that for the British Winchester bushel of 1697), i.e. 2 150.420∼ in^3 (35.239 07∼ L) if strictly striken.
See Table 8.

Table 8

US-C for dry goods					SI	BI
dry pint	...	...	...	...	551.∼ mL	0.969∼ pt
2	dry quart	...	...	...	1.10∼ L	1.94∼ pt
16	8	peck	...	...	8.81∼ L	1.94∼ gal
64	32	4	bushel	...	35.2∼ L	3.88∼ pk
			8	quarter	282.∼ L	7.75∼ bu

mass A unit †imputed from the preceding volume unit, especially for the measurement of grains, etc., the mass depending on the species and even variety of grain. These values are often set by statute, and reset from time to time. Current values in the USA are:

60 lb (27.216∼ kg) for wheat and soybeans,

56 lb (25.401∼ kg) for flour, rye, flaxseed, grain sorghum, and maize – the US or **International corn bushel**, but

70 lb (31.751~ kg) for husked maize,

48 lb (21.772~ kg) for barley,

45 lb (20.412~ kg) for rough rice,

38 lb (17.237~ kg) for oats.

The USA also has the bushel of cement of 80 else 100 lb.

The BI bushel was identically 80 lb for distilled water (*see* gallon).

butt (pipe) A large bulk-measure cask, with established volumes and quantities for various commodities in historic marketplaces, $= \frac{1}{2}$ tun.

BWG *UK* For steel wire, Birmingham Wire Gauge; *see* gauge.

byte *informatics* (also **octet**) 8 bits, though occasionally used for a grouping of 6 or 9 bits. The byte was coined as a term with the IBM '360' computer series in 1964, introducing a modular building block that replaced the varied long words of most earlier machines (and the simple chained characters of its predecessor, the 1401 computer). The growing use of computers for alphanumeric textual material brought a need for a character-sized entity. Initially this employed a 6-bit structure covering the alphabet only in upper case. While 7 bits would have sufficed for the set of typewriter characters, the 8-bit byte had the advantage of holding with reasonable efficiency either one alphanumeric character or two decimal digits, and for the alphanumeric set to be enhanced, all in a machine of compact modular design.

See also kibi-.

C

C, c [Lat: centum, the †Roman numeral for 100]
Metric As a lower-case prefix **c-**, *see* centi-.

   mathematics As **C**, *see* Euler's constant; as **C** or **c** for continuum, *see* infinity.

   physics As **C** or **c**, *see* speed of light (in a vacuum).

   electromagnetics *Metric* As **C**, *see* coulomb, *see also* SI alphabet for prefixes.

   informatics In †hexadecimal notation as **C** for 12, the third digit after 9.
   music *See* pitch.

c- *Metric* As prefix symbol, *see* C, c.

c_0 *sub-atomic physics* As a natural unit of velocity, $= c$, the †speed of light.

c_1, c_2 *radiation physics* first, second *See* radiation constant.

cable, cable length, cable's length *length* A nautical measure, $\approx$ 200 m.
Internat 1954 $\frac{1}{10}$ international nautical mile = 185.2 m (607.6115$\sim$ ft).
UK trad 100 fathoms = 600 ft m (182.88$\sim$ m), though defined by the
   Admiralty to equal $\frac{1}{10}$ †sea mile of 1 minute of latitude locally, hence
   somewhat variable about the value 608 ft (185.32$\sim$ m).
US-C trad 120 fathoms = 720 ft (219.456 m).

Cal., cal. *thermodynamics, nutrition* *See* calorie.

calendar *time* A scheme for grouping, labelling, and thereby
distinguishing the individual days. The Moon and the Sun provide
conspicuous natural groupings of very suitable size: the former the lunar
period or month of about 30 days, the latter the solar period or year of
about 365 days, i.e. about 12 of the lunar periods. (As discussed under
month and year, these periods exist in several forms; here we are
concerned with the obvious period for each, explicitly the †synodic
month and the †tropical year. The week as an entity is essentially
artificial, not necessarily of seven days, and of peripheral interest in this

context.) Unfortunately, neither grouping is an exact number of days, nor is the year an exact multiple of the natural month. While the familiar calendar adheres closely to the natural year, by incorporating months of relatively arbitrary size, the Jewish and Moslem calendars, for instance, have a natural month but a more erratic year. History has seen a great variety of renderings for these incompatibilities.

The lengths of the average natural (synodic) month and (tropical) year, in terms of the average natural (mean solar) day, are as follows:

1 month = 29.530 59~ days

1 year = 365.242 22~ days

hence

1 year = 12.368 27~ months.

With the recurrent pattern of the seasons of more functional significance than that of the Moon, there is an obvious advantage in giving primacy to fitting with the natural year. Since any reasonable calendar will be built on whole days, the fractional 0.242 2~ must be accommodated carefully if the calendar is to stay in step with the seasons. This, of course, is the cause of the **leap year** being in the familiar calendar. Introduced by Julius Caesar in the year 45 BCE by adding an extra day to February (the last full month of the year, which started at the spring equinox) every fourth year, this extra day (an †intercalary day) equated the year with 365.25 days, and hence overstated the need by about one day every 128 years. Thus, with this **Julian calendar**, the average period for the year exceeded that of the seasons, so they came progressively earlier as the centuries passed. By the 16th century this amounted to about 13 days. While the perturbation within a lifetime was insignificant, and the placement of equinoxes and solstices did not have naturally compelling points in the calendar, it was decided at that time both to alter the pattern so as to approximate yet more closely the true natural year and to reset the calendar to put the seasonal markers close to their traditional points in the calendar. Thus, Pope Gregory XIII adopted for the Roman Catholic Church what is known as the **Gregorian calendar**, with leap years not occurring in years that were multiples of 100 unless they were multiples of 400. Thus excluding three days per 400 years, this equated the year with $365\frac{97}{400} = 364.242\ 5\sim$ days – still overstated but only by about one day every 3 323 years. The finer adjustment, proposed but yet to apply, and potentially to be varied because of the minutely changing length of the average day and average year, is to exclude from being leap years those years that are multiples of

4 000. This would leave the year overstated by about one day per 20 000 years.

The occurrence of the northern spring equinox on or about 21 March is thus a deliberate feature of the Gregorian calendar, harking back to the Julian, but the original cause of any such arrangement is unclear. Since the tropical year equals 365.242 2~ days, equal seasons would be of 91.31~ days. However, because Earth's orbit is elliptical and the Sun at a focus thereof rather than the geometrical centre, the four seasons routinely differ in length from one another by a few days. The northern spring is nearly 93 days and its summer nearer to 94, but the southern spring is under 90 days and its summer under 89 (but includes the time of greatest proximity to the Sun, the perihelion). Hence, while the March equinox and the June solstice tend to be on the 20th or 21st of their month, the December solstice is more often on the 22nd and the September equinox on the 23rd (the offset being less than it might be because of the pattern of month sizes tending to correspond with the astronomical pattern). (Despite the summer solstice commonly being called 'Midsummer Day', the practice in many countries is to regard these dates as the starts of the respective seasons rather than their midpoints. This reflects the delayed effect of the seasonal travels of the Sun, but exaggerated. The Australian practice of regarding seasons as starting with the relevant month, e.g. 1 March instead of 21 March, is much more appropriate.)

This new (and still extant) calendar was implemented by the Roman church in October 1582, with the dates 5 to 14 October that year omitted, bringing the seasonal pattern not to the conditions of Caesar's time but to those of 325 CE, the time of the Council of Nicaea. The day following 4 October 1582 'old style' was thus 15 October 1582 'new style'. Though quickly adopted in many Catholic European countries, it was not widespread in Europe until 1700, when the continental Protestant areas adopted it, as did Scotland. England and her overseas possessions delayed introduction for 170 years, by then having to omit 11 days, jumping from 2 September 1752 'old style' to 14 September 1752 'new style'. The Orthodox world (Russia, Greece, etc., and Turkey) waited until after World War I, by which time the correction became 13 days, a change still not accepted by the Orthodox Church, which remains Julian.

As day, month, and year are all changing in size progressively, mainly due to the tidal effects induced by extra-terrestrial gravity, the precise challenge of the calendar will change significantly in due course.

Callier coefficient *photography* The ratio of the apparent densities of a photographic plate when viewed in parallel light to when viewed in diffuse light.

Calorie (kilogram-calorie, grand calorie, large calorie, major calorie) *See* calorie.

calorie (gram-calorie, petit calorie, small calorie) *heat energy* 4.1868 J ($3.968\,322\,93\sim \times 10^{-3}$ B.t.u.), but also **thermochemical calorie** $cal_{th} = 4.184$ J.

Although widely known as a unit for indicating the energy content of foodstuffs (and hence, to some extent, their 'fattening power'), the calorie is technically a measure of heat energy. Its use in nutrition corresponds to its use in physics, both being a measure of energy; however, the 'calorie' of the food packet is usually a **kilocalorie** or **kilogram-calorie** of 1 000 calories, reflecting a like ambiguous use in physiology. Sometimes distinguished by capitalization as **Calorie**, the latter is more clearly the **large calorie** or **grand calorie** with the true unit as the **small calorie**, **petit calorie**, or, more properly, the **gram-calorie**.

Defined in 1880 as the energy required to raise the temperature of 1 gram of water by 1 Celsius degree (i.e. 1 K) without further qualification, the term required an acknowledgement of the reference temperature for any precise use, since the amount is dependent on starting temperature. This is denoted by a suffix, e.g. cal_{15} denotes the rise from 14.5 to 15.5°C; cal_{mean} denotes the mean per degree over the range 0 to 100°C.

international steam calorie cal_{IT}	$= 4.186\,74\sim$ J,
mean calorie cal_{mean}	$= 4.190\,02\sim$ J
4°C calorie cal_4	$= 4.204\,5\sim$ J (water at its densest)
15°C calorie cal_{15}	$= 4.185\,5\sim$ J (15°C = cool indoors)
20°C calorie cal_{20}	$= 4.181\,90\sim$ J (20°C = warm indoors)

The cal_{15} is usually meant by the **mechanical equivalent of heat**. The international steam calorie $cal_{IT} = 4.186\,74\sim$, based on the international steam tables, prompted the rounded $= 4.186\,8$ J.

Generically also **heat unit** and from 1888 to 1896 the therm (*see also* B.t.u. and joule), the calorie proved difficult of laboratory verification.[29]

For foodstuffs, the calorific content is usually measured (in kcal, i.e. the heat energy that raises 1 kg one degree) by burning in an atmosphere of pure oxygen in a bomb calorimeter.

Cancer *astronomy See* zodiac; right ascension.

candela [candle] *luminous intensity*. Symbol cd. *Metric* The base photic unit.

SI, Metric-m.k.s.A. 1979 The intensity, in a given direction, of a source that emits monochromatic radiation of a frequency 540 THz and that has a radiant intensity in that direction of $\frac{1}{683}$ W·Sr^{-1} (1.11~ hefner, 0.981~ international candle). Hence:

cd·sr = lumen for luminous flux.

Metric-m.k.s. Defined, as discussed below, on a standard candle.

Metric-c.g.s. Defined, as discussed below, on a standard candle. The following are among the coherent derived units:

cd·cm^{-2} = stilb for luminance;

cd·sr·cm^{-2} = phot for illuminance.

History

The candela is a rationalization of a variety of units originally based on standard candles, and called **candle**. The pioneer was the British candle of 1860, defined by the Metropolitan Gas Act for gaslights in London;[30] it was recognized internationally in 1881, in an electrotechnical context, specified as a spermacetti candle of $\frac{1}{6}$ lb burning at 2 grains per minute, but without mention of wick or atmosphere. Replaced by a pentane lamp in 1898, it was agreed internationally in 1909, with the term **international candle**. Equivalent approximate values of other prior candles, all of which were rather roughly measured, include:

France	bougie-decimale	0.97 international candles
	carcel	9.7 international candles
USA	candle	1.06 international candles
Germany	hefner candle	0.903 international candles

the last continuing in use as a national standard for a considerable time.

The differing sensitivity of the eye to different wavelengths caused trouble in progressing from those lamps running at a temperature just below 2 000 K to electric tungsten filaments above 2 800 K. By 1924 appropriate conversion tables were agreed, and used until the adoption, in 1946, of a specification based on a black body set at 2 042 K, the freezing point of platinum (see below). Because of the round figure adopted for the area in the definition, this **new candle** was 1.9~ % less than its predecessor. The name candela followed in 1948, and a restatement of the black-body definition in 1967. However, results with the specification were proving to be too inconsistent, while related radiometric techniques offered superior precision and ease of realization. These circumstances led to the re-definition of 1979.

1946 CIPM: '4. The photometric units may be defined as follows: *New candle* (unit of luminous intensity) – The value of the new candle is such that the brightness of the full radiator at the temperature of solidification of platinum is 60 new candles per square centimetre'.

1948 9th CGPM 'candela' adopted in place of (new) candle.

1967–68 13th CGPM: '*decides* to express the definition of the candela as follows: The candela is the luminous intensity, in the perpendicular direction, of a surface of $\frac{1}{600000}$ square metre of a black body at the temperature of freezing platinum under a pressure of 101 325 newtons per square metre'.

1979 16th CGPM: '*decides*
1. The candela is the luminous intensity, in a given direction, of a source that emits monochromatic radiation of a frequency 540×10^{12} hertz and that has a radiant intensity in that direction of $\frac{1}{683}$ watt per steradian. 2. The definition of the candela (at the time called new candle) … is abrogated.'[8]

candle *luminous intensity* Of a different value in different countries, redefined with minor changes, first as a 'new candle', then as the candela. *See* candela.

candlepower *light-radiating capacity* The power of a light source of one candle, international candle, new candle or candela as appropriate.

Capricornus *astronomy See* zodiac; right ascension.

carat (**karat**) [Arabic: 'carob', the plant *Ceratonia siliqua*] *mass Internat 1907, UK 1913* As agreed by the 4th †CGPM and as **metric carat (CM)**, = 200 mg (3.086 5~ gr) – a deprecated unit under the SI. *Internat 1877, UK 1877* = 205.3 mg (3.168 3~ gr).
goldsmiths' (*USA* usually **karat** Measure of proportion, against a maximum of 24, notably to express purity of gold. Within practical tolerances, 24-karat gold is absolutely pure while 14-karat gold has 14 parts gold to 10 parts other metal(s). Terminology appears to relate to a gold/copper coin (the mark) that weighed 24 goldsmiths' carats of 12 gr = 288 gr (18.662~ mg).

carcel [B. G. Carcel; France] *flame luminosity France* The average luminosity of a carcel lamp (a clockwork-fed oil lamp used particularly in lighthouses) burning 42 g of calza oil per hour $\approx$ 1 cd, = 10 bougie decimal.

cardinal number A †number in the ordinary sense, e.g. 1 or one, 2 or two, as distinct from 1st or first, 2nd or second, etc., which are †ordinal numbers.

Cartier *See* relative volumic mass.

cascade unit (**radiation length, radiation unit**) *astrophysics* Applicable to cosmic rays, the mean path length required in a given medium to reduce the energy of a charged particle by 50%; examples are 50 mm in lead, 433 mm in water, and 332 m in air at †s.t.p. Cascade unit = $(\ln 2)^{-1}$ times the †shower length = 1.442~ shower length.

cat *See* katal.

catal *catalytic activity* *See* katal.

cc, c.c. *volume* A common abbreviation for cubic centimetre, but inherently of such meaning only within the English language; the correct expression is cm^3. The cc was for a long time the usual label, even on much scientific equipment, for a thousandth of a litre, i.e. the millilitre, symbolically mL or ml. While the two units can be and have officially been equated, problems with the definition and hence sizing of the litre make the equating of cm^3 with mL inappropriate, and the habitual use of cc for millilitre particularly unfortunate.

c.c.-atmos *energy* Metric-c.g.s. 1.01 3 ~ × 10^2erg (0.101 3~ J), being the energy of the atmospheric pressure per square centimetre of surface.[31]

cd *photics* SI *See* candela, also prefixed variously, as in kcd = kilocandela; *see* SI alphabet. (Could also mean centiday, i.e. $\frac{1}{100}$ day.)

CE [Common Era] *time* An equivalent to AD, used to avoid the religious connotations of AD while using what is effectively the universal †calendar. Written after the number, e.g. 1990CE or 1990 CE.

celestial latitude, celestial longitude *astronomy* *See* latitude.

celo *acceleration* BI-f.p.s. Very rarely used, identically ft·s^{-2} (0.3048 m·s^{-2}).

Celsius [A. Celsius; Sweden 1701–44] *temperature*. Symbol (deg C, degree C, °C) SI A scale and a unit of temperature, the scale having 100 at the freezing point of pure water and zero at its boiling point, expressed usually as °C for the scale readings and deg C or degree C for temperature intervals. Though long defined by those two values and points, Celsius is now a derived scale. In SI , the †kelvin is the defining unit and scale, with the degree C identical to the kelvin in size and the Celsius scale figure = $(T/K - 273.15)°C$ where T is the kelvin or thermodynamic temperature. *See* †temperature for comparative values and other conversions.

History
Apparently created early in Celsius's life (though initially with the zero at the freezing point of pure water and 100 at its boiling point, then reset into the familiar way by someone else in 1743, so somewhat erroneously attributed to him), the Celsius scale (widely called **centigrade** and also sometimes the **centesimal** scale) competed with its contemporary, the Fahrenheit scale, for acceptance until modern times. Its use in France and much of continental Europe during the devising of the metric system, and its appearance of having a decimal structure (though, of course, there is no scale of units, no unit equal to 10 degrees Celsius, for instance), and its lack of royal appellations, made it an easy choice for temperature in the new system. The progressive adoption of the metric system soon made it the usual scale outside the English-speaking world, and in most scientific work everywhere. The Celsius scale gave rise to the kelvin scale, which has the same unit size but its zero is set at the null point of thermodynamic activity $(-273.15 °C)$.

The Celsius scale has been called the **centesimal thermodynamic scale**, but, even when unqualified, that term is now more likely to mean the **absolute centesimal thermodynamic scale**, i.e. the kelvin scale.

See temperature for other scales and conversions between scales.

1948 9th CGPM Resolution 3: 'From the three names ("degree centigrade", "centesimal degree", "degree Celsius") proposed to denote the degree of temperature, the CIPM has chosen "degree Celsius". This name is adopted by the General Conference'
'To indicate a temperature interval or difference, rather than a temperature, the word "degree" in full, or the abbreviation "deg", must be used.'[8]

Celsius heat unit (**centigrade heat unit, c.h.u.**) *energy* The Fahrenheit-based B.t.u. translated to the Celsius scale, i.e. the energy required to raise the temperature of 1 lb of water by 1°C, thus = 1.8 B.t.u. (1.899 101~ kJ, 453.592 8~ calorie).

cent [Lat: hundred, also hundredth] The term is very familiar as a hundredth of a dollar and as 'centi', the prefix for the parallel division of the metric units. It is used in various circumstances for this purpose, and also for the reciprocal, i.e. a unit, such as the cental, that is 100 times another (for which 'hecto'is the metric prefix).

sub-atomic physics USA For reactivity, $\frac{1}{100}$ dollar.

music An †interval such that 100 cents = 1 semitone on the scale of equal temperament, so 1 200 cents = 1 octave and 1 cent represents a multiplication of frequency by $\sqrt[1200]{2} = 1.000\,578\sim$; for frequencies f_1 and f_2, the latter the higher, the difference in cents is:

$1\,200 \times \log_2 (f_2/f_1) = 3\,986.314\sim \times \log_{10} (f_2/f_1).$

Compare †centi-octave, which has 100 units to the octave.

cental *mass* Metric 100 kg (220.46∼ lb).
US-C 100 lb (45.359∼ kg) = (short) hundredweight.
BI 1859 (**cH, centner, kintal, new hundredweight**) 100 lb (45.359∼ kg), usually for grains; British (long) hundredweight of 112 lb, called also quintal. Removed from official measures by Statutory Instrument 2484 of 1978,[73] and by formal Act in 1985.[28]

centare [cent- + are] Symbol ca. *area.* Metric $\frac{1}{100}$ are = 1m^2.

centesimal With divisor/multiplier steps of 100. *Compare* decimal.

centesimal degree *temperature See* Celsius.
plane angle $\frac{1}{100}$ †revolution. *See* grade.

centesimal minute *plane angle* From centesimal degree, 1 centesimal minute = $\frac{1}{100}$ centesimal degree = $157.079\,6\sim \times 10^{-6}$ rad (32.4 arcsec).

centesimal second *plane angle* From centesimal minute, 1 centesimal second = $\frac{1}{100}$ centesimal minute = $1.570\,796\sim \times 10^{-6}$ rad (0.324 arcsec).

centesimal thermodynamic scale *temperature See* Celsius and kelvin.

centi- [Lat: 'hundred', hence 'hundredth'] Symbol c-. *Metric* The $\frac{1}{100}$ = 10^{-2} multiplier, e.g. 1 centigram = 1 cg = $\frac{1}{100}$ g; never contracted before a vowel.

Used extensively in Europe, the centigram and centilitre are little recognized elsewhere; the centigram along with the centiare (identically the square metre) are not legal in the UK for trade.[28] The centimetre, uniquely among non-†millesimal units, is universally recognized;

however, because of the risk of misinterpretation by a factor of 10, it is generally excluded in North America from formal drawings and records, the millimetre being the standard alternative to metre.

centigrade heat unit *See* Celsius heat unit.

Centigrade scale *See* Celsius.

centiHg *pressure* Centimetre of mercury (element Hg) = 10 mmHg.[115]

centimeter, centimetre [centi- + metre] *length* Symbol cm. *Metric* $\frac{1}{100}$ m (0.393 70~ in, i.e. $\frac{1}{2.54}$ in), the base unit for length in the †c.g.s. system.
   See also electromagnetic centimetre; electrostatic centimetre.

centimeter-gram-second system, centimetre-gram-second system *See* c.g.s. system.

centi-octave *music* An †interval such that 1 octave = 100 units, hence 1 centi-octave represents a multiplication of frequency by $\sqrt[100]{2} = 1.006\ 955\sim$; for frequencies f_1 and f_2, the latter the higher, the difference in centi-octaves is
   $$100 \times \log_2(f_2/f_1) = 332.193\sim \times \log_{10}(f_2/f_1).$$
Compare cent, which has 100 units to the semitone, 1 200 to the octave.

centner *mass BI See* cental.

centrad [centi-radian] *plane angle* $\frac{1}{100}$ rad, particularly for diffraction of light by a prism.

century [Lat: 'hundred'] *time* 100 years. The 'century' now generally means the 100 years varying only in the last two digits, e.g. 1900–1999, but classically meant the set one year later. *See* millennium.

cetane number [refers to the chemical used] *automotives* A measure of the 'knock' or compressive ignitability of a fuel, being the percentage by volume of the chemical cetane, which, when blended with methylnaphthalene, gives a like performance. Cetane rating applies to fuel for diesel engines, which, unlike gasoline engines, depend on compressive ignition of fuel. However, for adequate performance the ignition must not occur too readily; a cetane rating above 40 is typically required. *Compare* octane number.

CGPM [Conférence Génèrale des Poids et Mesures] The world conference, first convened in 1889 in Paris, and reconvened now every four years, that authorizes the metric measures, initially for the international version of the †metric system and its special subsets, like the †c.g.s. system, now solely for the †SI. *See also* CIPM.

c.g.s. system (**Metric-c.g.s.**, also **CGS system**) *Metric* A version of the general system that has its derived constants relating †coherently to the centimetre, the gram, and the second, in contrast with the metre, the kilogram, and the second of the †m.k.s. system and its contemporary form, the †SI system. In the c.g.s. system, for example, the unit of force is the dyne, which gives an acceleration of $1 \text{ cm} \cdot \text{s}^{-2}$ to a body of 1 g, so is a force just one hundred-thousandth of the newton of the SI, which gives an acceleration of $1 \text{ m} \cdot \text{s}^{-2}$ to a body of 1 kg. (The crucial factor favouring c.g.s. appears to have been that the gram was the mass of a cubic centimetre of water.[32]) Various formulations for the associated electrical units were introduced: initially the electromagnetic (†e.m.u. system) and the electrostatic (†e.s.u. system), which based all their units on mechanical equivalence, then, deriving from the former but with some base definitions of electric units, the †practical units and †international units. The **gaussian system** was a hybrid of the former pair. Agreed at the first International Electrical Conference in 1881,[33] the c.g.s. system is now obsolete. Base units were, for

- length: cm = centimetre;
- mass: g = gram;
- time: s = second;
- luminous intensity: cd = candela, originally new candle.

The derived coherent mechanical units common to all forms of the c.g.s. system were, for

- acceleration of free fall: gal = $\text{cm} \cdot \text{s}^{-2}$ (= $10^{-2} \text{ m} \cdot \text{s}^{-2}$);
- acoustics – specific impedance: rayl = $\text{dyn} \cdot \text{s} \cdot \text{cm}^{-3}$ (= $10^{-1}\text{Pa} \cdot \text{s} \cdot \text{m}^{-1}$ = $10 \text{ N} \cdot \text{s} \cdot \text{m}^{-3}$);
- acoustics – mechanical impedance: mechanical ohm = $\text{dyn} \cdot \text{s} \cdot \text{cm}^{-1}$ (= $10^{-3} \text{ N} \cdot \text{s} \cdot \text{m}^{-1}$);
- acoustics – mechanical mobility: mohm = $\text{cm} \cdot \text{dyn}^{-1} \cdot \text{s}^{-1}$ (= $10^{3} \text{ m} \cdot \text{N}^{-1} \cdot \text{s}^{-1}$);
- dynamic viscosity: P = poise = $\text{dyn} \cdot \text{s} \cdot \text{cm}^{-2}$ = $\text{g} \cdot \text{cm}^{-1} \cdot \text{s}^{-1}$ (= $10^{-1}\text{Pa} \cdot \text{s}$);
- energy, work, quantity of heat: erg = $\text{dyn} \cdot \text{cm}$ = $\text{cm}^2 \cdot \text{g} \cdot \text{s}^{-2}$ (= 10^{-7}J);
- force: dyn = dyne = $\text{cm} \cdot \text{g} \cdot \text{s}^{-2}$ (= 10^{-5}N);
- illuminance: ph = phot = $\text{cd} \cdot \text{sr} \cdot \text{cm}^{-2}$ (= 10^4lx);
- kinematic viscosity: St = stokes = $\text{cm}^2 \cdot \text{s}^{-1}$ (= $10^{-4} \text{ m}^2 \cdot \text{s}^{-1}$);
- luminance: L = lambert = $\text{cd} \cdot \text{cm}^{-2}$ (= 10^4 lx);
- luminous emission: sb = stilb = $\text{cd} \cdot \text{cm}^{-2}$ (= $10^4 \text{ cd} \cdot \text{m}^{-2}$);

- mechanical impedance: mechanical ohm = $dyn \cdot cm^{-1} \cdot s$ ($= 10^{-3}\ N \cdot s \cdot m^{-1}$);
- pressure: barad = $dyn \cdot cm^{-2} = cm^{-1} \cdot g \cdot s^{2}$ ($= 10^{-1} Pa$);
- quantity of light: lumerg = $cd \cdot sr \cdot s$ ($= 1 lm \cdot s$);
- specific acoustic impedance: rayl = $dyn \cdot s \cdot cm^{-3}$ ($= 10\ N \cdot s \cdot m^{-3}$).

For electromagnetic units, *see* electromagnetic unit, electrostatic unit, and Gaussian unit; *see also* practical unit.

Non-coherent units used included bar, calorie, darcy, langley, leo, pyron, and rad.

c.g.s. unit *Metric-c.g.s.* While various units of the †c.g.s. system have specific names, many are often referred to purely descriptively as the 'c.g.s. unit of capacitance', etc. Such terms usually imply the †electromagnetic units rather than the †electrostatic units.

cH *mass BI See* cental.

chain *length* The length of a standard physical chain.
BI , US-C (**surveyors' chain**, **Gunter's chain**) 66 ft = 22 yd = 20.1168~ m;
   see inch for greater details, including reference to **coast** or **survey chain** of 22 × 36/39.37 m = 20.116 840 2~ m.
   The chain was removed from official UK measures in 1985.[28]
USA **Rathborn chain** = 2 rod = 11yd = 33 ft (10.058 4 m). Divided into 10 primes, each of 10 links or seconds.
   Engineers' chain or **Ramsden's chain** = 100 ft (30.48 m) = 100 links.

char., character A single inscribed mark, symbol, etc., most usually in Western culture an alphabetic letter, a numeric digit or punctuation symbol. Hence by extension also the minimal explicit space that separates words, etc. The quantitative measure of a character depends on the size of the alphabet to be covered, provisions for unique versus dual representation, and any in-built self-checking capacity. In early telegraphy one character equalled five bits net, but one visible character could require two transmitted characters because of figure/letter duality (e.g. T and 5 were represented by the same pattern of 5 bits), differentiated by the last preceding letter-shift else figure-shift character. In modern computers and telecommunications one character was five then six bits (with upper/lower case duality hence shift characters), but now is usually eight bits (= 1 byte), allowing unambiguous coverage of numeric digits, the lower- and upper-case alphabet, plus punctuation and many other symbols.

characteristic *mathematics See* logarithm.

characteristic impedance of a vacuum Symbol Z_0.
376.730 313 461~ Ω.

charrière scale *length See* French.

chronon (tempon) *sub-atomic physics* 10^{-23} s, the decimally rounded time for electromagnetic radiation to span an electron.

c.h.u. *physics See* Celsius heat unit.

Ci *radiation physics See* curie.

CIE system [Commission International de l'Eclairage (i.e. Illumination)] *colorimetry* The key scientific system for describing and specifying †colour, based on the additive mixing of precisely defined coloured lights.

   The three colours red, green, and blue are sufficient to produce, by appropriate mixing, all colours, including white, the 'complete colour'. The three are therefore called the 'primary' colours, or, more particularly, the 'additive primary colours'. (Their initials RBG grace many computer monitors.) The common unit for measuring the quantity of each primary, and of the composite product, is the lumen; however, the quantity of the mixture is not the sum of that of the constituents, and the three primaries are not equi-potent, i.e. the formulation of white is not one of equal lumens of each primary. If we divide the quantity of each primary in a mixture by the respective number of lumens necessary to produce white, we have coefficients rationalized to white, i.e. in a sense corrected for the differential power of the different primaries. If we divide each of the three new coefficients by their sum, we get what are called normalized coefficients, i.e. three coefficients that add up to 1. If we regard these values as quantities in lumens of the respective primaries, then the quantity of produced colour, in lumens, is called the **trichromatic unit** or **T unit** of the product. If the coefficients are now divided by the number of lumens in the T unit, then we have the composition for one lumen of mixture, and the coefficients are termed the **trichromatic coefficients**. The calculations depend on what precisely is meant by red, green, and blue; the adopted standards for these as **cardinal illuminants** are monochromatic radiations of 700.0 nm, 546.1 nm, and 435.8 nm wavelengths, respectively.

   Any other three colours whose formulations are mathematically independent, i.e. such that no one can be obtained by multiplicative addition/subtraction of the other two, are equally viable as constructs for all colours. One standard set is:

the **X stimulus** $= + 2.3646$ red $- 0.5151$ green $+ 0.0052$ blue

the **Y stimulus** $= - 0.8965$ red $+ 1.4264$ green $- 0.0144$ blue

the **Z stimulus** $= - 0.4618$ red $+ 0.0887$ green $- 1.0092$ blue.

Related to these, the CIE in 1931 established three full-spectrum †standard illuminants.[34]

The three coefficients required to match any given sample with a base trio of independent reference colours are called the **tristimulus values**. These can be obtained by direct experimental comparison, but are usually obtained indirectly, by applying established tables of the values for 5-nm increments of wavelength to spectrographic data, else, for X, Y, and Z, similarly from experimental comparison using the RBG trio.[35]

CIPM [Comité International des Poids et Mesures] The world authority for metric 'weights and measures' (now as the SI), and supervisor of the †BIPM under the policy and dictates of the †CGPM.

circle *See* revolution.

circular *area* As applied to linear units of measure, a terminology for expressing the area of a circular section purely as the square of its diameter, hence avoiding incorporation of the †irrational factor †pi, i.e. ignoring the multiplier $\frac{1}{4\pi}$. Hence:

circular inch $= \frac{1}{4\pi}$ sq.in $= 0.7853982 \sim$ in^2 (506.7075$\sim$ mm^2),

circular mil $= \frac{1}{4\pi}$ sq.mil $= 0.7853982\sim$ sq.mil

$= 0.7853982\sim \times 10^{-6}$in^2 (506.7075$\sim$ μm^2).

circular mm $= \frac{1}{4}$ mm^2 $= 0.7853982\sim$ mm^2(0.001 211 737$\sim$ in^2).

Clark [T. Clark; UK 1801–67] *chemistry* With reference to water hardness, *see* degree.

classical electron radius *See* electron.

clausius [†R. J. E. Clausius; Germany 1822–88] (**rank**) *entropy Metric* The kilocalories divided by the kelvin temperature.

clay *geology* The smallest †particle size, typically less than 4μm or 0.002 in, but dependent on the scheme employed.

clo *textiles USA* For clothing, a measure of thermal insulation, one clo being nominally the amount required for maintenance of body temperature in typical sedentary indoor conditions, i.e. at 70°F, relative humidity $\leq$ 50%, air movement $\leq$ 10 ft per minute, the body temperature to be at 98.2°F with metabolism at 50 kilocalories per square metre per hour.[36] *Ipso facto*, this is equivalent to a normal business suit plus medium underwear and about a quarter of an inch of clothing. Hence:

$\quad$ 1 clo $\approx$ 0.18~ K·m^{2}W^{-1} (0.88~degF·h·ft^2·B.t.u.$^{-1}$, 1.54 tog).

clusec [cl per sec] *high-vacuum technology* In parallel with lusec, an evacuation rate of 1 centilitre per second, but usually used to express the power to extract at such rate when at a pressure of only 1 μm of mercury (i.e. at 0.0136 Pa) $\approx$ 1.3~ μW.

cmHg *pressure* Centimetre of mercury (element Hg) = 10 mmHg.

Co *rheology* *See* Cowling number.

cobble *geology* A †particle size, typically in the range 64 to 256 mm (2.5 to 10 in), but dependent on the scheme employed.

coherent Besides the base units metre, kilogram, second, etc., the †SI system has numerous simply named derived units, such as the newton and the joule. These two are, respectively, the force that causes one kilogram to be accelerated at 1 metre per (1) second per (1) second, and the work done by 1 newton acting over 1 metre. Being coherent refers to this consistent use of 1. In the old †c.g.s. system, with its base units the centimetre and the gram, the corresponding coherent units were the dyne and the erg, respectively the force that causes 1 gram to be accelerated at 1 centimetre per (1) second per (1) second, and the work done by 1 dyne acting over 1 centimetre. So 1 newton = 10^5 dyne, 1 joule = 10^7 erg, making each of the four compatible in a decimal sense within its respective other system, but not coherent therein.

$\quad$ The calorie is doubly a confusion. It raises the temperature of 1 gram of water 1 kelvin (degree Celsius), making it obviously not coherent within the SI, where the kilogram rather than the gram is the base unit for mass. But even within c.g.s., where the gram was the base unit, it was not truly coherent because of the intermediate water element; 1 calorie = 4.18~joule = 4.18~ m^2·kg·s^{-2} = 41.8~ cm^2·g·s^{-2} = 41.8~ erg.

$\quad$ A prime example of non-coherence within a single system would be the †horse power of the †f.p.s. system (with base units of foot, pound, and second), one horse power being 550 foot·pounds per second, but the distinctively metric units litre (1 cubic decimetre = $\frac{1}{1000}$ m^3) and †are (1 square decametre = 100 m^2) are not coherent in any version of metric except, in the case of the litre, a barely touched system using the decimetre, gram, and second as its base units.

colour, color What is commonly referred to as colour is technically **hue**, as distinct from **chroma** (the purity of a colour) and **colour value** (the component greyness amount).[37] All three must be considered together for complete specification of 'colour'. Among pertinent methodical schemes[38] are the †Munsell system of 1907 (primarily for art and used heavily in the printing industry), the **Ridgway** system of 1912 (for ornithology),[39] the †Ostwald system of 1918, the †CIE system and †ISCC-NBS system. Both the names and the significance of colours have been subject to standardization.[40]

combining weight *mass See* equivalent weight.

common logarithm (**Briggsian logarithm**) The †logarithm to the base 10, i.e. the power to which 10 must be raised to equal the given number, written for some number x as $\log_{10} x$ but more usually as just $\log x$.

compass point *plane angle* $\frac{1}{32}$ revolution ($= \frac{1/4}{16\,\text{rad}}$, $11° \ 15'$); *see* point.

complex number *See* number.

Compton wavelength [A. H. Compton; USA 1892–1962] *fundamental constant*. Symbol λ_c. A critical distance below which certain quantum-mechanical effects take place for any particle. For the electron as a reference constant, $= h/m_e c = 2.426\,310\,215(18) \times 10^{-12}$ m with †relative standard uncertainty 7.3×10^{-9}.[4]

The crossed symbol λ denotes the **Compton wavelength over 2 pi** (i.e. 2π) $= 386.159\,264\,2(28) \times 10^{-15}$ m, which is the †natural unit of length.

conductance quantum *fundamental constant*. Symbol G_0. $7.748\,091\,696(28) \times 10^{-5}$ S, with †relative standard uncertainty 3.7×10^{-9}.[4]

constant of gravitation *See* Newtonian gravitational constant.

Coordinated Universal Time *See* Universal Time.

1975　15th CGPM: '*considering* that the system called "Coordinated Universal Time" is widely used, that it is broadcast in most radio transmissions of time signals, that this wide diffusion makes available to the users not only frequency standards but also International Atomic Time and an approximation to Universal Time (or, if one prefers, mean solar time), *notes* that this Coordinated Universal Time provides the basis of civil time, the use of which is legal in most countries, *judges* that this usage can be strongly endorsed.'[8]

cord *volume UK, North America* For firewood, 128 ft³ (3.625~ m³), being a stack 8 ft long × 4 ft high of 4 ft logs; rare now in UK (where it was also applied to quarried material), but common in North America. A **cord foot** is one foot along such a stack.

The term **face-cord** applies more generally, for logs of any specific length, to a stack that has the same length and height. Technically, any fraction of a cord can indicate any appropriate fractioning of one or more of the pertinent dimensions; however, a third of a cord would normally imply a face-cord of 16 in logs.

coul (informal), **coulomb** [C. A. de Coulomb; France 1736–1806] *quantity of electricity, electric charge.* Symbol C. The coulombs of electricity transported or of charge delivered by a †steady current equal the product of current in amperes and time in seconds. But the Coulomb was defined as a base unit for the †e.s.u. system.

SI, Metric-m.k.s. = s·A. The following are among the coherent derived units:
- C·m⁻² for electric flux density, surface density of charge, electric polarization;
- C·m⁻³ for volumic charge, charge density;
- C·kg⁻¹ for exposure to X- or gamma-rays;
- C·V = joule for energy, work, quantity of heat;
- C·V⁻¹ = farad for electric capacitance.

This coulomb is equatable with 6.241 45~ × 10¹⁸ electrons.

Metric-c.g.s. See abcoulomb; statcoulomb. *See also* practical unit.

History
The name 'coulomb' was agreed, along with related units and the use of the †c.g.s. system, in 1881 at the first International Electrical Conference,[33] as the 'quantity of electricity defined by the condition that an ampère gives one coulomb per second', with the implication that there be both an absolute form and a corresponding †practical unit. The former, later discriminated as the abcoulomb, falls within the †e.m.u. system, and is fundamentally definable in terms of purely mechanical units. The **practical coulomb** = 10⁻¹ abcoulomb.

The creation of explicit laboratory specifications of the ampere, ohm, and volt, which were subsequently found to be slightly discrepant from what was intended, gave a slightly altered coulomb as a unit derived from them. The IEC of 1908 covered the discrepancy by adopting the distinct name **international coulomb**. Because of experimental vagaries, the value for conversions is normally referred to as the mean international coulomb, $= 0.99985\sim$ C. There is also the **US international coulomb**, $= 0.999835\sim$ C.

With the implementation of the Metric-m.k.s.A. system in 1948, and its basing of electrical units on an ampere compatible with the original absolute units, the modern coulomb became essentially the old practical coulomb. Sometimes called the **absolute coulomb**, this became identically the coulomb of the SI.

Since quantity of electricity or electric charge is inherently a count of electrons (or equivalently charged particles), the logical practice would be to define the coulomb from the electron, then the volt from coulomb, rather than the existing reverse practice. Were the electromagnetic units being created afresh today, the coulomb would likely be the charge of 10^{18} electrons, i.e. $0.1602\sim$ of the extant coulomb.

countable *See* infinity.

cow-calf unit *agriculture See* livestock unit.

Cowling number [T. G. Cowling; UK 1906–90] *rheology*. Symbol *Co*. Characterizing the steady flow past a fixed object of a conducting fluid (such as a physical plasma) subject to a magnetic field parallel to the direction of flow, the dimensionless ratio of the square of the magnetic flux density to the product of magnetic permeability, volumic mass and the square of the speed of flow. Called also the **second Cowling number** (the first being its product with the magnetic †Reynolds number), it is identically the reciprocal of the square of the †Alfvén number.[61]

c.p.s. *electromagnetics* Cycles per second; *see* cycle.
　　informatics Characters per second; *see* character.

crinal *acceleration* Metric-d.k.s. $1\,\text{dm}\cdot\text{s}^{-2} = 0.1\text{m}\cdot\text{s}^{-2}$ $(0.32808\sim\text{ft}\cdot\text{s}^{-2})$.

crith [Gk: 'barleycorn'] *mass* The mass of 1 litre of hydrogen at †s.t.p., $= 90.6\sim$ mg $(1.40\sim\text{gr})$. Since the mass of any gas is proportional to its †relative molecular mass, which for hydrogen gas is virtually 2, the mass in crith of 1 litre of any gas under standard conditions is essentially half its relative molecular mass ('molecular weight').

crocodile *electric potential* UK 1 MV.

C Series, C4, etc., *paper and printing See* paper size.

cu., cubic The three-dimensional derivative of a linear measure. See relevant latter. *See* note under square; *see also* cc.

cubit *length* A very ancient unit, equated with the length of the human forearm from elbow to outstretched finger-tips, and the most versatile body measure. While very variable between persons, for one person it is consistently close to a quarter of the fathom, i.e. the reach finger-tip to finger-tip of outstretched arms, and close to two foot lengths. However, in most formal systems it has been equated with $1\frac{1}{2}$ feet.[41] The common cubit contained 24 †digits, but the Royal cubit four extra. *See also* ell.

The Great Pyramid of Ghiza was measured in the 1880s by Flinders Petrie as having sides of length 230.25 to 230.4 m, with a mean of 230.355 m. Postulating them as being 250 cubits points to a cubit of about 461 mm and a fathom of about 1.84 m, i.e. very close to today's international †nautical mile, the length of 1 minute of latitude. In turn this indicates that the Ancient Egyptians, nearly 5 000 years ago, measured Earth's radius.

cumec *flow rate* Cubic metre per second $= 1\,\mathrm{m^3 \cdot s^{-1}}$. *Compare* cusec.

cup *volume* Traditionally half a pint, often regarded as 250 mL but, if it is to equal sixteen tablespoons of 15 mL, more appropriately 240 mL (see Table 9).

Table 9

Metric				SI	US-C liq
teaspoon	...	...	...	5 mL	0.169~ oz
3		tablespoon	...	15 mL	0.507~ oz
48		16	cup	240 mL	8.12 ~ oz

curie [P. Curie; France 1859–1906] *radiation physics*. Symbol Ci. The most widely known but now obsolescent unit of radioactivity, originally defined as the amount of radioactivity deriving from 1 gram of radium, but divorced from that element and generalized in 1953 to be the quantity of any radionuclide that had 3.7×10^{10} disintegrations per second (the rounded number of the original definition).[42] The name was originally accepted at the 1910 International Congress of Radiology and Electricity in Brussels,[43] its matching standard developed under the control of Mme. Sklodowska Curie. Succeeded in 1975 by the becquerel,

1 curie = 37 GBq, but the 1978 decision of the †CIPM considering it acceptable to continue to use the curie with the SI still stands.

The curie, as agreed in 1910, was defined technically as the radioactivity of the amount of radon gas in equilibrium with 1 gram of radium.[44] Such radiation is mixed in type, so the number of disintegrations is not the number of particles produced, nor does the generalized definition imply the number of particles produced; *see* becquerel for further discussion.

Although not †coherent with the †metric system in any way, the curie was widely subject to the metric prefixes. In particular, because of the curie being a large unit for very many uses, millicurie (mCi) and microcurie (μCi) were common expressions. The more rounded and appropriately sized rutherford of 10^6 disintegrations per second was accepted in 1949, but has rarely been used.

The curie was not applicable to X-rays and other electromagnetic radiations, the roentgen being the parallel unit for those, but one subsequently overlapping the curie by being applied generally to ionization. Nor was the curie a measure of absorbed dose, i.e. of the amount of energy imparted to matter by the indicated radiation; this was the rem.

For **intensity millicurie** *see* sievert.

1964   12th CGPM: '*accepts* that the curie be still retained, outside SI, as the unit of activity, with the value $3.7 \times 10^{10} \text{s}^{-1}$. The symbol for this unit is Ci.'[8]

Curie temperature scale *temperature* (also *magnetic temperature*) A scale for use close to the †thermodynamic null ('absolute zero'), below the lower limit of the †international temperature scale. It depends on the proportionality of the susceptibility of a paramagnetic material to the †absolute temperature (Curie's law).

cusec *flow rate* Cubic foot per second = $1 \text{ ft}^3 \cdot \text{s}^{-1}$, = $0.028\,317\sim \text{m}^3 \cdot \text{s}^{-1}$.

cwt [Lat: centum + weight] *UK* Hundredweight.

cycle [Gk: 'circle'] One set of ordered events or phenomena that recur without change in their essentials; hence, like one lap of a circle, the passing of one set leaves circumstances apparently unchanged except for the passage of time. For a literal lap of a circle, the terms 'revolution' and 'turn' are usual, e.g. in revolutions per minute for the rotation of an engine, and in ampere·turns for magnetomotive force from the windings of an electric motor. (The expressions two-cycle and four-cycle applied to engines, in contrast, refers to the logical sequence of valve action.)

In much of science the series of phenomena is a continuum of change, e.g. in amplitude of a signal, as with electric alternating current. The standard for electric power in North America is for 60 cycles per second, the household supply cycling from 0 up to $+165$ V, progressively down through 0 to -165 V, then back to 0 to deliver the mean 110 V. (Elsewhere 50 cycles per second is usual, with a higher voltage; *see* r.m.s.) Radio and other electromagnetic radiations, and related features of particle physics, have a similar oscillation or waveform, usually of vastly greater frequency. While such oscillations might be seen as different from the circle, they are usually expressible as trigonometric functions of angular variables that recurrently lap the circle.

The number of cycles per second is the most obvious applicable measure; abbreviated to c.p.s. and vernacularly just 'cycles', this is the hertz. For a travelling entity the quotient speed over hertz gives wavelength; expressed reciprocally to distance, this gives †wave number.

D

D, d As **D**, the †Roman numeral for 500, particularly in post-Roman use, derived graphically from the left half of the letter M (= thousand).
Metric As a lower-case prefix, **d-**, *See* deci-.
Metric Until 1960, as an upper-case prefix, **D-**, *see* deca-.

   time SI **d** *See* day.

   electromagnetics **D** *See* debye; *see also* Q factor for **d** as dissipation factor.

   radiation physics As **D**, a unit of X-ray dosage, $= 10^2$ roentgen.

   astronomy As **D**, *see* astronomical unit.

   informatics In †hexadecimal notation as **D** for 13, the 4th digit after 9.

   biochemistry SI Unofficially but commonly **D**, *see* dalton (also Da).

   music See pitch.

   woodworking For sizes of nails, *see* pennysize.

d- *Metric* As symbol, *see* D, d.

Da *area Metric* To 1960, decare. *Compare* da. Now daa.
   biochemistry SI Unofficially, *see* dalton (also D).

da *area Metric* Deciare. *Compare* Da.

da- *SI 1960 See* Deca-. Was D-.

daa [da- + a] *area SI 1960 See* Decare.

dag *mass SI 1960* Decagram. Was Dg.

daL, dal *volume SI 1960* Decalitre. Was Dl.

dalton [J. Dalton; UK 1766–1844] *mass.* Symbol D. Identically the †atomic mass unit, used mainly for biochemical molecular weights, expressed typically in kilodaltons. *See also* mole.

dam *length SI 1960* Decametre. Was Dm.

daraf [reverse spelling of farad] *electric elastance Metric* The elastance of a circuit in darafs equals the reciprocal of its capacitance in farads, so, for a †steady current, equals the ratio of the rise in the potential produced between its plates in volts to the applied charge in coulombs.

Metric-m.k.s. (symbol usually D) $F^{-1} = V \cdot C^{-1}$, identically $V \cdot (A \cdot s)^{-1}$ and $J \cdot C^{-2}$ ($= m^2 \cdot kg^1 \cdot s^{-4} \cdot A^{-2}$ in base terms).

Metric-c.g.s. See abdaraf; statdaraf. *See also* practical unit.

darcy [H. Darcy; France 1803–58] *physical permeability Metric-c.g.s.* Identically the volume, in cubic centimetres, of liquid having viscosity 1 centipoise passing through 1 square-centimetre surface of porous material in 1 second, under a pressure gradient of one †atmosphere per centimetre of travel.[45] A coarse sand-bed can be over 100 darcy, a very fine one under 10; sandstone is about 1 darcy.

date *See* calendar; Julian date.

day *time*. Symbol d. Any of the various periods associated with one rotation of Earth. As explained under †time, there are various lengths of day. The obvious day is the time between successive high noons or other sundial readings, termed the **apparent solar day**. Because of the parameters of Earth's elliptical orbit, the length of such period varies up to 30 seconds from the nominal 24 hours, the greatest discrepancy being reached as Earth approaches perihelion (its closest point to the Sun, currently near the beginning of January, but nearly half an hour earlier each orbital period; *see* year). The averaged value, the **mean solar day**, has long been defined as 24 hours.

The routine offset value of a sundial from a regular clock for any particular day is computed by the †equation of time. The cumulative effect currently brings the sundial a maximal 16 minutes 23 seconds fast in early November and a maximal 14 minutes 20 seconds slow in mid-February. Around those dates and in mid-May and late July the length of the apparent solar day is the average 24 hours. Sundial readings are correct in mid-April and mid-June, near the start of September and in late December. In June, and more so in December, the solar days are longer than the mean as the solstice is passed, causing the apparent paradox of sunrise being later as the daylight lengthens for the winter hemisphere, and of sunset being later as the daylight shortens for the summer hemisphere (an anomaly that can persist for a week), and is mirrored in a like period preceding the solstice.

Because of the friction of tidal effects (applicable to land as well as water), the rotational speed of Earth is progressively slowing, lengthening the mean day by nearly 15μs per year, using the second now fixed by the atomic clock. The size of that second was set to the second of ephemeris time, related to the year 1900 but equated with $\frac{1}{86400}$ of the mean solar day as observed over the period 1750–1892. Thus it is effectively $\frac{1}{86400}$ of the mean solar day around the year 1820, making the like day of 2001, by accumulation of all those microseconds, about 86 400.0028 s long. Over 365 days, the excess amounts to about 0.75 s, which prompts the inclusion of †leap seconds in †Universal Time.

Since Earth is simultaneously travelling about the Sun, with the same anti-clockwise turning viewed from the (North) Pole Star as its rotation, it has to turn fractionally more than one rotation to bring the Sun back to a corresponding position each day, by the amount that offsets over a year the effective one counter-rotation of orbiting the Sun. Hence the †sidereal day, the time for one revolution relative to the stars and of which there are $366\frac{1}{4}$ in a year, is less than the apparent solar day by about $\frac{1}{366}$ day, i.e. almost 4 minutes. The variation of the sidereal day itself is a very minor fraction of a second, so the prefix 'mean' is usually omitted. Current values are as follows:[10]

> 1 sidereal day = 0.997 269 566 33~ mean solar day
>
> 1 mean solar day = 1.002 737 909 35~ sidereal day
>
> hence
>
> 1 sidereal day = 23.934 470~ mean solar hours
>
> = 23 hours 56 minutes 4.090 53~ seconds (mean solar)
>
> 1 mean solar day = 24.065 710~ sidereal hours
>
> = 24 hours 3 minutes 56.555 37~ seconds (sidereal)

The long-established scale on day is as shown in Table 10. This scale still holds, except that the number of seconds in a minute, hour, and calendrical day can vary slightly; *see* Universal Time.

Table 10

second						
60*	minute					
3 600	60	hour				
86 400′	1440	24	day			
			7	week		
			28	4	vernacular 'lunar month'	
			30±		solar calendar month	
			365±	52±	12	solar calendar year

The ± indicates approximate values; * *see* second for exceptions.

In European tradition, the day starts at midnight, i.e. such that high noon is its midpoint. However, this demarcation is not universal, even within European practice. Astronomers, for obvious reasons, favoured noon as the start (*see* †astronomical day system as well as †calendar for the overall sequential identification of days), and such was also long the practice at sea, the former a half-day behind, the latter the same ahead. (The dating of Capt. James Cook's initial activities in Australasia presents an interesting example. His log changed date at noon, normally putting it ahead of the familiar date. However, having travelled westward from Britain without advancement for any dateline, he was, by then, a day behind overall. Thus his afternoons corresponded to the familiar date, his mornings were behind.)

By natural extension, the term day is used, appropriately qualified for the equivalent periods, for other planets, e.g. the Martian day.

See also degree-day.

dB *See* decibel usually, but *see also* deciboyle.

DBA or **dB(A)**, **dBB** or **dB(B)**, etc. *See* decibel.

Dean number *rheology* A dimensionless quantity relating to momentum transport along a curved pipe, being half the †Reynolds number multiplied by the square root of the ratio of pipe width to radius of curvature.

debye [P. J. W. Debije; Netherlands, Germany, USA 1884–1966] *electric dipole moment*. Symbol D. 10^{-18} statcoulomb·cm,[46] $= 3.335\ 6\sim \times 10^{-30}$ C·m, but the name has been used for the comparably sized

elementary charge × angstrom $= 16.022\sim \times 10^{-30}$ C·m;
elementary charge × Bohr radius $= 8.478\ 4\sim \times 10^{-30}$ C·m.

dec- *Metric* Contracted form of †deca- as in †decare (never of †deci-).

deca- [Gk: 'ten'] E.g. **decagon** = ten angles (hence ten sides). *Metric* (now **da-**, until 1960 **D-**) The 10 multiplier, e.g. 1 decagram = 1 dag = 10 g; contractable to dec- before a vowel, e.g. decare = 1 daa = 10 a.

Though used extensively in Europe, the decametre, decagram, decalitre, and decare are little recognized, so undesirable elsewhere; among them, only decare is legal in the UK.[28]

decade A group of ten items.

time Ten years. As 'the decade' now generally means the ten years varying only in the last digit, e.g. 1990–1999, but classically meant the set one year later. *See* millennium.

physics Sometimes used to mean a range spanning a multiplication by 10, notably for a span of frequencies, especially a range spanning between consecutive powers of 10.

decare [deca- + are] (also **dekare**; **dam** since 1960, **Dm** previously)
area Metric $10 \text{ a} = 1\ 000\text{m}^2$ ($1\ 308.0{\sim}\text{yd}^2$) $= \frac{1}{10}$ hectare. An undesirable term. A legal unit in the UK, the only country using the deca-prefix.

deci- [Lat: 'ten'] Symbol d-. *Metric* The $\frac{1}{10} = 10^{-1}$ multiplier, e.g. 1 decigram = 1dg $= 10^{-1}$g; never contracted before a vowel.

Used extensively in Europe, the decimetre, decigram, and decilitre are little recognized elsewhere; the decigram along with deciare are not legal in the UK.[28]

decibel *electrics, acoustics*, etc. Symbol dB. A comparative measure of power levels, applied to electric signals, sound, etc., introduced in 1924 as the **transmission unit**[47] measured, for powers P_1 and P_2, as
$$= 10 \log_{10} P_1/P_2$$
which means that an increment of 1 unit represents a multiplication of power ratio by $10^{0.1}$, an increase of nearly 26%. The name decibel was in regular use by 1929[48] and agreed, along with the phon, in 1937 at the first International Acoustical Conference[136] as a tenth of the bel defined as
$$= \log_{10} P_1/P_2.$$
Either provides thereby a †geometric scale.

The decibel is very much more usual than the bel because its integer scale is so often appropriate to the precision desired; a change in sound of 1 dB is very close to the discriminatory sense of the human ear.

Where power is proportional to the square of amplitude (or voltage in electrical circuits with the same resistance), the dB figure is identically
$$10 \log_{10} P_1/P_2 = 10 \log_{10} (A_1/A_2)^2 = 20 \log_{10} A_1/A_2$$
where A_1 and A_2 are the respective amplitudes. (This holds exactly only for a pure harmonic signal of a single frequency, but is sufficiently close to true for electrical signals generally to be applied collectively to their mixture of frequencies.)

Although decibel is a relative scale, it is often expressed as though it were absolute, with an implied reference signal. For sound, using the distinguishing label **dBA** but often just as dB, the reference is usually the threshold of human audibility, taken as sound of an amplitude (pressure) of 20 μPa and a frequency of 1 kHz. Sound sufficient to cause pain is of the order of 10^{12} times as strong as the threshold, i.e. increased by 120 dB (to 120 dBA);[49] chronic exposure to anything over 85 dBA (e.g. near non-sound-deadened pneumatic drills or within 1 000 m of a jet aircraft) can cause permanent hearing damage. If either low or high frequencies are attenuated, giving emphasis to mid-range frequencies and beyond, the labels **dBB** and **dBC**, respectively, are used. (Such selective attenuation provides measures corresponding better to human sensitivities, with comparable readings for sounds of different pitches sensed by the ear as equally loud.) In electric communication circuits, **dBW** is relative to a signal of power of 1 watt, **dBm** to a power of 1 mW (hence 0 dBm = −30 dBW). Whatever the setting, good professional practice requires clear specification of the reference signal employed whenever the decibel and similar units are used absolutely. All labels can be written with the final, qualifying letter in brackets.

The decibel has been applied widely to ratios of logarithmic progress, but with considerable confusion when the distinct power versus amplitude situation does not apply, hence with inconsistency as to using 10 versus 20 as the multiplier of the logarithm. Various efforts have been made to adopt a more general term, the **boyle**, brig, **decilog**, and neper being examples. For both the deciboyle (used for pressure) and the completely general decilog,[50] the formula

$$10 \log_{10} p_1/p_2$$

is used for any comparable variables p_1 and p_2 (without consideration of the power-to-amplitude factor, which is notable for deciboyle since amplitude is electrical pressure). The boyle scheme uses the †torr or mm of mercury as its unit, but the characteristic atmospheric pressure of 1 bar (100 kPa, 750.062∼ torr) as its reference.[51] The number of **deciboyles** (coincidentally labelled dB) for pressure p is

$$10 \log (p/750.062\sim) = 10 \log p - 10 \log 750.062\sim = (10 \log p) - 28.751\sim.$$

More fundamentally, since $10 \log_{10} x$ is identically $\log_a x$ where $a = {}^{10}(10) = 1.258\ 925\sim$, it has been proposed that this number be recognized as special, and be called **logit**,[52] the name reflecting the fact that it provides a parallel to the unit in the normal arithmetic sense. Other suggestions include decilu, decomlog, and decilit.[53]

As $10 \log_{10} p = \log_b p$ for $b = 10^{0.1} = 1.258\ 925\ 4\sim$, this number has been called the **standard ratio**, and proposed as a unit.

deciboyle [deci- + boyle] *pressure*. Symbol dB. A rare †geometric scale for expressing pressure;[51] *see* decibel.

decile *statistics* A form of †quantile; divides the range into tenths.

decillion See Table 55 under thousand.

decilog *engineering* A rare †geometric scale of general applicability.[54] *See* decibel.

decimal With divisor/multiplier steps of 10 between units, in contrast with the steps of 2 and its powers for †binary and its compounds, of 12 for †duodecimal, of 16 for †hexadecimal, of 20 for †vigesimal, and of 60 for †sexagesimal, etc. The original †metric system of the 1790s made decimal scaling usual in Europe, though it had then been usual in China for a thousand years. The †SI system, although nominally decimal through its metric heritage, is largely †millesimal, i.e. in steps of 1 000. Steps of 100 are †centesimal.

decimal notation, decimal number, decimal scale The familiar numbering system using the graphic characters 0, 1, ... , 9, deriving from the Indo-Arabic numerals which, employing the zero character as a cypher, allowed the positional scheme that is its essence. Thus 301 represents

$$= 3 \times 100 + 0 \times 10 \ + 1 \times 1$$
$$= 3 \times 10^2 + 0 \times 10^1 + 1 \times 10^0$$

Alternative schemes in use include †binary, †octal, and †hexadecimal.

decimetre-kilogram-second system *See* d.k.s. system.

deg *temperature* Abbreviation for †degree, used notably for expressing a span of degrees rather than a temperature scale reading; *see* Celsius.

degree [Lat: 'step'] Symbol ° generally, also deg. A step in any scale that is actually or figuratively discrete. When used in an ordinal manner, e.g. 'second degree', the degree often relates to gradations of authority, dignity, proficiency, or rank, with higher numeric values being superior. In other contexts, higher values can be seen as inferior. The common stance is that the terminology is open-ended at the numerically high end, leaving first degree tied to the anchor point from which to measure ascent or descent. For questioning, etc., the third degree is the implicit limit.

   length $\frac{1}{360}$ of a great circle of Earth, being 1 degree of longitude along the Equator

$$= 111.296\sim \text{km} \ (69.156\sim \text{mi})$$

else the similar but slightly variable 1° of latitude along a meridian

near the Equator = 110.551~ km (68.693~ mi)
near a Pole   = 111.669~ km (69.388~ mi)
at latitude ϕ = (111.413 cos ϕ −0.094 cos 3ϕ) km.

Also the very variable degree of longitude along an identified parallel of latitude, which ranges progressively down in size from the 111.296~ km at the Equator to zero at the Poles, closely proportional to cos ϕ.

See minute for some derived units; *see also* geographic mile.

plane angle. Symbol also arcdeg, degree of arc. The traditional measure, by definition = $\frac{1}{360}$ of one †revolution = $\frac{2\pi}{360}$ rad = 0.017 453~ rad. Values are unlimited, but any value outside the range 0 to 360 represents identically the angle having the value within that range that differs by an integer multiple of 360. Thus the right angle, being 90°, is identically 450°, 810°, et seq., also −270°, −630°, et seq.

The figure 360 for degrees in the revolution appears to derive from the number of days in the year. The Chinese traditionally have 365$\frac{1}{4}$ degrees in a revolution, one degree representing the average daily change in the rotating celestial scene, making this definition consistent with other aspirations to have 'natural units'. Sumer, Babylon, else their precursors likely had the same initially, but then changed it to be a round number in their terms, specifically to an integer multiple of their number base, 60. The ready division of the circle into sixths, producing the regular hexagon of six equilateral triangles, would have encouraged this six-fold picture, each of the angles of these perfectly symmetric triangles being the base 60°. The 60° angle might be called a 'hexangle', in contrast to the right angle with its peculiar figure of 90°. Since the Babylonian-based scheme of 12 hours to the half-day persists for the clock (despite attempts to decimalize the clock along with length and mass), the use of 360° to girdle Earth is highly convenient, yielding 15° of longitude per hour of solar change. (Lest one think that 360 or 365$\frac{1}{4}$ or the more accurate 365.242~ degrees per revolution is awkward, it should be noted that the SI standard, the radian, has 6.283 2~ units per revolution. This last is, of course, 2π, which places the radian close to the hexangle, which might be regarded as a primitive radian, using the crude but sometime used approximation $\pi \approx 3$; *see* pi and Table 11.)

solid angle *See* square degree.

mathematics The power to which a variable is raised. For multi-term expressions, the maximal aggregate power of the variable(s) in a single term. Thus for the single-variable polynomial

$x^6 + 2x^5 + 3x + 4$

Table 11

					SI	
second	...	...	...	...	4.85~ µrad	
60	minute	...	...	...	291.~ µrad	
3600	60	degree	...	...	17.5~ mrad	
		90	right angle	...	1.57~ rad	
		180	2	straight line	3.14~ rad	$= \pi$ rad
		360	4	2 revolution	6.28~ rad	$= 2\pi$ rad

the degree is 6; for the multi-variable expression

$$x^6 + 5x^4y^3z^2 + 7x^3y^2 + 15y^4z^2$$

the degree is $4 + 3 + 2 = 9$ (being greater than 6, $3 + 2$, and $4 + 2$).

statistics See degrees of freedom.

physics A unit in many scales for measuring temperature, e.g. †Celsius (centigrade), †Fahrenheit, †Rankine (but not now the kelvin).

hydrometry 1912 As universal hydrometer degree, = 100 times the specific gravity. For petroleum products *see* API gravity.

viscometry The unit in the †Engler and †MacMichael systems for viscosity.

hardness of water The proportion of calcium salt, usually calcium carbonate ($CaCO_3$):

English or **Clark degree** = grains of $CaCO_3$ per gallon (1:70 000);

French degree = grams of $CaCO_3$ per hectolitre (1:100 000);

German degree = grams of calcium oxide per hectolitre (1:100 000).

The approximate relations are

5° English = 7° French = 4° German = 70 p.p.m. $CaCO_3$,

these figures being a commonly acknowledged ceiling for softness. Twice this concentration would make it definitely hard; three times is very hard.

photography The unit in the DIN and Scheiner systems for photographic emulsion speed, i.e. †film speed.

geography As degrees of latitude and longitude, the angular offset of a point respectively from the plane of the Equator and the plane of the Greenwich meridian. *See* latitude.

medicine For burns, the respective degrees relate first to the epidermis, the second to the deeper skin tissues, the third to the underlying tissues.

See also Engler degree; proof.

degree-day, degree-hour A measure of cumulative warmth, indicating the heating/cooling requirements of buildings and the growth potential of (botanical) plants, being the cumulative count of degree difference from a defined base temperature. The typical reference base for plants in temperate climates is 15°C; with this, a single period with a maximum of 25°C would give 10 units, as would a run of five periods with a maximum of 17°C. For plants, days below the base count zero, while the greater the value the better. For buildings, where a temperature close to 20°C is typically optimal in moderate climates, days below the base count negatively and indicate energy needed for supplemental heating, while greater positive figures point to energy for cooling. As occupant activity contributes heat, the typical reference base for buildings is also close to 15°C. In the UK it is 15.5°C and, because the focus is heating needs, degree-days are expressed inverted to the above, i.e. positive values for temperatures below reference value.

For the designated period, the value is usually obtained using a single observed temperature during the period. Clearly it thereby provides a crude measure in that hours of from-base temperatures can vary considerably for a given single figure, but finer measurement requires more detailed recording. A continuous record obviously provides a better picture for any period length.

degrees of freedom *statistics* Effectively the count of observations, within a set of such, that could be altered independently without changing one or more statistics relating to the set. For a set of n observations with a known mean, the degrees of freedom $= n - 1$, since all but one could be changed with compensating change only to one other to produce the same mean. If the set is an array with m columns and n rows, and column totals are known, then the degrees of freedom $= (m - 1) \times n$; if both column and row totals are known, the degrees of freedom $= (m - 1) \times (n - 1)$.

deka- [Gk: 'ten'] *Metric See* deca-.

demal *chemistry*. Symbol D. A scarcely used term for a solution with 1 gram-equivalent of solute per cubic decimetre, coined to accommodate the accepted discrepancy between litre and cubic decimetre.[55] *Compare* normal solution.

denier [Lat: denarius] *textiles*. Symbol Td. *See* yarn units.

density The amount of some quantity per unit volume, notably of mass per unit volume, termed distinctively **mass density** but more properly **volumic mass**. Used similarly for amount of substance per unit volume (*see* amagat unit) and for electromagnetic quantities (*see* SI unit for examples). Used also per unit area in the form 'surface density', e.g. $C \cdot m^{-2}$ for surface density of charge.

deuterium unit *sub-atomic physics* The †deuteron energy or the corresponding mass, = 12 prout.[56]

deuteron *sub-atomic physics* The nucleus of the deuterium atom. Values[4] of associated fundamental constants, with †relative standard uncertainty, are:

deuteron magnetic moment (μ_d)	$0.433\,073\,457(18) \times 10^{-26}\,J \cdot T^{-1}$	4.2×10^{-8}
deuteron mass (m_d)	$3.343\,583\,09(26) \times 10^{-27}\,kg$	7.9×10^{-8}
	$= 3\,670.482\,9550(78)\,m_e$	2.1×10^{-9}

dex [decimal exponent] For a pure number, including the ratio of two quantities expressed in the same unit, the exponent of the power of ten that most closely approximates it, hence effectively its (decimal) †order of magnitude.

d factor [dissipation factor] *electromagnetics* See Q factor.

Dg *mass* Metric To 1960, decagram. Now dag. *Compare* dg.

dg *mass* Metric Decigram. *Compare* Dg.

dhrystone [a play on words, whetstone being a predecessor] *informatics* A computer program package representative of 'systems programming' in the 1980s, used to benchmark effective speed of computers.

diamond hardness dynamic test *See* hardness numbers.

diem Day, hence **per diem** = per day, = d^{-1}.

digit [Lat: 'finger'] A numeric character: 1, 2, … , 9, also 0 and, in †hexadecimal notation, the letters A, B, … , F standing beyond 9. Hence the measure for 'length' of a number, e.g. 1 357 has a length of four digits and is a four-digit number. In computer contexts the length may include leading zeros.

All computers work fundamentally in †binary mode, with information represented by a succession of on/off or equivalent pulses. Mathematically these are represented, in binary notation, by strings of 0's and 1's, referred to as binary digits or bits. Various computers of the 1960s and 1970s had their storage designed with bits grouped into decimal or other digits, some designs even allowing for program-specified mixing to deal with the problems of non-decimal currencies and measurement units. Partly through that and partly through the rigours of methodical translation of binary to decimal for output, the word 'digit' came from obscurity into relatively common usage. (However, many people, unfortunately, still refer to a seven-digit telephone number as having seven 'numbers', and to its first digit as being its 'first number'.)

Most modern computers allow numbers to be stored and arithmetically processed internally in a purely binary form, in a decimal form using four bits per digit, and in an 8-bit †character form (the byte) that encompasses alphabetic and punctuation characters along with the familiar decimal digits. (*See also* real number; floating-point number.)

   length Classically a sixteenth of a foot, or a twenty-fourth of a cubit.

   astrophysics As used in expressing the amount of overlap of the Sun and the Moon in an eclipse, a digit represents half the length of the apparent diameter (i.e. a radius length), the two objects being virtually the same apparent size from Earth.

DIN [Ger: Deutsche Industrie-Normen] A general term relating to the official standards of Germany. *photography See* film speed.

Dionysian period [Dionysius Exiguus] (**Great Paschal period**, **Victorian period**) *time* 532 years, being the least common multiple of 19 and 28, the years respectively over which the lunar pattern repeats (*see* Metonic cycle) and over which the days of the week generally repeat within the Gregorian †calendar. Though thus ostensibly being the repetition pattern of the Christian Easter, it fails to be exactly so because the Metonic cycle is imprecise by over an hour and because the 28-year factor is disturbed by the exception to leap years established for most centenary years.

diopter, dioptre *photics* A unit for expressing the focusing or diverging strength of a lens or mirror, being the reciprocal of the focal length measured in metres; for a diverging optical element it is expressed as a negative value, for a converging one as a positive value.

displacement ton The ton used to express the mass of ships, traditionally the long ton of 2 240 lb (1 016.0~ kg), but typically equated with a volume of 35 ft^3 (0.991 09~ m^3), the approximate volume of 1 long ton of salt water.

   Since, by Archimedes' principle, the mass of a floating object equals the mass of water displaced, the mass of a ship can be deduced, as its displacement, from the three-dimensional plans of its hull, cut at the level of flotation. Clearly that level depends on the load added to the fixtures (from hull to engines to gun-turrets). Its full displacement tonnage corresponds to its being fully laden, including with fuel, the level of immersion then according with the Plimsoll line for the pertinent sea. Its displacement when empty of fuel and freight, and of any temporary ballast, is called the lightweight displacement; the difference between that and the full displacement is termed the deadweight tonnage, which is used to express the capacity of freight ships, but is of little relevance for warships. However, a distinct †measurement ton is widely used for assessing the commercial size of a ship for dues and similar purposes. *See also* shipping ton.

dissipation factor *electromagnetics*. Symbol d. *See* Q factor.

distance modulus *astronomy* A simple expression for the distance from Earth of an astronomical object, being the difference between its observed apparent magnitude and its computed absolute magnitude (*see* stellar †magnitude). If m and M are those respective magnitudes, then the modulus is $(m - M)$ and the distance d in parsecs is given by

$$\log_{10}(d) = ((m - M) + 5)/5$$

Used by professional astronomers for distant star clusters and galaxies.

d.k.s. system (Metric-d.k.s., also **D.K.S. system**) *Metric* A version of the general system that has its derived constants relating †coherently to the decimetre, the kilogram, and the second, in contrast with the centimetre, the gram, and the second of the †c.g.s. system and with the metre, the kilogram, and the second of the †m.k.s. system and the SI system. In the d.k.s. system, for example, the unit of force is the crinal, which gives an acceleration of $1\,\mathrm{dm \cdot s^{-2}}$ to a body of 1 kg, just a tenth of the newton of the SI, which gives an acceleration of $1\,\mathrm{m \cdot s^{-2}}$ to a body of 1 kg. The d.k.s. system had little acceptance and is now essentially defunct; the contemporary standard is the SI system using the m.k.s. scheme.

DL *paper and printing See* paper size.

Dl *volume Metric* To 1960, decalitre. Now daL, else dal. *Compare* dl.

dL, dl *volume Metric* Decilitre. *Compare* Dl.

Dm *length Metric* To 1960, decametre. Now dam. *Compare* dm.

dm *length* Metric Decimetre. *Compare* Dm.

Dobson number, Dobson unit [G. M. B. Dobson; UK 1889–1970] *physics* For measurement of areal concentration of a very rarified gas, coined for ozone in the upper atmosphere, the length in multiples of 10 μm of a column of the gas when forced to be at †s.t.p.; numerically the partial pressure of the gas relative to that exerted by a 10 μm column at s.t.p. Typical upper atmospheric ozone values are a few hundred Dobson units, indicating, via †Loschmidt's number, of the order of 5×10^{27} molecules per square metre of Earth's surface. 1 Dobson unit $= 10^{-5}$ atmo-metre.

dollar *sub-atomic physics* USA The degree of departure from critical condition of a nuclear reactor, the value 1 being the threshold of reactivity for a self-sustaining chain reaction. *See also* inhour; nile.

double prime A pronunciation of the symbolic ″ that is essentially two parallel strokes, but usually typed as a single symbol. *See* second.

dr., drachm *mass See* dram.

draconic month [Draco, the dragon, a constellation] *See* month.

dram *mass.* Symbol dr. Distinctively the dram or drachm avoirdupois: $= \frac{1}{16}$ oz av $= 27\frac{11}{32}$ gr $= 1.7718\sim$ g. *See* pound ($= 256$ dr) for more precise values; *see* ounce for scale. The dram was removed from official UK measures in 1985.[28]

    See apothecaries' scale for the distinct dram used for medications.
    weight, force See gravitational system.
    volume See fluid dram.

drex *textiles See* yarn units.

drill size *engineering* The effective diameters in millimetres and inches of 'numbered' drill sizes (labelled #) are given in the folded Table 12.

drop *volume* Essentially the smallest amount of liquid that can occur naturally in near-spherical form. It is usually stylized in pointed pear-shaped form, the shape it tends to have as it breaks free from some surface holding it by surface tension, hence its smallness depending on gravitational circumstances. Modern research indicates that the smallest feasible three-dimensional drop or **droplet** of pure water is of 6 molecules hence a volume of about 18 zeptolitres, $= 18$ zL $= 18 \times 10^{-21}$ L.

dry pint *See* pint.

Table 12

mm	#	in	mm	#	in	mm	#	in
0.3429	80	0.0135	2.0828	45	0.082	4.9149	10	0.1935
0.3683	79	0.0145	2.1844	44	0.086	4.9784	9	0.196
0.4064	78	0.016	2.2606	43	0.089	5.0546	8	0.199
0.4572	77	0.018	2.3749	42	0.0935	5.1054	7	0.201
0.5080	76	0.020	2.4384	41	0.096	5.1816	6	0.204
0.5334	75	0.021	2.4892	40	0.098	5.2197	5	0.2055
0.5715	74	0.0225	2.5273	39	0.0995	5.3086	4	0.209
0.6096	73	0.024	2.5761	38	0.1015	5.4102	3	0.213
0.6350	72	0.025	2.6416	37	0.104	5.6134	2	0.221
0.6604	71	0.026	2.7051	36	0.1055	5.7912	1	0.228
0.7112	70	0.028	2.7940	35	0.110	5.9436	A	0.234
0.7417	69	0.0292	2.8194	34	0.111	6.0452	B	0.238
0.7874	68	0.031	2.8702	33	0.113	6.1468	C	0.242
0.8128	67	0.032	2.9464	32	0.116	6.2484	D	0.246
0.8382	66	0.033	3.0480	31	0.120	6.3500	E	0.250
0.8890	65	0.035	3.2639	30	0.1285	6.5278	F	0.257
0.9144	64	0.036	3.4544	29	0.136	6.6294	G	0.261
0.9398	63	0.037	3.5687	28	0.1405	6.7564	H	0.266
0.9652	62	0.038	3.6576	27	0.144	6.9088	I	0.272
0.9906	61	0.039	3.7338	26	0.147	7.0358	J	0.277
1.0160	60	0.040	3.7973	25	0.1495	7.1374	K	0.281
1.0414	59	0.041	3.8608	24	0.152	7.3660	L	0.290
1.0668	58	0.042	3.9116	23	0.154	7.4930	M	0.295
1.0922	57	0.043	3.9878	22	0.157	7.6708	N	0.302
1.1811	56	0.0465	4.0386	21	0.159	8.0254	O	0.316
1.3208	55	0.052	4.0894	20	0.161	8.2042	P	0.323
1.3970	54	0.055	4.2164	19	0.166	8.4328	Q	0.332
1.5113	53	0.0595	4.3053	18	0.1695	8.6106	R	0.339
1.6129	52	0.0635	4.3942	17	0.173	8.8392	S	0.348
1.7018	51	0.067	4.4958	16	0.177	9.0932	T	0.358
1.7780	50	0.070	4.5720	15	0.18	9.3472	U	0.368
1.8542	49	0.073	4.6228	14	0.182	9.5758	V	0.377
1.9304	48	0.076	4.6990	13	0.185	9.8044	W	0.386
1.9939	47	0.0785	4.8006	12	0.189	10.0838	X	0.397
2.0574	46	0.081	4.8514	11	0.191	10.2616	Y	0.404

dry quart *See* quart.

D Series *paper and printing See* paper size.

D unit *radiation physics* An obsolete term equalling 102 roentgen.

duodecimal [Lat: 'two' + 'ten'] With divisor/multiplier steps of 12, in contrast with the steps of 2 for †binary, 10 for †decimal, 16 for †hexadecimal, etc. Not used by any significant modern methodology, but widely inherited in weights, measures, and currency from the Romans (for whom uncia, the etymological source for 'inch' and 'ounce', came to mean a twelfth), and before that a constituent of the composite †Babylonian numerals.

duty *engineering* BI-f.p.s. An early name for foot·pound.

dwt. *mass* Pennyweight. *See* troy scale.

dyn *mechanics* Metric-c.g.s. *See* dyne. Also prefixed, as in cdyn = centidyne.

dynamical time [calculated by the laws of dynamics] *See* ephemeris time.

dyne [Gk: 'power'] *force*. Symbol dyn. *Metric-c.g.s.* Identically gram·gal, i.e. the force that gives to a mass of 1 gram an acceleration of 1 centimetre per second squared, $= 10^{-5}$ N $(2.248\ 089\ 4\sim \times 10^{-6}$ lb-f) $(= \text{cm·g·s}^{-2}$ in c.g.s. base terms). The following are among the coherent derived units:

- dyn·s·cm^{-2} = poise for dynamic viscosity
- dyn·cm = erg for energy, work, quantity of heat
- dyn·cm^{-2} = barad for pressure
- dyn·s·cm^{-3} = rayl for specific acoustic impedance
- $\text{dyn·cm}^{-1}\text{·s}$ = mechanical ohm for mechanical impedance
- $\text{dyn}^{-1}\text{·cm·s}^{-1}$ = mohm for mechanical mobility.

As the pioneer †coherent unit of force, in place of a weight, introduction of the dyne met learned opposition.[57] The name **large dyne** was proposed for the m.k.s. equivalent, which became the newton.

E, e *Metric* As an upper-case prefix, **E**, *see* exa-, e.g. Eg = exagram.

   mathematics As **E**, initial letter of exponent, *see* floating-point number. As **e**, the base of 'natural' logarithms (*compare* common logarithms). More definitively, the unique transcendental †number such that the derivative of the function e^x relative to x is identically e^x, hence likewise the integral. Expressed in simple graphical terms, the uniqueness is that the graph of e^x is identically the graph of its own slope values, and of its own accumulating area covered. A precise valuation is given by the series

   $$1 + \tfrac{1}{1!} + \tfrac{1}{2!} + \tfrac{1}{3!} + \cdots$$

where ! represents the mathematical factorial expression, i.e. 2! equals the product 2×1, $3! = 3 \times 2 \times 1$, etc., giving

   $e = 2.718\,281\,828\,459\,045\,235\,360\,287\,471\,352\,662\,497\,757\,247\,093\sim$

with reciprocal

   $e^{-1} = 0.367\,879\,441\,171\,442\,321\,595\,523\,770\,161\,460\,867\,445\,811\,131\sim$.

   sub-atomic physics As **e**, *see* elementary charge.

   radiation physics See n unit.

   informatics In †hexadecimal notation as **E** for 14, the 5th digit after 9. For use such as 0.9876E+12, *see* floating-point number.

   music See pitch.

E- *Metric* As symbol, *see* E, e.

Earth mass *astrophysics* The usual reference mass for comparative indication of planetary masses, being the mass of Earth, $= 5.974\,2\sim \times 10^{24}$ kg ($6.585\,4\sim \times 10^{21}$ short ton). A new, more elaborate, but unconfirmed evaluation[127] gives $5.972\,245(82) \times 10^{24}$ kg.

eclipse year *astronomy* The period over which the intersection line with the ecliptic of the tilted orbit of the Moon passes, in the one bearing, through the Sun, i.e. akin to the equinoctial †year of the tilted equatorial plane. However, the gravitational influence of the Sun causes the lunar orbital plane to gyrate relatively quickly, making this period only $346.620\,05 \sim$ mean solar †days. Since eclipses, of the Sun by the intrusive Moon and of the Moon by shadowing Earth, depend on near alignment of the three bodies, the (half) eclipse year is crucial, along with the synodic †month, for recurrence of eclipses. *See also* saros.

Eddington number [A. S. Eddington; UK 1882–1944] *physics* 10^{79}, the estimate by Eddington of the number of particles in the Universe. Somewhat oddly, this is only 1 away in exponent terms from the square of 10^{40}, which Paul Dirac[58] subsequently showed to be the characteristic range of size in the Universe, e.g. of the radius of the Universe to that of the electron, the Coulomb force between proton and electron to the gravitational force between them.[59]

E_h *sub-atomic physics See* Hartree.

Ei- *informatics See* exbi- , e.g. as **EiB** = exbibytes, **Eib** = exbibits.

eighth *music See* interval.

eighth gram, eighth metre *Metric See* tenth gram.

eighth-mile *length* Usually called simply eighth in North America, furlong in other English-speaking contexts (for both *see* furlong).

8va *music See* octave.

8vo *paper and printing See* octavo; *see also* paper size.

einstein [A. Einstein; Germany, USA 1879–1955] *physics* Sometimes used for speed relative to the †speed of light in vacuum.

   chemistry (**einstein unit**) A unit of energy absorbed in a photochemical reaction, being †Avogadro's number times the energy of a quantum of the radiation (i.e. its frequency multiplied by the †Planck constant).[60] One einstein unit is the amount absorbed by one gram-molecule of target material.

electric constant *physics.* Symbol ε_0. *SI* The adopted value for electric †permittivity of vacuum $= (4\pi \cdot c^2)^{-1} \times 10^7 \text{F} \cdot \text{m}^{-1} = 8.854\,2\sim \times 10^{-12}$ $\text{C}^2 \cdot \text{m}^{-1} \cdot \text{J}^{-1}$.

electromagnetic centimetre *electromagnetic inductance* Metric-c.g.s.-e.m.u. An old expression for what is the abhenry, $= 1$ nH.

electromagnetic spectrum *physics* Electromagnetic waves (light, radio waves, electricity, x-rays, etc.) cover a continuum of frequencies, increasingly exploited technologically by mankind and detected in nature. The main types, with indications of usage, are shown broadly in Table 13, grouped by †decades. Naturally occurring waves of many

Table 13

Frequency	Wavelength	Nomenclature	Source
Hz	m		
10^{23}	$2.99792458 \times 10^{-15}$		cosmic photons, outer space
10^{22}	$2.99792458 \times 10^{-14}$		gamma rays, radioactive nuclei
10^{21}	$2.99792458 \times 10^{-13}$		gamma rays, x-rays
10^{20}	$2.99792458 \times 10^{-12}$		x-rays, inner shell of atoms
10^{19}	$2.99792458 \times 10^{-11}$		x-rays, positron-electron annihilation
10^{18}	$2.99792458 \times 10^{-10}$		ultraviolet, x-rays, atoms in sparks
10^{17}	$2.99792458 \times 10^{-9}$	ultraviolet	atoms in sparks and arcs
10^{16}	$2.99792458 \times 10^{-8}$	ultraviolet	atoms in sparks and arcs
10^{15}	$2.99792458 \times 10^{-7}$	visible light	atoms, hot bodies, molecules
10^{14}	$2.99792458 \times 10^{-6}$	infrared	hot bodies, molecules
10^{13}	$2.99792458 \times 10^{-5}$	infrared	hot bodies, molecules
10^{12}	$2.99792458 \times 10^{-4}$	far-infrared	hot bodies, molecules
10^{11}	$2.99792458 \times 10^{-3}$	microwaves	electronic devices
10^{10}	$2.99792458 \times 10^{-2}$	microwaves, radar	electronic devices
UHF 10^{9}	$2.99792458 \times 10^{-1}$	radar, radio (UHF t.v.)	electronic devices, 'space'
VHF 10^{8}	2.99792458	radio (FM and VHF t.v.)	electronic devices
HF 10^{7}	29.9792458×10^{1}	shortwave radio	electronic devices
MF 10^{6}	299.792458×10^{2}	radio (AM)	electronic devices
LF 10^{5}	2.99792458×10^{3}	long-wave radio	electronic devices
VLF 10^{4}	2.99792458×10^{4}	induction heating	electronic devices
10^{3}	2.99792458×10^{5}		
10^{2}	2.99792458×10^{6}	power	rotating machinery
10^{1}	2.99792458×10^{7}	power	rotating machinery
1	2.99792458×10^{8}		commutated direct current
0	infinite	direct current	storage batteries

frequencies reach Earth's surface, from outer space, in the solar wind, through the interaction of such waves with Earth's magnetosphere, and from atmospheric storms; natural waves below 1 kHz have been recorded. *See* UHF, VHF, HF, MF, LF, and VLF; *see also* visible light.

electromagnetic unit (EM unit, e.m. unit) *Metric-c.g.s.* While various units of the †e.m.u. system have specific names obtained by prefixing more familiar names with 'ab', most are often referred to purely descriptively as the 'e.m. unit of capacitance', etc. The relevant units are, by subject and showing both distinct name and the SI equivalent of value, one e.m. unit of:

capacitance	abfarad	$= 1\,GF$
electric charge	abcoulomb	$= 10\,C$
electric conductance	abmho	$= 1\,GS$
electric current	strength abampere	$= 10\,A$
electric potential	abvolt	$= 10\,nV$
electric resistance	abohm	$= 1\,n\Omega$
inductance	abhenry	$= 1\,nH$
magnetic field strength	oersted	$= \frac{1}{4\pi}\,abampere \cdot turn \cdot cm^{-1}$
magnetic flux	maxwell	$= 10\,nWb$
magnetomotive force	gilbert	$= \frac{1}{4\pi}\,abampere \cdot turn$

electron *sub-atomic physics* Values[4] of associated fundamental constants, with †relative standard uncertainty, are:

electron charge $1.602\,176\,462(63) \times 10^{-19}$ C 3.9×10^{-8}
                                  $= -1.0 \times$ elementary charge,

electron gyromagnetic ratio (γ_e) $1.760\,859\,794(71)$
                                  $\times 10^{11}\ s^{-1} \cdot T^{-1}$ 4.1×10^{-8}

electron magnetic moment (μ_e) $-9.284\,763\,62(37)$
                                  $\times 10^{-24}\ J \cdot T^{-1}$ 4.0×10^{-8}

electron mass (m_e) $9.109\,381\,88(72) \times 10^{-31}$ kg 7.9×10^{-8}

classical electron radius $2.817\,940\,285(31) \times 10^{-15} \cdot$ m 1.1×10^{-8}

For relative masses, *see* neutron; proton.

electron volt (originally **equivalent volt**) *fundamental constant.* Symbol eV. The kinetic energy received by 1 electron or other †elementary charge moving through a potential of 1 volt, $= 1.602\,176\,462(63) \times 10^{-19}$ J with †relative standard uncertainty 3.9×10^{-8}.[4]

electrostatic centimetre *electromagnetic capacitance Metric-c.g.s.-*

e.s.u. An old expression for what is the the stafarad, $= 1.112\,6\sim$ pF $= 10^{-3}$ jar.

electrostatic unit (ES unit, e.s. unit) *Metric-c.g.s.* The various units of the †e.s.u. system have specific names obtained by prefixing more familiar names with 'stat', but are often referred to purely descriptively as the 'e.s. unit of capacitance', etc. The relevant units are, by subject and showing both distinct name and the SI equivalent of value, one e.s. unit of:

capacitance	statfarad	$= 1.112\,650\sim$ pF
electric charge	statcoulomb	$= 333.5641\sim$ pC
electric conductance	statmho	$= 1.112\,650\sim$ pS
electric current	statampere	$= 333.5641\sim$ pA
electric potential	statvolt	$= 299.7925\sim$ V
electric resistance	statohm	$= 898.7552\sim$ TΩ
inductance	stathenry	$= 898.7552\sim$ TH
magnetomotive force	statampere·turn	$= 3.335\,641\sim \times 10^{-10}$ A·turn

For the numeric values deriving from c, the speed of light: *see* e.s.u. system.

elementary charge (also **proton charge**) *fundamental constant.* Symbol e. The charge of the proton (and identical, except for having positive sign, to that of the electron), $= 1.602\,176\,462(63) \times 10^{-19}$ C with †relative standard uncertainty 3.9×10^{-8}.[4] The †atomic unit of electric charge and the fundamental unit for the electromagnetic domain; *see* base unit.

elite *paper and printing* Applied to a typewriter of fixed pitch, describes a type size equivalent to 10 point that has 12 characters per inch, and which looks superior to the standard 10 characters per inch of †pica (but retains the 12-point height of pica, giving the same 6 lines per inch).

ell [ulna, one of the two long bones of the forearm, a bone also called the cubitus, which gives †cubit] *length* An old body measure based on the human arm. Although suggestive of the forearm alone, British practice corresponded to the whole arm plus some fraction of the chest, hence a yard or more. It was the reference unit for the old measure in Scotland. Modern usage is primarily with textiles.

See also cubit.

textiles UK Three named sizes, each a multiple of the †quarter unit of 9 in.

Flemish ell = 27 in (0.685 8 m) = outstretched hand to near shoulder,
English ell = 45 in (1.143 m) = outstretched hand to opposite shoulder,

French ell = 54 in (1.371 6 m) = outstretched hand to opposite elbow.

em *paper and printing* A measure of type body size, = 2 en = 12 points, being the width of the letter m in a standard reference text. The em, technically the †pica em in the British scheme and the cicero in the general European scheme, is minutely less than a sixth of an inch in the relevant scale.

EM unit, e.m. unit *electromagnetics* Metric-c.g.s. *See* electromagnetic unit.

e.m.u. system *electromagnetics* Metric-c.g.s. One of two widely used schemes of electrical units developed in the 1860s and 1870s within the defunct †c.g.s. system, the other being the †e.s.u. system. The unitary base of the e.m.u. system is defined by a theoretical magnetic pole that exerts a force of 1 dyne on an identical pole 1 centimetre distant in a vacuum, with a magnetic permeability value set at 1. Thus all the electrical units are based on purely mechanical units; none are base units, such as the ampere is in the SI. All units are derived by established relational equations, with the adoption of value 1 for magnetic permeability; most are labelled with familiar metric names prefixed by 'ab' (from absolute); all can be referred to as '†e.m. unit of . . . ' or '†electromagnetic unit of . . . '.

The e.s.u. units have the same definitions and relationships but with dielectric permittivity set at 1. Since the product of dielectric permittivity and magnetic permeability inherently equals the square of c, the speed of light, the corresponding units of the respective systems are related by various powers of c (established during the c.g.s. era as $2.997\,930 \sim \times 10^{10}$ cm·s^{-1}). *See* e.s.u. system for the cross-system numeric factors.

Reference should be made to electromagnetic unit or to particular unit names for relevant definitions and SI equivalences.

These absolute units were mostly far from the amounts occurring in everyday work, so were augmented by decimal-power multipliers to become †practical units, introducing the now-familiar unqualified names:

1 ampere	= 10^{-1} abampere
1 volt	= 10^{8} abvolt
1 ohm	= 10^{9} abohm
1 farad	= 10^{9} abfarad
1 henry	= 10^{9} abhenry

The basis of definition was changed to an ohm expressed in terms of the resistance of a specified column of mercury (Hg ohm), with an equivalent ampere (Ag ampere) expressed as a depositional rate of silver. In 1901 it was also noted that, were permeability set at 10^{-7} rather than 1, the whole array of terms fitted an m.k.s. scheme without those decimal multipliers. However, at the same time, the new laboratory definitions were found to be at minor variance with the previous ones. The consequence was agreement in 1908 on distinctively labelled †international units, with

1 international ohm = 1.000 49~ abohm
1 international volt = 1.000 34~ abvolt

These were in turn discarded in 1948 for the new definitions within the †m.k.s.A. system, which became those of the SI.

EN [European Norm] A general term relating to official standards of the European Union.

en *paper and printing* A measure for type body size, = 6 points = $\frac{1}{2}$ em.

engineering mass unit (hyl, metric slug, mug, par, TME)
engineering Metric-m.k.s. Within the †gravitational system, the mass accelerated 1 metre per second per second by 1 kilogram-force, = g kg, where g is the acceleration of free fall, in $m \cdot s^{-2}$; 1 TME = 9.806 65 kg (21.633~ lb).

Engler degree [C. O. V. Engler; Germany 1842–1925] *viscosity Europe*
The ratio of the time for 200 mL of an oil to flow through a specified orifice, relative to the time for distilled water to do the same, expressed as degrees Engler; 6° Engler is roughly 1 centipoise (1.~ mPa·s).

enzyme unit (International Union of Biochemistry unit)
biochemistry The amount of an enzyme that catalyses the transformation of 1 micromole of substrate per minute (under defined conditions of pH, concentration, and temperature) interpreted as one microequivalent for catalysis involving multiple bonds. If the enzyme is identically the substrate compound, the appropriate amount is two micromoles. 14.28 enzyme units = 1 Sumner unit.

eon *geology* The largest unit of the †geochronologic scale, being often thought of as a billion (10^9) years (and proposed by Gamow to be exactly that[81]). Deriving etymologically from 'life', eon, in the form of the current eon (the Phanerozoic, lasting about 570 million years), encompassed the whole fossil record when it was introduced. Preceding time was lumped together as the undifferentiated lifeless 'pre-Cambrian' eon of 4 billion years. Now that too is seen as including life, and is divided. The next smaller unit is era.

eötvös [R., Baron von Eötvös; Hungary 1848–1919] *geology*. Symbol E. Used for expressing horizontal gravitational field gradient, i.e. the change in gravitational acceleration, being defined originally as 10^{-9} gal per cm of horizontal traverse, $= 10^{-9}$ $m \cdot s^{-2} \cdot m^{-1}$. With gravitational acceleration at sea level ranging from $9.78 \sim m \cdot s^{-2}$ at the Equator to $9.83 \sim m \cdot s^{-2}$ at the Poles, over a distance of 10 000 km, the average is about 5 E along a meridian. The variation averages zero along a parallel. Local geological conditions can vary these figures by ± 5 E.

epact *time* The (number of) days by which a full solar year exceeds a year of 12 full lunar months, else, particularly in calculations relating to Easter, the age of the Moon (i.e. the number of days after the new moon) at the start of the calendrical solar year. More generally, any length of †intercalary days.

Ephemeris Time (dynamical time, ET, gravitational time, newtonian time) [ephemeris, Gk: 'diary', the almanac *American Ephemerides* in its 1960 edition having first publicized this idea] *time* A scheme for measuring time with a standardized day length (as distinct from the variable length of the natural day), based on an idealized motion for Earth and the Moon. Though introduced in 1960, it used the year 1900 as its reference standard, defined as equalling 31 556 925.974 7 seconds. The scheme uses only the Newtonian laws of dynamics and the theory of gravity, as modified by relativity, hence its synonymous names. The scheme was displaced in 1984 by †Terrestrial Dynamic Time; *see* Universal Time. *See also* day; time.

epoch [Gk: 'pause'] *time* A specific point in time or the interval between two such points, but not as such a measure of time. In astronomy it is usually the single point, otherwise it generally has the extended meaning; in geology it can be millions of years. By contrast, an epoch within the initial cosmic 'big bang' is less than a microsecond.

   geology The fourth-largest unit (following period) of the †geochronologic scale, and the tertiary one into which it, excluding the pre-Cambrian, is divided: examples include Pliocene, Pleistocene, and Holocene. Typical size is about 15 million years.

equation of time The traditional name for the relationship between

Table 14

Day	Jan	Feb	Mar	Apr	May	Jun	Day
1	+3m12s	+13m33s	+12m34s	+4m08s	−2m51s	−2m25s	1
2	+3m40s	+13m41s	+12m23s	+3m40s	−2m59s	−2m16s	2
3	+4m08s	+13m48s	+12m11s	+3m32s	−3m06s	−2m06s	3
4	+4m36s	+13m55s	+11m58s	+3m14s	−3m12s	−1m56s	4
5	+5m03s	+14m01s	+11m45s	+2m57s	−3m18s	−1m46s	5
6	+5m30s	+14m06s	+11m31s	+2m40s	−3m23s	−1m36s	6
7	+5m57s	+14m10s	+11m17s	+2m23s	−3m27s	−1m25s	7
8	+6m23s	+14m14s	+11m03s	+2m06s	−3m31s	−1m14s	8
9	+6m49s	+14m16s	+10m48s	+1m49s	−3m35s	−1m03s	9
10	+7m14s	+14m18s	+10m33s	+1m32s	−3m38s	−51s	10
11	+7m38s	+14m19s	+10m18s	+1m16s	−3m40s	−39s	11
12	+8m02s	+14m20s	+10m02s	+1m00s	−3m42s	−27s	12
13	+8m25s	+14m19s	+9m46s	+44s	−3m44s	−15s	13
14	+8m48s	+14m18s	+9m30s	+29s	−3m44s	−3s	14
15	+9m10s	+14m16s	+9m13s	+14s	−3m44s	+10s	15
16	+9m32s	+14m13s	+8m56s	−01s	−3m44s	+23s	16
17	+9m52s	+14m10s	+8m39s	−15s	−3m43s	+36s	17
18	+10m12s	+14m06s	+8m22s	−29s	−3m41s	+49s	18
19	+10m32s	+14m01s	+8m04s	−43s	−3m39s	+1m02s	19
20	+10m50s	+13m55s	+7m46s	−56s	−3m37s	+1m15s	20
21	+11m08s	+13m49s	+7m28s	−1m00s	−3m34s	+1m28s	21
22	+11m25s	+13m42s	+7m10s	−1m21s	−3m30s	+1m41s	22
23	+11m41s	+13m35s	+6m52s	−1m33s	−3m24s	+1m54s	23
24	+11m57s	+13m27s	+6m34s	−1m45s	−3m21s	+2m07s	24
25	+12m12s	+13m18s	+6m16s	−1m56s	−3m16s	+2m20s	25
26	+12m26s	+13m09s	+5m58s	−2m06s	−3m10s	+2m33s	26
27	+12m39s	+12m59s	+5m40s	−2m16s	−3m03s	+2m45s	27
28	+12m51s	+12m48s	+5m21s	−2m26s	−2m56s	+2m57s	28
29	+13m03s	+12m42s	+5m02s	−2m35s	−2m49s	+3m09s	29
30	+13m14s		+4m44s	−2m43s	−2m41s	+3m21s	30
31	+13m24s		+4m26s		−2m33s		31

Day	Jan	Feb	Mar	Apr	May	Jun	Day

natural solar time, as exhibited by the sundial, and regular clock time, accommodating the variation in the length of the solar day. Table 14

Day	Jul	Aug	Sep	Oct	Nov	Dec	Day
1	+3m33s	+6m16s	+12s	−10m05s	−16m20s	−11m11s	1
2	+3m45s	+6m13s	−7s	−10m24s	−16m22s	−10m49s	2
3	+3m57s	+6m09s	−26s	−10m43s	−16m23s	−10m26s	3
4	+4m08s	+6m04s	−45s	−11m02s	−16m23s	−10m02s	4
5	+4m19s	+5m59s	−1m05s	−11m20s	−16m22s	−9m38s	5
6	+4m29s	+5m53s	−1m25s	−11m38s	−16m20s	−9m13s	6
7	+4m39s	+5m46s	−1m45s	−11m56s	−16m18s	−8m48s	7
8	+4m49s	+5m39s	−2m05s	−12m13s	−16m15s	−8m22s	8
9	+4m58s	+5m31s	−2m26s	−12m30s	−16m11s	−7m56s	9
10	+5m07s	+5m23s	−2m47s	−12m46s	−16m06s	−7m29s	10
11	+5m16s	+5m14s	−3m08s	−13m02s	−16m00s	−7m02s	11
12	+5m24s	+5m05s	−3m29s	−13m18s	−15m53s	−6m34s	12
13	+5m32s	+4m55s	−3m50s	−13m33s	−15m46s	−6m06s	13
14	+5m39s	+4m44s	−4m11s	−13m47s	−15m37s	−5m38s	14
15	+5m46s	+4m33s	−4m32s	−14m01s	−15m28s	−5m09s	15
16	+5m52s	+4m21s	−4m53s	−14m14s	−15m18s	−4m40s	16
17	+5m58s	+4m09s	−5m14s	−14m27s	−15m07s	−4m11s	17
18	+6m03s	+3m57s	−5m35s	−14m39s	−14m56s	−3m42s	18
19	+6m08s	+3m44s	−5m56s	−14m51s	−14m43s	−3m13s	19
20	+6m12s	+3m30s	−6m18s	−15m02s	−14m30s	−2m43s	20
21	+6m15s	+3m16s	−6m40s	−15m12s	−14m16s	−2m13s	21
22	+6m18s	+3m01s	−7m01s	−15m22s	−14m01s	−1m43s	22
23	+6m20s	+2m46s	−7m22s	−15m31s	−13m45s	−1m13s	23
24	+6m22s	+2m30s	−7m43s	−15m40s	−13m28s	−43s	24
25	+6m24s	+2m14s	−8m04s	−15m47s	−13m11s	−13s	25
26	+6m25s	+1m58s	−8m25s	−15m54s	−12m53s	+17s	26
27	+6m25s	+1m41s	−8m46s	−16m01s	−12m34s	+47s	27
28	+6m24s	+1m24s	−9m06s	−16m06s	−12m14s	+1m16s	28
29	+6m23s	+1m07s	−9m26s	−16m11s	−11m54s	+1m45s	29
30	+6m21s	+49s	−9m46s	−16m15s	−11m33s	+2m14s	30
31	+6m19s	+31s		−16m18s		+2m43s	31
Day	**Jul**	**Aug**	**Sep**	**Oct**	**Nov**	**Dec**	**Day**

shows the typical discrepancy of apparent high noon each day of the year relative to 12:00 noon on the true clock for actual longitude, e.g. for 1 July, the Sun is typically overhead at 3 minutes 33 seconds past noon on the true local clock. (The use of time zones means that people living across a band of about 15° of longitude share one standard clock time; correction to true local clock time for the longitude involves adding four minutes to the clock for every degree of longitude that the locality is west of the standard meridian for the time zone, four seconds to the clock for every minute of longitude, else similarly but subtracting if east of the meridian.)

The use of 'daylight saving time' or 'summer time' necessitates a further addition, usually of one hour. Thus for any point on longitude 98°W in the Central time zone of North America, which has standard meridian 90°W, there must be an addition of 32 minutes generally, but an hour and 32 minutes if 'daylight saving' applies: the Sun will be at its zenith there on 1 July at 12:35:33 for places that abjure the 'saving', and at 13:35:33 standard time for places that adopt it.

Because of minor variations in Earth's behaviour, and the exigencies of the calendar, the actual discrepancy for any date can be up to 15 s from the value shown, especially around perihelion (currently near the beginning of January, but nearly half an hour earlier each orbital period – *see* year).

A discrepancy of 16 minutes, almost reached in mid-February and exceeded in November, represents an angular discrepancy of 4° from true south for the Sun at local clock noon.

equinoctial Relating to an equinox, i.e. one of the two points in the year when Earth's axis is perpendicular to the direction of the Sun, making day and night equal in length. For **equinoctial year** *see* year.

equivalent volt *See* electron volt.

equivalent weight *chemistry* The mass in grams of an element else compound that could combine with, else displace, 8 g of oxygen. Thus the equivalent weight of any ion equals the aggregate of the †relative atomic masses of its constituents divided by the valence of the ion, e.g. to three decimal places, 1.008 for H^+, 22.990 for Na^+, 35.453 for Cl^- but $\frac{96.062}{2} = 48.031$ for SO_4^{--}. The definition can be applied with little loss of precision to more rounded figures for other such ions, notably to 1 g of hydrogen and 35.5 g of chlorine. Ions combine, and are produced by ionization, in equal amounts in equivalent terms, prompting the synonym **combining weight**.

If, in place of grams, milligrams are used as the unit of mass, the result is called the milliequivalent (**mEq**). Likewise micrograms gives the

microequivalent or **μEq**. Employing the kilogram gives **kilogram-equivalent**, while **gram-equivalent** can be used to distinguish the original definition. Each is usually expressed as per litre or other volume; the milliequivalent per litre equals the †molality times the valence.

er *radiation physics* [equivalent roentgen] *See* rep.[44]

era *geology* The second-largest unit (following eon) of the †geochronologic scale, and the primary one into which it was originally divided, being, excluding the pre-Cambrian, one of the Cenozoic, the Mesozoic, or the Palaeozoic. These range from about 65 (incomplete) to 330 million years.

   The next smaller unit is period.

erg [Gk: 'work'] *energy, work, quantity of heat* Metric-c.g.s. Identically dyne·centimetre, i.e. work done by 1 dyne acting over 1 centimetre, $= 10^{-7}$J$(= $ cm^2·g·s^{-2} in c.g.s. base terms). Erg can be prefixed, as in cerg $=$ centierg.

erlang (sometimes **traffic unit**) [A. K. Erlang; Denmark 1879–1924] *telecommunications.* Symbol E. A measure of telephone traffic load on a multi-channel route, effectively the mean number of simultaneous calls. Used equivalently with estimated demand for planning, a value of 0.7 per channel often being regarded as maximal for tolerable service quality. Sometimes considered as total call minutes per hour on a route, with 42 ($= 60 \times 0.7$) erlangs per channel maximal for fair likelihood of a caller finding the circuit free. For data transmission over a leased circuit the erlangs of load equal the seconds of activity per hour divided by 3 600.

ES unit, e.s. unit *electromagnetics* Metric-c.g.s.-e.s.u. *See* electrostatic unit.

e.s.u. system (**electrostatic units**) *electromagnetics* Metric-c.g.s. One of two widely used schemes of electrical units within the defunct †c.g.s. system, the other and more used being the †e.m.u. system. The unitary base of the e.s.u. system is defined as the charge that exerts a force of 1 dyne on an identical charge 1 cm distant in a vacuum. Thus all the electrical units are based on purely mechanical units; none are base units, such as the ampere is in the SI. All units are derived by established relational equations, with the adoption of value 1 for dielectric permittivity; most are labelled with familiar metric names prefixed by 'stat' (from electrostatic), though such a prefix may be omitted within a clear context; all can be referred to as 'e.s. unit of ...' or '†electrostatic unit of ...'.

The e.m.u. system adopts the same basic definitions and relational equations, but has magnetic permeability set at 1. Since the product of magnetic permeability and dielectric permittivity inherently equals the square of c, the speed of light, the corresponding units of the respective systems are related by various powers of c (established during the c.g.s. era as $2.997\,930 \sim \times 10^{10}$ cm·s^{-1}). Thus for:

volt	ab- = c^{-1} stat-	stat- = c^{+1} ab-
so	ab- = $33.356\,35 \sim \times 10^{-12}$ stat-	stat- = $29.979\,30 \sim \times 10^9$ ab-
ampere,		
coulomb	ab- = c^{+1} stat-	stat- = c^{-1} ab-
so	ab- = $29.979\,30 \sim \times 10^9$ stat-	stat- = $33.356\,35 \sim \times 10^{-12}$ ab-
farad,		
mho	ab- = c^{+2} stat-	stat- = c^{-2} ab-
so	ab- = $898.758\,4 \sim \times 10^{18}$ stat-	stat- = $1.112\,646 \sim \times 10^{-21}$ ab-
henry,		
ohm	ab- = c^{-2} stat-	stat- = c^{+2} ab-
so	ab- = $1.112\,646 \sim \times 10^{-21}$ stat-	stat- = $898.758\,4 \sim \times 10^{18}$ ab-

Reference should be made to electrostatic unit or to particular unit names for relevant definitions and SI equivalences, and to e.m.u. system for comparative notes.

ET *astrophysics See* Ephemeris Time.

E unit, e unit *radiation physics See* n unit.

***Eu*, Euler number** [L. Euler; Switzerland 1707–83] *rheology* A dimensionless quantity relating to momentum transport. The ratio of the pressure difference to the product of volumic mass and the square of a representative speed.[61]

Euler's constant *mathematics* Symbol C. The limit as n goes to infinity of the difference between the sum Σr^{-1} for $r = 1, 2, \dots, n$ and $\log n$,
$$= 0.577\,215\,664\,901\,532\,860\,61 \sim.$$

eV *sub-atomic physics See* electron volt.

EVT *viscosity Europe* Equi-viscous temperature.

ex- *SI* Contracted form of †exa-.

exa- [hexa, Lat: 'six'] Symbol E-. *SI* The $10^{18} = 1000^6$ multiplier, e.g. 1 exagram = 1 Eg = 10^{18} g; contractable to ex- before a vowel.

informatics Sometimes 2^{60}, but *see* exbi-.

exbi- [exa- binary] Symbol Ei-. *informatics* The 1 152 921 504 606 846 976 = 2^{60} multiplier, as exbibytes (EiB) and exbibits (Eib). *See* kibi-, exa-.

exponent, exponential *mathematics* The exponent is the part of an expression indicating the power to which a term is raised, i.e. the x in a^x, whether x is an integer, some other number, or some elaborate expression itself. An **exponential function** involves one or more such terms, the members of a †geometric series being simple examples.

The term **exponential series** is also used within mathematics to describe the series that converges on the special number †e.

F, f *Metric* As a lower-case prefix, **f-** , *see* femto-, e.g. fg = femtogram.
   length As **f**, foot in such constructs as †f.p.s. system.
   physics As **F**, *see* Faraday constant.
   electromagnetics As **F**, *see* farad; *see* SI alphabet for prefixes.
   informatics In †hexadecimal notation as **F** for 15, the 6th digit after 9.
Since this is the highest digit in hexadecimal, a string of F's represents a
string (four times as long) of binary 1's, and hence a maximal absolute
value in a given context.
   music See pitch.

f- *Metric* As symbol, *see* F, f.

face-cord *volume* See cord.

Fahrenheit [G. D. Fahrenheit; Poland, Netherlands 1686–1736]
temperature. Symbols deg F, degree F, °F. A scale and a unit of
temperature, its defining points being 32 at the freezing point of pure
water and 212 at its boiling point, thus with 180 degrees between the
two. Readings on the scale are expressed usually as °F, temperature
intervals preferably as deg F or degree F, sometimes F°. The equivalent
'absolute' scale, with identically sized units but its zero at the
†thermodynamic null, is the †Rankine scale.
   See temperature for other scales and conversions between scales.

History
Created by its namesake in about 1712 (perhaps exactly
contemporaneously with the creation of the †Celsius or Centigrade
scale), the Fahrenheit scale was used widely in the English-speaking
world until recently. With the extensive adoption of the †SI system, it is
now only a relic except in the USA, where it remains the prevailing
customary scale.
   The original reference points appear to have been zero at the freezing
point of heavy brine and 96° (nominally 100) at normal human body
temperature. This unusual pattern derives from Newton's duodecimal
mind recommending 12 subdivisions from ordinary freezing point to

body temperature, then †Roemer, with a compatible sexagesimal mind, adopting 60 subdivisions from the freezing of brine to boiling point. Fahrenheit's greater precision prompted a further subdivision on a binary basis, four-fold relative to Roemer and more so relative to Newton.

Fahrenheit, seeking the lowest possible practical temperature, doped his brine with ammonium chloride, so he had his zero clearly below Roemer's. With the four-fold increase over Roemer, the scale became 240 for boiling point, something like 100 for body temperature and 32 for the freezing of normal water. A scale set as described was then further modified to make the boiling point of water exactly 212°, making the freezing-to-boiling point 180 degrees for ordinary water, a number often seen as a coincidence but more likely to have been deliberate, being the number of degrees (of another type) in two right angles or the straight line, and completely in tune with the sexagesimal beginnings. This shift also meant that normal body temperature is somewhat higher than the original 96°, being subsequently regarded as 98.4 else 98.6°. The outcome was a scale that had, essentially by design, the points 0° and 100° corresponding closely to the lower and upper limits of human comfort, an approach which made the scale inherently preferable in many everyday contexts, and would have been as equally effective as the Celsius scale relative to science. However, Celsius was the more familiar to the creators of the metric system; 'the rest is history'.

Fanning friction number [J. T. Fanning; USA 1837–1911] *rheology*. Symbol f. A dimensionless quantity characterizing turbulent isothermal fluid flow in pipes; assuming smooth pipe, typical ranges have $f = 0.046/Re$, where Re is the †Reynolds' number.

farad [M. Faraday; UK 1791–1867] *electric capacitance* Symbol F. The farads of a capacitor equal the ratio of charge in coulombs to the rise in the potential across its plates in volts, identically the number of ampere·seconds per volt of rise in the potential across the plates of the capacitor. For a †steady current $= C \cdot V^{-1}$, and identically $s \cdot A \cdot V^{-1}$. *SI, Metric-m.k.s.A. 1948* ($= m^{-2} \cdot kg^{-1} \cdot s^4 \cdot A^2$ in base terms). The following is among the coherent derived units:

• $F \cdot m^{-1}$ for dielectric permittivity.

The capacitance of a typical piece of equipment is less than 1 μF, resulting in the nF and the pF being common, the latter sometimes called vernacularly a '**puff**'. Because of MF (now correctly the symbol for the massive megafarad) and mf having at some time been common symbols for microfarad, the millifarad (correctly mF) is rarely referred to; quantities that would merit the unit are expressed in microfarads, e.g. 1 200 μF rather than 12 mF.

Metric-c.g.s. See abfarad; statfarad. *See also* practical unit.

History
Sometimes distinguished as the **Latimer–Clark farad**, such a unit had been in use in the UK since 1867 at the size of the current μF, being about the capacitance of 2 000 ft of submarine cable of the day. International agreement came, along with that for related units and the use of the †c.g.s.-system, in 1881 at the first International Electrical Conference,[33] as the 'the capacity defined by the condition that a coulomb in a farad gives a volt', with the implication that there be both an absolute form and a corresponding †practical unit. The absolute form, later discriminated as the abfarad, falls within the †e.m.u. system, and is fundamentally definable in terms of purely mechanical units. The **practical farad** $= 10^{-9}$ abfarad.

The creation of explicit laboratory definitions of the ampere, ohm, and volt, which were subsequently found to be slightly discrepant from those intended, gives a slightly altered farad as a unit derived from them. The IEC of 1908 covered the discrepancy by adopting the distinct name **international farad**. Because of experimental vagaries, the value for conversions is normally referred to as the mean international farad; $= 0.999\,510\sim$ F. There is also the **US international farad**, $= 0.999\,505\sim$ F, defined by Congress in an Act of 1894.

With the implementation of the Metric-m.k.s.A. system in 1948, and its basing of electrical units on an ampere compatible with the original absolute units, the modern farad became essentially the old practical farad. Sometimes called the **absolute farad**, it became identically the farad of the SI.

1946 CIPM '*Farad* (unit of capacitance) The farad is the capacitance of a capacitor between the plates of which there appears a potential difference of 1 volt when it is charged by a quantity of electricity of 1 coulomb.'[8]

Faraday constant *fundamental constant*. Symbol F. The product of the †Avogadro constant and †elementary charge, $= 96\,485.3415(39)$ C·mol^{-1} with †relative standard uncertainty 4.0×10^{-8}.[4]

fathom *length* Originally a body measure, being the distance spanned finger-tip to finger-tip by the outstretched arms and hands.
British/American Usually, and distinctively as the **warship fathom**, 6 ft (1.828 8 m, 72 in), but sometimes $= \frac{1}{100}$ cable's length. Despite the long

history of being 6 feet, the fathom had other meanings even in 19th-century England, being traditionally 5.5 ft (1.676 4 m, 65 in) on merchant ships and both 5 ft (1.524 m, 60 in) and 7 ft (2.136 6 m, 84 in) on fishing vessels. *See* inch for precise sizing.

History

A 'fathom' was used in ancient Egypt, then in ancient Greece, but lost place to the 5-foot pace in Roman practice. As a natural body measure it would have been fairly close to 6 ft or 1.8 m. By chance, such a size produces a figure of a little over 1 000 fathoms for the †geographic mile, i.e. the distance between two points on a meridian separated by 1 minute of latitude. The actual value of this distance is very close to the modern standard for the international †nautical mile (INM), of 1852 m (6076.~ ft), so about 1 013 BI fathoms. The French, measuring Earth in the 1790s to establish the size of the metre, used their fathom (the toise) as their measuring stick.

FBM, f.b.m. [foot board measure] *See* board foot. This abbreviation occurs particularly in North America, also with the multiplicative prefixes m- for thousand, mm- for million.

f·c, f.c. *illumination See* foot·candle.

femt- *SI* Contracted form of †femto-.

femto- [Danish: 'fifteen'] Symbol f-. *SI* The 10^{-15} multiplier, e.g. 1 femtogram = 1 fg = 10^{-15} g; contractable to femt- before a vowel.

fermi [E. Fermi; Italy, USA 1901–54] *sub-atomic physics*. Symbol fm. 10^{15} m, i.e. the femtometre (with which its symbol coincides), appropriate for the scale of the atomic nucleus. An SI-deprecated name.

Fermi coupling constant *fundamental constant*. Symbol G_F. $G_F/(\hbar c)^3 = 1.166\,39(1) \times 10^{-5}$ Ge·V^{-2} with †relative standard uncertainty 8.6×10^{-6}.[4]

f.g.s. system [foot grain second] Lord Kelvin is reputed to have worked in this predecessor to the †f.p.s. system.

Fibonacci numbers [Leonardo Fibonacci of Pisa; c.1170–c.1250] *mathematics* Numbers in a sequence such that, following 1 and 1 (else 0 and 1), each is the sum of its two immediate predecessors, thus:

1, 1, 2, 3, 5, 8, 13, 21, 34, 55, 89, 144, 233, 377, 610, 987, 1597, 2584, ...
The ratio of one to its immediate predecessor progressively

approximates, ever more closely, to the †golden ratio. The sequential members occur in various areas of nature, e.g. the numbers of ancestors of a male bee, perhaps the branching of certain plants, and the spacing of our planets.[62]

Fibonacci, in a book of 1202, brought the Indo-Arabic †numerals, with their zero cypher and decimal point, into European culture.

fictitious year *astronomy See* Besselian year.

fifth *music See* interval.

fifth gram, fifth metre *Metric See* tenth gram.

film speed *photography* The **ASA**, **DIN**, and other ratings on films indicate the rapidity of response of the photographic emulsion to light. All have values increasing with speed, the ASA values being linearly proportional to speed.

The DIN and **Scheiner** schemes are †exponential, expressed in degrees with 6° representing a quadrupling of speed; but the British Scheiner scheme has values 10 above the DIN, the American Scheiner 5 above DIN. The alphabetical **Ilford** scheme is also exponential, with successive letters being doublings (upward from A). The first widely used scheme was the †H and D, which was identical to the ASA except for being multiplied by 25. Table 15 shows comparative values, but the method of assessing differs: ASA and H and D indicate maximal sensitivity, DIN and Scheiner the minimal exposure for forming a detectable image.

fineness *goldsmiths* A measure of proportion, against a maximum of

Table 15

ASA	DIN	H and D	Ilford
1	1°		
4	7°	100	A
16	13°	400	C
64	19°	1 600	E
100	21°		
200	24°		
256	25°	6 400	G
400	27°		
512	28°	12 800	H
1 000	31°		

1 000, notably to express purity of gold. Thus a fineness of 995 indicates 0.5% impurities.

fines *engineering* For ore processing, material sieving < 35 mm.

fine-structure constant *fundamental constant*. Symbol α.
A dimensionless quantity, being half the ratio of the square of the †elementary charge to the product of the permittivity of vacuum, the †Planck constant, and the †speed of light in vacuum, = 7.297 352 533(27) $\times 10^{-3}$ with †relative standard uncertainty 3.7×10^{-9}.[4] The inverse = 137.035 999 76(50).

finite number Any †number in the normal sense. *Compare* infinity.

finsen unit [N. R. Finsen; Denmark 1860–1904] *physics*. Symbol FU.
A measure of the intensity of †ultraviolet rays, pertinent to natural and artificial sun-tanning, 1 FU corresponding to an energy intensity of $10 \, \mu W \cdot m^{-2}$ from rays of wavelength 296.7 nm (1010.$\sim$ THz).

firkin A small cask, for ale or beer = $\frac{1}{2}$ barrel = $\frac{1}{32}$ tun; for wine = $\frac{1}{3}$ tun.

first *music See* interval.
  other For **first degree** *see* degree.

First Point of Aries *astronomy* The celestial point on the extended line passing from Earth through the Sun at the moment of the northern spring equinox, and the accepted starting moment of the tropical year and of the †zodiac (q.v. for further explanation), hence the zero point for expressing †right ascension and celestial longitude (*see* latitude). It is a slowly moving point relative to the stars, originally in the star pattern Aries, which in classical times signified the opening of the year that began at that equinox.

five *Metric* For use applied as a suffix to units, *see* tenth gram.

fixed-point number A number written in the usual way, with the explicit or implicit decimal point at the true break between integer and fractional parts of its value; *compare* floating-point number.

fl. dr. *See* fluid dram.

fl. oz. *See* fluid ounce.

flask *mass* The traditional unit for mercury, = $76\frac{1}{2}$ lb (34.700$\sim$ kg).

floating-point number A †number expressed as a mantissa value qualified by an exponent value, e.g. $(0.123\,456, -4)$ to mean $0.123\,456 \times 10^{-4}$ else $0.123\,456 \times 16^{-4}$, etc., depending on the implied radix, i.e. numbers in which the true value is obtained by floating the decimal (or equivalent) point the indicated number of places, four to the left in the above example, producing $0.000\,0123\,456$ if with a decimal radix. With †normalization of the mantissas to an appropriate set range (typically to the highest †arithmetic value less than 1, as illustrated here), this notation allows retention, within a fixed amount of space, of comparable numeric significance over a wide range of numeric size. In external display, it is usual to adopt decimal radix and the convention of using E preceding the exponent, e.g. $0.987\,6\,E + 12$ to mean $0.987\,6 \times 10^{12}$.

floor (**storey**, **story**) *length* A vernacular measure of height, using the floor-to-floor distance of familiar buildings. Ten feet or 3 metres would be fair equivalents, though the figures for scientific laboratory buildings and for lower floors of hotels are usually distinctly greater.

FLOPS *informatics* [FLoating-point Operations Per Second] A measure, using a standard mix of pertinent instructions, of the effective calculating speed of a computer working with floating-point numbers. Usually expressed as megaflops, MFLOPS.

fluid drachm, fluid dram *volume*. Symbol fl. dr. For apothecaries, $\frac{1}{8}$ fluid ounce, hence the volume of $\frac{1}{8}$ oz avoirdupois or 2 dr avoirdupois of water, but slightly less than the volume of 1 dr apothecaries of water. 1 fl. dr. $= 60$ minims.
BI $3.551\,6\sim$ mL. Removed from official UK measures in 1970.[28]
US-C Solely for liquids, $3.696\,7\sim$ mL.

fluid ounce *volume*. Symbol fl oz, also liq oz. Volume of 1 oz avoirdupois of water.
BI $\frac{1}{20}$ BI pt $= \frac{1}{160}$ BI gallon $= 28.413\sim$ mL – see Table 16(a).
US-C (Symbol also liquid ounce, liq oz) Solely for liquids, $\frac{1}{16}$ US pt $= \frac{1}{128}$ US gallon $= 29.574\sim$ mL; see Table 16(b).

flux unit (**jansky**) *astrophysics* For spectral energy flux in radio-astronomy, $= 10^{-26}$ W·m^{-2}·Hz^{-1}.

fm Femtometre; *see* femto- and metre, also the identically sized fermi.

f number *photography* Originally written f/8, etc., usually pronounced

Table 16(a)

BI				SI	US-C liq
minim	...	...	...	59.2~ μL	0.0200~ oz
60	fluid dram	...	...	3.55~ mL	0.120~ oz
480	8	fl oz	...	28.4~ mL	0.961~ oz
		20	pint	568.~ mL	19.2~ oz

See gallon for upward extension.

Table 16(b)

US-C liq				SI
minim	...	...	...	61.6~ μL
60	fluid dram	...	...	3.70~ mL
480	8	fl oz, liq oz	...	29.6~ mL
		16	pint	473.~ mL

See gallon for upward extension.

just 'eff eight', etc., and now usually written accordingly as f8, etc., a measure of aperture size, hence of light-gathering capacity. Technically it is a measure of relative aperture, the ratio of focal length to diameter of the entrance pupil. Since light-gathering capacity is proportional to area rather than diameter, it is quadrupled by a doubling of f number. Standard f numbers increase progressively by $\sqrt{2} \approx 1.4$, representing a doubling of aperture area. The British range starts at 1, giving values including 2, 2.8, 4, 5.6, et seq.; the traditional European range is offset from this. In contrast, **US number** starts with 1 equalling $\frac{f}{4}$ and proceeds proportional to area, so every doubling of area gives a whole number.

Fo, Fo* *rheology* See Fourier number.

folio *paper and printing* A †paper size, being half of a full sheet, the actual size depending on that of the full sheet.

foolscap *paper and printing* British A †paper size, being that with the full sheet of 13.5 in × 17 in (342.9 mm × 431.8 mm), but used in UK also to mean specifically the †folio (i.e. half) of this, being 8.5 in × 13.5 in, else 13 in (215.9 mm × 342.9 mm, else 330.2 mm), termed **legal size** in N. America.

foot *length*. Symbol ft. *Internat, BI, US-C* 12 in = $\frac{1}{3}$ yd = 304.8~ mm (precisely that since 1959); *see* inch for greater details, including reference to **coast** or **survey foot** of $\frac{12}{39.37}$ m = 304.800 610~ mm. *Canada See also* perche.

volume BI 1985 The cu. ft was removed from official UK measures. *See also* board foot, cord foot, Hoppus foot.

foot board measure *See* board foot.

foot·candle *illuminance*. Symbol fc. *BI* Conceived as the illumination received 1 foot away from an international [†]candle, later re-phrased as lumen per square foot; 1 fc = 1 lumen ft^{-2} (10.763 9~ lux, i.e. lumen·m^{-2}). The composite name should not be punctuated by a hyphen.

foot-grain-second *See* f.g.s. system.

foot·lambert *luminance*. Symbol ft·L. The average luminance of a surface producing 1 foot·candle, i.e. 1 lumen ft^{-2} (3.426 259~ cd·m^{-2}, $\frac{1}{\pi}$ cd·ft^{-2}, 10.763 91~ m·lambert). If this is via reflection rather than emission, the proportion of reflected to received light is called the surface reflection factor. The composite name punctuated by a hyphen is misleading. *See also* lambert.

foot of water *See* head of liquid.

foot·pound (**foot·pound-force**, originally **duty**) *engineering*. Symbol ft·lb. *BI-f.p.s.* The unit of work in the f.p.s. [†]gravitational system, being the work done by 1 pound-force acting over a distance of 1 foot, = 1.355 817 95~ J (32.174 048 9~ ft·pdl). (The corresponding unit in the non-gravitational form is the [†]foot·poundal.) The composite name punctuated by a hyphen is misleading.

The [†]pound·foot, although having the same units and thereby interchangeable with foot·pound, is preferably used distinctively for torque. With torque, the applicable foot is measured perpendicular to the line of the force; with work it is along that line.

foot·poundal *physics*. Symbol ft·pdl *BI-f.p.s.* The unit of work in the non-gravitational form of the system, being the work done by 1 poundal acting over a distance of 1 foot, = 42.140 173 7~ mJ (0.031 080 950 2~ ft·lb-f). (The corresponding unit in the f.p.s. [†]gravitational system is the foot·pound or foot·pound-force.) The composite name punctuated by a hyphen is misleading.

foot·pound-force *See* foot·pound.

foot-pound-second *See* f.p.s. system.

force 8 gale, etc., *See* Beaufort scale.

four *Metric* For use applied as a suffix to units, *see* tenth gram.

Fourier number [J. B. J. Fourier; France 1768–1830] *rheology*
As *Fo*, relating to heat transport, the dimensionless ratio of the product
of thermal diffusivity and a representative time interval to the square
of a representative length.[61]

As *Fo**, the **Fourier number for mass transfer**, relating to transport of
matter in a binary mixture, the dimensionless ratio of the product
of thermal diffusion coefficient and a representative time interval to
the square of a representative length.[61] Identically the ratio of *Fo*
to the †Lewis number.

fourth *music See* interval.

fourth gram, fourth metre *Metric See* tenth gram.

f.p.s. system (FPS system) *BI* A system of units that uses the foot, the
pound, and the second as base units. (*Compare* the m.k.s. system and its
contemporary form, the SI system, with its base units the metre, the
kilogram, and the second. See also †f.g.s. system.)

In a physics context, the f.p.s. system has the pound as a unit of mass;
the distinctive †coherent units include:

- length: ft = foot (= 0.3048 m);
- mass: lb or lb-m = pound-mass (= 0.453 592∼ kg);
- time: s = second;
- force: pdl = poundal = ft·lb-m·s^{-2} (= 0.138 255∼ N, 0.031 081 0∼ lb-f);
- dynamic viscosity: poundal·second per sq. foot = ft·lb-m·s^{-2}
  (= 0.671 968∼ N·s·m^{-2}).

In an engineering context, the f.p.s. system has the pound as a unit of
†weight, i.e. a force (called the †pound-force, lb-f); hence it is a
†gravitational system. The coherent unit of mass is the slug or gee pound
(= 32.174 lb-m = 14.593 880 9∼ kg); all coherent units involving mass are
correspondingly different from the physicists' units.

Fr rheology See Froude number.

frame *photography* A single picture, particularly on a movie film,
where the number of frames shown per second is crucial to the quality of
the movement.

informatics A defined package of information; originally the set of several bits across a magnetic tape, akin to the use in photography, but now any equivalent package, through to the megabyte package needed to represent a graphic image, i.e. frame in the original photographic sense. For telecommunications, frame is used to indicate any collection of bits subject to error detection and other transmission control, in contrast to a pre-set collection of 8-bit bytes or other fixed module.

franklin [B. Franklin; USA 1706–90] *electric charge* Metric-c.g.s.-e.s.u. A name for statcoulomb.[63]

fraunhofer [J. von Fraunhofer; Germany 1787–1826] *electromagnetics* For spectral width, 10^6 times the ratio of line width to wavelength.

French *length* $\frac{1}{3}$ cm, a unit used for shoe sizes in bygone times, but now sometimes used for the width of optical telecommunication fibres.[64] Hence also **French scale** and **charrière scale**.

   music As **French vibration** or **French frequency**, a notation for vibrational frequency, defined in terms of time, mass, and inertia, effectively twice the usual expression of frequency (pitch).

fresnel [A. J. Fresnel; France 1788–1827] *photics* For frequency, identically THz, i.e. 10^{12} Hz.[65]

frigorie *mechanics* Europe For the rate of extracting heat during refrigeration, 1 kilogram · calorie per hour, a trivially small unit for most purposes.

fringe value *optics* See brewster.

frog unit *pharmacology* The amount fatal to a frog.[66]

Froude number (also **Reech number**) [W. Froude; UK 1810–79] *rheology*. Symbol *Fr*. Relating to momentum transport, the dimensionless ratio of the speed of the fluid to the square root of the gravitational acceleration times a length element.[61] Used particularly in naval architecture, it characterizes fluid flow with an open surface. (The name has been used for the square of the above, the ratio of inertia force to gravity force.[67])

ft, ft. *see* foot; for **ft · lb**, *see* foot · pound.

funal *mechanics* Metric-m.t.s. The original name for the sthene.

furlong ['furrow's length'] *length* Internat, BI, US-C (**eighth, eight-**

mile) $\frac{1}{8}$ mile $= 220$ yd $= 5.0292\sim$ m; *see* inch for greater details, including reference to **coast** or **survey eighth** of $220 \times \frac{36}{39.37}$ m $= 201.168\,402\sim$ m.

The furlong was removed from official UK measures in 1985.[28]

G, g *Metric* As an upper-case prefix, **G-**, *see* giga-, e.g. GJ = gigajoule.

mass Metric As **g**, *see* gram, also prefixed variously, as in mg = milligram(s); *see* SI alphabet. Strictly lower-case.

acceleration As **g**, gravity, a representative value of Earth's gravitational acceleration at sea level, often taken as 10 m·s^{-2} (*see* standard gravity). Used, with appropriate positive and negative multipliers, for acceleration generally, especially as experienced in accelerating, braking, turning, and colliding vehicles. Hence g scale. *See also* G; G scale.

It should be noted that the (usually italicized) symbol g stands for the acceleration due to gravity in a general sense, too, i.e. the value particular to a location and altitude.

plane angle France As **ᵍ**, but used in an elevated position, i.e. **ᵍ**, *see* grade.

fundamental constant As **G**, *see* Newtonian constant of gravitation.
electromagnetics As **G**, *see* gauss.
music See pitch.

G- *Metric* As symbol, *see* G, g.

G₀ *electromagnetics see* conductance quantum.

gage *See* gauge.

Gal *physics Metric-c.g.s. See* gal.

gal (also **Gal**) [Galileo Galilei; Pisa, Florence 1564–1642] *acceleration Metric-c.g.s.* cm·s^{-2},[11] (= 10^{-2} m·s^{-2}). Also occurs prefixed, as in cgal = centigal.

There is no equivalent term in the SI; the 1978 decision of the †CIPM considering it acceptable to continue to use the gal with the SI still stands (though it very rarely appears).

gal. *volume See* gallon.

galactic latitude, galactic longitude *astronomy See* latitude.

galactic year (**cosmic year**) *astronomy* About 250 million ordinary years, being the time for the Sun (more correctly the average star in its neighbourhood) to complete one revolution around the centre of our Galaxy, the Milky Way. The computed value depends on †galactocentric distance as well as rotational speed of the Galaxy.

galactocentric distance *astronomy*. Symbol R_0. The radial distance of the Sun from the centre of our Galaxy, the Milky Way, $\approx 8.5 \times 10^3$ parsecs (1.7×10^9 AU, 250×10^{15} km, 28×10^3 ly, 163×10^{15} mi).

gale *See* Beaufort scale.

galilei, galileo [Galileo Galilei; Pisa, Florence 1564–1642] *See* gal; leo.

gallon *volume* The reference unit for most customary volumetric measurement, except for the US dry system based on the bushel (in British tradition equal to eight gallons).

BI (**Imperial gallon**) *1985* 4.546 09 L (1.200 950~ US gal, 277.419 4~ in^3).
1963 4.546 092~ L (1.200 950~ US gal, 277.419 6~ in^3). The base unit for volume, the volume of 10 lb of distilled water of volumic mass 0.998 859 g/mL weighed in air of volumic mass 0.001 217 g/mL against weights of volumic mass 8.136 g/mL, and formally interpreted as equating with 4.545 964 591 L, all referring to the original litre defined by the kilogram of water of 1.000 028 dm^3. See Table 17(a).
1825 The base unit for volume, the volume of 10 lb of water under specific conditions equivalent to but less precise than those above.

US-C liq The base unit for volume, = 231 in^3 (identically the British wine gallon of 1706) = 3.785 411 784 L (0.832 674 2~ BI gal). See Table 17(b).
US-C dry There is no gallon for dry goods; *see* bushel.

Table 17(a)

BI, Australia, Canada, New Zealand, etc.,						SI	US-C liq
fl oz	...	...	...	...	...	28.4~ mL	0.961~ oz
5	gill	...	...	...	...	142~ mL	4.80~ oz
20	4	pint	...	...	...	568~ mL	1.20~ pt
40	8	2	quart	...	...	1.14~ L	1.20~ qt
160	32	8	4	gallon	...	4.55~ L	1.20~ gal
				2	peck	9.09~ L	2.40~ gal
				8	4 bushel	36.4~ L	9.61~ gal

For downward extension, *see* fluid ounce.

Table 17(b)

US-C for liquids						SI
liq oz	...	...	...	...	...	29.6~ mL
4	gill	...	...	...	...	118.~ mL
16	4	pint	...	...	...	473.~ mL
32	8	2	quart	...	...	946.~ mL
128	32	8	4	gallon	...	3.79~ L
				63:2	barrel	119.~ L
				42	petroleum barrel	159.~ L
				63	3:2 hogshead	238.~ L

For downward extension, *see* fluid ounce.

gamma [Anglicized name of the Greek letter 'g'] *mass*. Symbol γ. A one-time popular, but now obsolete and SI-deprecated name for the microgram (μg).

mechanics (more usually **G**) *See* Newtonian constant of gravitation.

electromagnetics Metric-c.g.s.-Gaussian A very small unit of magnetic flux density, appropriate for describing variations in the intensity of Earth's magnetic field,[69] but obsolete; 10^{-5} gauss, = 1nT.

optics, electronics A measure of amplification, usually the ratio of the logarithms of output and input, akin to the decibel.

photography For contrast enhancement, as above for optics. Also a measure of development of an emulsion, dependent on the developer and increasing with time to a limiting value termed 'gamma infinity'. Hence the slope of a response curve for an electronic photomultiplier, relating, for instance, brightness to excitation, e.g. for specifying CRTs.

gas constant *physics* The constant of proportionality in the universal gas equation, = 8.314 472(15)~ J·(K·mole)$^{-1}$ with †relative standard uncertainty 1.7×10^{-6}.[6]

gauge (gage) *railways* The standard gauge for rails is 4 ft 8½ in, a value apparently deriving from the spacing of Roman (and perhaps earlier) chariot wheels, which needed to be consistent for practical management, as their use set tracks literally in stone. Various narrower gauges have been used in steep terrain, to reduce the effort in terracing and other earth works, and in temporary situations, e.g. during the harvesting of sugar cane. A broad gauge, typically of 5 ft 3 in, has been used extensively in flat terrain.

iron and steel A multitude of wire and sheet gauges has been used in the industrial world, each manufacturer having a unique scale in earlier years. Several that still survive are listed in Table 18 with dimensions in inches. (For mass apply 480 lb·ft^{-3} for iron, 481.2 lb·ft^{-3} for steel generally, but US Manufacturers' standards use 41.82 lb·in^{-1}·ft^{-2} for rolled sheet.)

SWG = Standard Wire Gauge, UK official in 1883, standard in 1893
  (**Imperial Wire Gauge**, **British Standard Wire Gauge**);
BWG = **Birmingham Wire Gauge** (**Stubbs Iron Wire Gauge**) of 1884;
(**British**) **Stubbs Steel Wire Gauge** (continues to gauge 80 for 0.013 in);
US Steel Wire Gage (**Roebling Wire Gage**, **Brown and Moen Wire Gage**, **American Steel and Wire Gage**);
AWG = **American Wire Gage**, from the **Brown and Sharp** of the 1850s;
W. & M. Music Wire Gage (USA);
US Standard Steel Plate Gage, authorized in 1893;
U.S. Manufacturers' Steel Sheet Gage;
BG = **Birmingham Gauge** for steel sheet, from 1884.

Modern practice is to use direct measures of thickness, based on the R10″ series of †preferred numbers expressed in millimetres.[70]

Table 18

Gauge	SWG	BWG	StSW	US SW	AWG	W&M	USStd	ManSSBG
15/0								1.000
14/0								0.9583
13/0								0.9167
12/0								0.8750
11/0								0.8333
10/0								0.7917
9/0								0.7500
8/0								0.7083
7/0	0.500		0.4900		0.0087	0.5000		0.6666
6/0	0.464		0.4615	0.5800	0.0095	0.4687		0.6250
5/0	0.432		0.4300	0.5165	0.0100	0.4375		0.5883
4/0	0.011	0.454	0.3938	0.4600	0.0110	0.4062		0.5416
3/0	0.372	0.425	0.3625	0.4096	0.0120	0.3750		0.5000
2/0	0.348	0.380	0.3310	0.3648	0.0133	0.3437		0.4452
0	0.324	0.340	0.3065	0.3249	0.0144	0.3125		0.3964
1	0.300	0.300	0.227	0.2830	0.2893	0.0156	0.2812		0.3532

(contd.)

Table 18 (*contd.*)

Gauge		SWG	BWG	StSW	US SW	AWG	W&M	USStd	ManSSBG
2	0.276	0.284	0.219	0.2625	0.2576	0.0166	0.2656		0.3147
3	0.252	0.259	0.212	0.2437	0.2294	0.0178	0.2500	0.2391	0.2804
4	0.232	0.238	0.207	0.2253	0.2043	0.0188	0.2344	0.2242	0.2500
5	0.212	0.220	0.204	0.2070	0.1819	0.0202	0.2187	0.2092	0.2225
6	0.192	0.203	0.201	0.1920	0.1620	0.0215	0.2031	0.1943	0.1981
7	0.176	0.180	0.199	0.1770	0.1443	0.0230	0.1875	0.1793	0.1764
8	0.160	0.165	0.197	0.1620	0.1285	0.0243	0.1719	0.1644	0.1570
9	0.144	0.148	0.194	0.1483	0.1144	0.0256	0.1562	0.1495	0.1398
10	0.128	0.134	0.191	0.1350	0.1019	0.0270	0.1406	0.1345	0.1250
11	0.116	0.120	0.188	0.1205	0.09074	0.0284	0.1250	0.1196	0.1113
12	0.104	0.109	0.185	0.1055	0.08081	0.0296	0.1094	0.1046	0.0991
13	0.092	0.095	0.182	0.0915	0.07196	0.0314	0.0937	0.0897	0.0882
14	0.080	0.083	0.180	0.0800	0.06408	0.0326	0.0781	0.0747	0.0785
15	0.072	0.072	0.178	0.0720	0.05707	0.0345	0.0703	0.0673	0.0699
16	0.064	0.065	0.175	0.0625	0.05082	0.0360	0.0625	0.0598	0.0625
17	0.056	0.058	0.172	0.0540	0.04526	0.0377	0.0562	0.0538	0.0556
18	0.048	0.049	0.168	0.0475	0.04030	0.0395	0.0500	0.0478	0.0495
19	0.040	0.042	0.164	0.0410	0.03589	0.0414	0.0437	0.0418	0.0440
20	0.036	0.035	0.161	0.0348	0.03196	0.0434	0.0375	0.0359	0.0392
21	0.032	0.032	0.157	0.03175	0.02846	0.0460	0.0344	0.0329	0.0349
22	0.028	0.028	0.155	0.0286	0.02535	0.0483	0.0312	0.0299	0.03125
23	0.024	0.025	0.153	0.0258	0.02257	0.0510	0.0281	0.0269	0.02782
24	0.022	0.022	0.151	0.0230	0.02010	0.0550	0.0250	0.0239	0.02476
25	0.020	0.020	0.148	0.0204	0.01790	0.0586	0.0219	0.0209	0.02204
26	0.018	0.018	0.146	0.0181	0.01594	0.0626	0.0187	0.0179	0.01961
27	0.0164	0.016	0.143	0.0173	0.01419	0.0658	0.0172	0.0164	0.01745
28	0.0148	0.014	0.139	0.0162	0.01264	0.0720	0.0156	0.0149	0.01562
29	0.0136	0.013	0.134	0.0150	0.01126	0.0760	0.0141	0.0135	0.01390
30	0.0124	0.012	0.127	0.0140	0.01002	0.0800	0.0125	0.0120	0.01230
31	0.0116	0.010	0.120	0.0132	0.00893		0.0109	0.0105	0.01100
32	0.0108	0.009	0.115	0.0128	0.00795		0.0102	0.0097	0.00980
33	0.0100	0.008	0.112	0.0118	0.00708		0.0094	0.0090	0.00870
34	0.0092	0.007	0.110	0.0104	0.00630		0.0086	0.0082	0.00770
35	0.0084	0.005	0.108	0.0095	0.00561		0.0078	0.0075	0.00690
36	0.0076	0.004	0.106	0.0090	0.00500		0.0070	0.0067	0.00610

37	0.0068	0.103	0.0085	0.00445	0.0066	0.0064	0.00540
38	0.0060	0.101	0.0080	0.00396	0.0062	0.0060	0.00480
39	0.0052	0.099	0.0075	0.00353	0.0059		0.00430
40	0.0048	0.097	0.0070	0.00314	0.0053		0.00386
41	0.0044	0.095	0.0066	0.00280	0.0051		0.00343
42	0.0040	0.092	0.0062	0.00249	0.0049		0.00306
43	0.0036	0.088	0.0060	0.00222	0.0047		0.00272
44	0.0032	0.085	0.0058	0.00198			0.00242
45	0.0028	0.082	0.0055	0.00176			0.00215
46	0.0024	0.079	0.0052	0.00157			0.00192
47	0.0020	0.077	0.0050	0.00140			0.00170
48	0.0016	0.075	0.0048	0.00124			0.00152
49	0.0012	0.072	0.0046	0.00111			0.00135
50	0.0010	0.069	0.0044	0.00099			0.00107

gauss [K. F. Gauss; Germany 1777–1855] *magnetic flux density*. Symbol Gs, sometimes G. *Metric-c.g.s.-Gaussian 1930* As agreed by the International Electrotechnical Committee in 1930,[71] identically maxwells per square centimetre; technically defined in a three-dimensional system, it corresponds in the SI, with its extra base unit the ampere, to 100 μT. The gauss is quite small by earthly standards, 1 Gs being only about four times Earth's flux density, but it is subdivided, with 1 gauss = 10^5 gamma.

    magnetomotive force UK 1895 Briefly the unit that became the gilbert.

    magnetic field strength Metric-c.g.s. 1900 As agreed at the International Electrical Congress,[72] briefly the unit that became the oersted.

Gaussian gravitational constant *astronomy*. Symbol k.
0.017 202 098 95, used in defining the astronomical unit of length from the mass of the Sun, using Kepler's third law and a day of 86 400 s, its dimensionality being such that k^2 has the dimensions of the †Newtonian constant of gravitation.[10]

Gaussian system (**Metric-c.g.s.-Gaussian**) *Metric-c.g.s.* A hybrid version of the general †c.g.s. system incorporating as electric units those of the †e.s.u. system but as magnetic units those of the †e.m.u. system. Since those two subsidiary systems were based on an assumed dimensionless unit value for respectively the dielectric permittivity and the magnetic permeability of free space, and the product of those two attributes necessarily equals the reciprocal of the square of the speed of light, the Gaussian system has an inbuilt contradiction. However, as long as this scale discrepancy, of approximately $10^{21}\,\text{cm}^2 \cdot \text{s}^{-2}$, was kept in mind, the system was of value for the respective realms of electrical and magnetic units. The composite set is shown under †Gaussian unit.

Note
The term has also been used for the millimetre–milligram–second system used by Gauss for the pioneer †absolute system.

Gaussian unit *Metric-c.g.s.* The units of the †Gaussian system, each, except for the gauss, identically an †electrostatic unit else, if involving magnetism, †electromagnetic unit. While this meant them having specific names tied to those systems, in a Gaussian context they would usually be referred to as the 'Gaussian unit of capacitance', etc. The relevant units are, showing distinct name and SI equivalent value, one Gaussian unit of:

capacitance	statfarad	$= 1.112\,650\sim \text{pF}$
electric charge	statcoulomb	$= 333.564\,1\sim \text{pC}$
electric conductance	statmho	$= 1.112\,650\sim \text{pS}$
electric current	statampere	$= 333.564\,1\sim \text{pA}$
inductance	stathenry	$= 898.755\,2\sim \text{TH}$
electric potential	statvolt	$= 299.792\,5\sim \text{V}$
electric resistance	statohm	$= 898.755\,2\sim \text{T}\Omega$
magnetic field strength	oersted	$= 79.577\,47\sim \text{m}^{-1}\,\text{A}$
magnetic flux	maxwell	$= 10\,\text{nWb}$
magnetic flux density	gauss	$= 100\,\mu\text{T}$

See electrostatic unit for an explanation of the elaborate numeric factors.

Gaussian year *time See* year.

Gay-Lussac [J. L. Gay-Lussac; France 1778–1850] *See* relative volumic mass.

gee pound, geepound *See* slug.

Gemini *astronomy See* zodiac; right ascension.

gemmho [reverse spelling of 'meg' then of 'ohm', from 'megohm'] *electric conductance* Effectively the micromho, i.e. 10^{-6} mho. A circuit is said to have a conductance of n gemmho if it has a resistance of n^{-1} megohm.

generation *time* The typical interval from one generation to the next, ranging from hours for bacteria through days to weeks for very many organisms, to decades. For the humans it is taken to be 25 to 35 years, though procreation can occur from 10 years old to 50 and now 60 for the female, to over 80 for the male.

geochronologic scale (**geologic time scale**) *geology* The time scale, spanning millions of years, deduced from the records of rocks and their embraced fossils. The listing in Table 19 gives an indication of the current consensus, including estimates of starting and finishing times, in years before the present, for each named period and an indication of biological evolution (and resulting extinctions).

Table 19

Eon Era Period Epoch	Years BP	
	4 550 000 000	
Hadean		
	3 500 000 000	
Pre-Cambrian		
Eozoic		First cellular life – marine cyanobacteria
	2 450 000 000	
Archaeozoic		Green algae, freshwater cyanobacteria
	1 500 000 000	
Proterozoic		Terrestrial algae and fungi
	570 000 000	
Phanerozoic		
Palaeozoic		
Cambrian		Fishes
	508 000 000	
Ordovician		Spore-producing plants
	436 000 000	
Silurian		Vascular plants, centi/millipedes, amphibia
	404 000 000	
Devonian		Seed plants, non-flying insects, spiders
	363 000 000	
Carboniferous		
Mississippian		Flying insects
	312 000 000	
Pennsylvanian		Reptiles
	288 000 000	

(*contd.*)

Table 19 (*contd.*)

Eon Era Period Epoch	Years BP	
Permian		
	244 000 000	Extinction of trilobites, rugose corals, etc.
Mesozoic		
Triassic		Mammals
	208 000 000	Extinction of much marine life
Jurassic		Birds
	145 000 000	
Cretaceous ('K')		Flowering plants
	65 000 000	Extinction of dinosaurs, ichthyosaurs, etc.
Cenozoic		
Tertiary ('T')		
Palaeocene		
	57 500 000	
Eocene		Bats, horses, and whales
	36 000 000	
Oligocene		
	23 500 000	
Miocene		Many mammals of modern appearance
	5 000 000	
Pliocene		First *Homo*
	1 800 000	
Quaternary		
Pleistocene		First *Homo sapiens*
	11 000	
Holocene		
	Today	

While well established and widely accepted, the scale undergoes continual revision and subdivision, as a result of on-going research and reinterpretation, and does not exist in a universally accepted detailed form.

The more recent the time, the more detailed the scale becomes, from a billion (10^9) years at the most distant subdivision to barely 10 000 at the most recent. The various subdivisions – eon, era, period, epoch, and age (then sub-age) – are the **geologic time units**. The eon is nominally a billion years, actually a half to two billion; virtually the whole of the detailed time scale lies in the one (current) eon, the Phanerozoic. The other units have no nominal size.

The period prior to the Cambrian, representing 80% of the total time of Earth's existence, is subject to much variety in its division and subdivision. Initially it was collectively the Pre-Cambrian, one undifferentiated period bereft of discernible markers of life or of geological activity. Increasing sophistication of techniques in recent years has changed matters, and progressively this huge span of time is being subdivided, but its subdivisional names are erratic; the above gives but one example, another example is shown in Table 20.

Table 20

Eon Era Period	Years BP
	4 550 000 000
Priscoan	
	4 000 000 000
Archaean	
	2 500 000 000
Proterozoic	
	570 000 000
Phanerozoic	
Palaeozoic	
Cambrian	

geodesic (adj.), **geodesically** (adv.) [Gk: 'Earth' + 'divide'] Applied to units of length, indicates the setting of its size to accord with value obtained by geodesy (the division and measurement of Earth's surface). The †geographic mile and its relatives, notably the †nautical mile, are key examples as is the metre.

See also latitude.

geographic (adj.), **geographically** (adv.) [Gk: 'Earth' + 'writing'] Pertaining to geography, the description of Earth's surface and adjacent atmosphere, both natural and man-made. *See* geographic mile.

geographic mile (meridian mile) *length* Distinct from the statute mile (the familiar mile of English), the geographical mile is a longer unit. Though the common word 'mile' derives from the Latin for one thousand, and the statute mile derives from the Roman mile of a thousand paces, the geographic mile has a very much earlier and different provenance, belonging to dynastic Egypt if not earlier, where it was sized †geodesically. It is the distance along a meridian of 1 minute of latitude, or, more generally, the distance along any great circle that subtends 1 minute of angle at Earth's centre. Because Earth is an oblate spheroid, with equatorial radius 0.34% greater than polar radius, this real distance varies from 1 842.∼ m to 1 862.∼ m. The locally specific size is usually called a **sea mile** in navigation. The geographic mile and the nautical mile are mean values. The latter has long been officially standardized, since 1954 internationally at 1 852 m (6 076.∼ ft). The geographic mile has usually been interpreted as equal to the nautical mile of the map-maker, though the Central Bureau of Longitude promulgated the figure of 1 852 m for it a hundred years ago.[11]

Historically, the geographic mile was often divided into 1 000 fathoms (q.v. for extended discussion). Three miles routinely make 1 league.

The kilometre, when created in 1792, was, within its context, effectively a geographic mile. The metre was defined as $\frac{1}{10000000}$ of the meridional distance from Equator to Pole, making the kilometre equal to $\frac{1}{10000}$ of that distance. Since the decimalization efforts of the time were also applied to angles, but in centesimal rather than truly decimal steps, the 90 degrees of latitude from Equator to Pole became 100 grad, each divided into 100 centesimal minutes, each of 100 centesimal seconds. Thus, within the nascent metric system, there were 10 000 minutes of latitude from Equator to Pole, identical with the number of kilometres.

geologic time scale, geologic time unit *See* geochronologic scale.

geometric [Gk: 'Earth' + 'measure'] *mathematics* The **geometric mean** of *n* numbers is the *n*th root of their collective product; *compare* arithmetic.

Applied to a series of numbers, 'geometric' indicates that adjacent members differ by a constant multiplier, the 'common ratio' (any finite number). The geometric series with common ratio *b* has the form

$$a, \quad a \cdot b^1, \quad a \cdot b^2, \quad a \cdot b^3, \quad \ldots$$

for some value *a*. *Compare* arithmetic.

For measurement scales, geometric means that a step of any one size in the scale value represents a given amount of multiplicative change in the measured item, regardless of place on the scale. The scale for musical †pitch is geometric relative to frequency, essentially so for just intonation but precisely so for the scale of equal temperament, in which a rise of 1 semitone anywhere involves a multiplication of frequency (pitch) by $\sqrt[12]{10} = 1.0595\sim$, a rise of 1 full tone multiplication by the square of this factor, i.e. $1.1225\sim$, a rise of three semitones multiplication by the cube and so on.

Since

$$\log(a \cdot b^n) = \log a + n \cdot \log b$$

†logarithms transform multiplication into addition. Taking logarithms of the above series converts it into the †arithmetic series

$$\log a + 0\log b, \quad \log a + 1 \cdot \log b, \quad \log a + 2 \cdot \log b, \quad \log a + 3 \cdot \log b, \quad \ldots$$

with common difference $\log b$. Thus the modified †savart scale for musical †interval, expressed as

$$\text{number of savarts} = 996.578\sim \times \log_{10}(f_2/f_1)$$

for the interval between frequencies f_1 and f_2, gives 25 savarts for the semitone, everywhere. A rise of 1 full tone equals $25 + 25 = 50$, a rise of 3 semitones equals 75 savarts and so on. Thus, relative to any initial note, the value in savarts for a sequence of notes separated by 1 semitone form an arithmetic series. However, because of what its values represent, the savart scale is referred to as being a geometric scale.

As illustrated by the familiar piano, the notes of music are discrete, i.e. there is some relevant space between them (minor physically, but distinct in frequency terms). However, the range of frequencies that can be generated, e.g. by a violin, the human voice, or a machine, forms a continuum. The logarithmic transformation of the savart could be applied to any interval, very small to very large, were it required, with consistent results. The Ancient Grecian scale for †stellar magnitude had the discrete values 1, 2, 3, 4, 5, and 6. Though an arithmetic series of itself, each step forward in the scale involved a diminishing of brightness by about 60%, hence it was, roughly, a geometric scale. Adoption of a logarithmic basis has allowed consistent representation of any magnitude. The geometric bel scale, in contrast, was initiated on a logarithmic basis; each increase of 1 represents a multiplication of the measured power level by $1.26\sim = \sqrt[10]{10}$.

Most scales for measurement using named units are essentially geometric, though the multiplier might not be constant; many volume scales have successive units doubling, while the traditional division of the various foot units of Europe was by 12, down to inch, then to †line, then to †point. The original †metric system went seven decimal steps, from milli- up to kilo-, then †myria-. Many scales are essentially geometric but not precisely so, convenience of manufacture often demanding compromise (see †preferred numbers).

The various proposed **logarithmic scales of pressure** (logarithmic in this sense being synonymous with geometric) are particular examples, addressing the fact that a change in pressure of a given absolute magnitude depends for its significance on the absolute pressure to which it is applied. These are all akin to the decibel scheme, varying in the units for expressing absolute pressure and the comparative reference pressure. The boyle scheme uses the †torr or mm of mercury as its unit, but the characteristic atmospheric pressure of 1 bar (100 kPa, 750.062~ torr) as its reference. The number of **deciboyles** (coincidentally labelled dB) for pressure p is:

$$10 \log(p/750.062\sim) = 10 \log p - 10 \log 750.062\sim = (10 \log p) - 28.751\sim.$$

For a generic scheme, *see* †preferred numbers.

Gerlach *liquor and food processing See* Baumé.

German R unit *radiation physics See* r unit.

GeV *sub-atomic physics* Giga electron-volt, $= 10^9 eV$.

G$_F$ *sub-atomic physics See* Fermi coupling constant.

gf, g-f *engineering See* gram-force.

GHA [Greenwich Hour Angle] *astronomy See* right ascension.

gibi- [giga-binary] *informatics* Symbol Gi-. The $1\,073\,741\,824 = 2^{30}$ multiplier, as **GiB** = gibibytes and **Gib** = gibibits. *See* kibi-.

gig- *Metric* Contracted form of †giga-, as in gigohm = GΩ.

giga- [Gk: 'giant'] Symbol G-. *SI* The 10^9 multiplier, e.g. 1 gigajoule = 1 GJ $= 10^9$ J; contractable to gig- before a vowel, e.g. 1 gigohm = 1 GΩ = $10^9 \Omega$.
informatics Sometimes $1\,073\,741\,824 = 2^{30}$, but *see* gibi- then kibi-.

gilbert [W. Gilbert; England 1540–1603] *magnetomotive force* Metric-c.g.s.-e.m.u. *1930* the analog of volt of electromotive force, $= \frac{1}{4\pi}$ abampere·turn, $= \frac{10}{4\pi}$ A·turn = 0.795 774 7~ A·turn.

The corresponding †practical unit was the pragilbert, $= 10^{-1}$ gilbert. The mean **international gilbert** = 0.999 85~ gilbert = 0.794 58~ A·turn.

Originally termed the gauss, it became the gilbert soon after 1900, but officially so only by agreement by the International Electrotechnical Committee in 1930.[71] There is no equivalent unit in the SI.

gill *volume* BI, US-C liq $\frac{1}{4}$ pint $= \frac{1}{32}$ gallon.

Giorgi system [G. Giorgi; Italy 1871–1950] *See* m.k.s. system.

positive

gm *mass* Metric A common, but deprecated representation for gram; g is the standard symbol throughout the history of the †metric system.

GMT *time See* Greenwich Mean Time.

G number *engineering See* Wobbe index.

golden number, golden ratio The number 1.618~, unique in that its value equals the ratio of its integer part to its fractional part, i.e.

$1.618\sim = 1 + 0.618\sim = \frac{1}{0.618\sim}$

Meeting this condition requires, for fractional part f, that

$1 + f = \frac{1}{f}$, hence that $f + f^2 = 1$, hence that $f^2 + f - 1 = 0$

which has the sole real solution

$2f = -1 + \sqrt{5}$, or

$f = 0.618\,033\,988\,75\sim.$

Thus the golden ratio, usually represented by the Greek letter phi, is

$\phi = 1.618\,033\,988\,75\sim$

and its reciprocal equals f. Known to Pythagoras and his followers over 2 000 years ago, the ratio is much vaunted as being the perfect proportion for very many facets of art and architecture.

It is inherent in the opening definition that f is the reciprocal of ϕ, i.e. $\phi^{-1} = f$, and that the ratio 1:0.618~ is identically 1.618~:1. Thus, if a line is cut into two sections having mutual proportions equal to the golden ratio, the ratio of the whole line to the larger section is also the golden ratio. Such a cutting is called a **golden section**, with the larger section (obviously 61.8~% of the whole) called the **golden mean**. A rectangle having its length-to-breadth ratio equal to the golden ratio is called a **golden rectangle**. Because of the $\sqrt{5}$ factor, both the golden rectangle and the golden section can be constructed by simple geometric processes, using just a ruler and compasses. The golden section occurs inherently in various geometric situations, e.g. the golden mean of the radius of any circle equals the length of each side of the regular decagon inscribed within the circle.

If a full square is inscribed at one end of a golden rectangle, the residual rectangle is also golden, e.g. if originally a rectangle of 1.618~ × 1 allowing a square of 1 × 1, the residual rectangle would be 0.618~ × 1. Since the residual is golden, the same would apply to it, and ever onwards without limit. This comes from an inherent consequence of the basic definition, for, writing ϕ for $1 + f$ and $\phi - 1$ for f, the original condition translates to

$\phi = 1/(\phi - 1)$, hence that $\phi^2 - \phi = 1$, hence that $\phi^2 - \phi - 1 = 0$.

If we multiply this by ϕ^n for any value of n we get

$$\phi^{n+2} - \phi^{n+1} - \phi^n = 0$$

hence

$$\phi^{n+2} = \phi^{n+1} + \phi^n \qquad \phi^{n+2} - \phi^n = \phi^{n+1} \qquad \phi^{n+2} - \phi^{n+1} = \phi^n.$$

Thus the sum of adjacent powers of ϕ equals the next higher power, the value of any one is the difference between the adjoining ones, and the difference of adjacent powers equals the next smaller; this is unique to the multiplier 1.618~ in the world of †geometric series. For the line, the process is likewise endless in that the lesser section at one step equals the length of the golden mean of its associate.

gon [Gk: 'angle'] *plane angle* See grade.

Gr *rheology* See Grashof number.

gr. *mass* See grain.

grad *plane angle* See grade.

grade *plane angle*. Symbols ^{g}, gon, grad. $\frac{1}{100}$ right angle $= \frac{1}{400}$ turn, $= \frac{\pi}{200}$ rad $= 0.015\,707\,96\sim$ rad (54 arcmin).

Like degrees, grade has a distinct symbol: its initial letter elevated to superscript, e.g. 50^g. It was introduced along with the general metric units following the revolution in France in the 1790s, as part of the decimalization of the time, where it influenced the size of the metre (*see* geographic mile). The unit has never succeeded in displacing the familiar hexagesimal degree outside France, where it is still used. It is subdivided centesimally into †centesimal minutes then centesimal seconds.

Graetz number [L. P. Graetz; USA 1856–1941] *rheology* Symbol Gz. *USA 1941* A dimensionless quantity characterizing the transfer of heat by streamline fluid flow in a pipe. It is the product of the †Reynolds number, the †Prandtl number, and the ratio of the circumference to the length of the pipe.

grain Symbol gr. Adverting to the seed of wheat, barley, etc., the grain is a long-standing elementary unit for measuring mass, and has been used for length too. In Britain the initial entity was the barleycorn, the seed grain of barley, specifically two-rowed barley.

mass The grain, originally the UK troy grain, is the common elemental unit of the †avoirdupois, †troy and †apothecaries' scales in the UK and the USA. Defined originally by barley grains 'neither small nor large and taken from the middle of the ear'[73] the unit of modern times differed minutely between the UK and the USA until the agreed international standard[142] was implemented in 1959 (though in UK not exclusively until 1964).

The grain was officially abolished for the UK in 1985.[28]

Internat, BI, US-C, Australia, Canada, New Zealand, etc.,

    1959 64.798 91 mg (0.323 994 55 metric †carat of 200 mg).

    prev Within 0.01% of the above for centuries; *see* pound for details.

History

The avoirdupois scale (*see* ounce and hundredweight) has a pound of 7 000 grains = 16 oz of 437.5 gr; both the troy and the apothecaries' scales have an ounce of 480 grains, but a pound of only 12 ounces, giving 5 760 grains to the pound. The latter two scales differ in their intermediate subdivision. Comparative values across the three scales are shown in Table 21.

Table 21

gr	av, avdp	troy	ap, apoth	g
1	1 grain	1 grain	1 grain	0.0648~
20			20 gr = 1 scruple	1.30~
24		24 gr = 1 pennyweight		1.56~
27.3~	27$\frac{11}{32}$ gr = 1 dram			1.77~
60			3 scr = 1 drachm, dram	3.89~
437$\frac{1}{2}$	16 dr = 1 oz av			28.3~
480		20 dwt = 1 oz troy	8 dr = 1 oz ap	31.1~
5 760		12 oz = 1 lb troy	12 oz ap = 1 lb ap	373.~
7 000	16 oz = 1 lb av			454.~

weight, force See gravitational system.

gram *mass.* Symbol g. *Metric* The central unit of mass, with all masses being expressed as grams alone else with a standard decimal multiplier else divisor prefixes. Originally defined as the mass of 1 cubic centimetre of water, the gram has, since 1799, been defined as 1 thousandth of the †prototype kilogram in the archives. For Metric-c.g.s. the gram was the base unit, but with the SI the kilogram is the base unit (i.e. the unit with which pertinent other units are †coherent.)

gram-calorie *See* calorie.

gram-equivalent *See* equivalent weight.

gram-force *mechanics*. Symbols gf, g-f, pond. *Metric* The force represented by 1 gram of mass subject to †standard gravity, i.e.

$$1\,\text{g-f} = 1\,\text{g} \times 980.665\,\text{cm}\cdot\text{s}^{-2} = 980.665\,\text{dyn} = 9.806\,65\,\text{mN}.$$

This was a base unit of the †gravitational system created from the normal †c.g.s. system (in which the gram is mass and the dyne the unit of force).

gramme, gramme- *See* gram, gram-, e.g. for **gramme-rad** *see* gram-rad.

gram-mole, gram-molecular weight, gram-molecule *See* mole.

gram-rad *radiation physics* For absorbed dose of radiation, the same amount of absorbed energy as that of the rad, but concentrated in the gram instead of the kilogram, $= 1\,000\,\text{rad} = 10^5\,\text{erg}\cdot\text{g}^{-1} = 10\,\text{Gy}$.

gram-roentgen *radiation physics* For absorbed energy, the amount absorbed by 1 gram of air when receiving 1 roentgen of radiation, $= 8.38\sim\,\mu\text{J}$. The unit is trivially small, absorbed energies being typically of the order of a million such (i.e. 1 megagram-roentgen).

granule *geology* A †particle size, typically 2 to 4 mm (0.787 to 0.157 4 in).

Grashof number [F. Grashof; Germany 1826–93] *rheology*
As *Gr*, relating to momentum transport, the dimensionless ratio of the product of the acceleration of free fall, the cubic expansion coefficient, a representative temperature difference, and the cube of a representative length to the square of the kinematic viscosity.[61] It characterizes free convection in a fluid.
As *Gr**, the **Grashof number for mass transfer**, relating to transport of matter in a binary mixture, the dimensionless ratio of the product of the acceleration of free fall, the cubic expansion coefficient, the difference in mole fraction, the negative of the partial differential of volumic mass relative to positional coordinate, and the cube of a representative length to the product of the volumic mass and the square of the kinematic viscosity.[61]

grave *mass Metric* 1792 The original name for kilogram.

gravel *geology* A †particle size, typically 2 to 64 mm (0.787 to 2.5 in).

gravimetric system (**gravitational metric system**) *See* gravitational system.

gravitational constant *See* Newtonian gravitational constant; Gaussian gravitational constant.

gravitational system (technical system) Any system that uses †weight, in its true sense, rather than mass, as its pertinent base unit is called a gravitational or technical system.[1] Weight is a force, being the product of mass and the effective acceleration of gravity (represented by the newton in the SI). The gram, kilogram, etc., are measures of mass, but the pound was properly a weight, so the British system is thus inherently a gravitational one. However, the pound has come to be regarded as a unit of mass, equalling close to 454 g, but it continues also to be interpreted, competingly, as that mass subjected to the acceleration of gravity. Physicists have long regarded the pound as a mass, which should be called the pound-mass if there is any ambiguity, but engineers have traditionally seen it as a weight (i.e. a force, distinguished where necessary as pound-force or lb-f). Similarly, engineers recognize a kilogram-force (kg-f) and have a gravitational version of the inherently non- gravitational SI.

The actual gravitational acceleration at Earth's surface, routinely represented by g, is about $9.81 \text{m} \cdot \text{s}^{-2}$ or $32.2 \text{ft} \cdot \text{s}^{-2}$, but varies with locale, by about $\pm 0.5\%$. In some cases the exact acceleration for a locale is pertinent, but any established system requires a fixed standard. For the gravitational SI system (the Potsdam or **international gravimetric system**) this is †standard gravity, set in 1901 at $= 9.806\,65$ $\text{m} \cdot \text{s}^{-2}$ ($32.174\,0\sim \text{ft} \cdot \text{s}^{-2}$). For the BI system, it has long been set at $32.174 \text{ft} \cdot \text{s}^{-2}$ ($9.806\,64\sim \text{m} \cdot \text{s}^{-2}$). Adding a precautionary suffix to the units of mass gives

kilogram-force = kilogram-mass $\times g$
pound-force   = pound-mass $\times g$

The inclusion of the numeric component of g in each makes these relationships not †coherent, for a gravitational system or for a non-gravitational one. The relevant units that give coherence and the relationships are:

- non-gravitational systems:

newton	= kilogram-mass	$\times 1 \text{m} \cdot \text{s}^{-2}$
poundal	= pound-mass	$\times 1 \text{ft} \cdot \text{s}^{-2}$

- gravitational systems:

kilogram-force (kg-force, kg-f)	= engineering mass unit	$\times 1 \text{m} \cdot \text{s}^{-2}$
pound-force (lb-force, lb-f)	= slug	$\times 1 \text{ft} \cdot \text{s}^{-2}$

These matching units, each related to the more familiar by the respective numeric component, are thus valued in the SI as follows:

- force units:

newton, the SI derived unit for force	$= 1\,N = 1\,m \cdot kg \cdot s^{-2}$
poundal	$= 0.031\,080\,997\,1\sim lb\text{-}f$
	$= 0.138\,255\,163\sim N$
kilogram-force (kg-force, kg-f)	$= 9.806\,65\,N$
pound-force (lb-force, lb-f) $= 32.174\,lb\text{-}$	
mass $\cdot ft \cdot s^{-2}$	$= 4.448\,214\,90\sim N$

- mass units:

kilogram-mass	$= 1\,kg$
pound-mass	$= 0.453\,592\,37\,kg$
engineering mass unit (hyl, metric slug, etc.)	$= 9.806\,65\,kg$
slug (g pound, gee pound)	$= 32.174\,lb\text{-}mass$
	$= 14.593\,880\,9\sim kg$

The kilogram-force is also called the kilopond, the term pond having been adopted in the gravitational version of the †c.g.s. system for the gram-force. The pound-mass has no common distinctive name.

Corresponding weight versions of related units exist, e.g. milligram-force, tonne-force, etc., with the SI scene; ounce-force, ton-force, etc., with the British.

gravity *physics* *See* standard gravity.

petroleum processing *See* API gravity.

brewing Adverting to specific gravity (i.e. †relative density) but actually a reciprocal of specific gravity measured, in its usual manner, relative to water. It is basically expressed as 1 000 times that reciprocal, but commonly expressed less 1 000. Thus a specific gravity of 0.975 gives a 'gravity' of $\frac{1000}{0.975} = 1\,025.64\sim$, expressed rounded as 1 026 or just 26.

gray [L. H. Gray; UK 1905–65] *radiation physics*. Symbol Gy. *SI* The derived unit for absorbed dose of ionizing radiation, specific energy imparted, †kerma, absorbed dose index, being energy imparted to a dosed material per unit mass, identically $J \cdot kg^{-1}$ ($= m^2 \cdot s^{-2}$ in base terms). Hence

$Gy \cdot s^{-1}$ for absorbed dose rate.

Though the effect of any radiation depends on its amount of energy, it also depends on the type of radiation and on the energy levels of particular particles. The effect on biological tissue depends on the type of tissue and other factors; the unit applying to the net effect is the sievert.

The sievert is dimensionally identical with the gray, but the two are distinct as regards context, i.e. the gray should be used only for absorbed dose, the sievert only for dose equivalent. Otherwise, it is accepted that the gray can be used, within the field of ionizing radiations, with other physical quantities also expressed in joules per kilogram.

The gray was defined only by the 15th †CGPM of 1975, to succeed the rad as the measure of absorbed dose; 1 Gy = 100 rad.

1975   15th CGPM: re ionizing radiations '*adopts* the following special name for the
SI unit of ionizing radiation: *gray*, symbol Gy, equal to one joule per
kilogram.

Note. – The gray is the SI unit of absorbed dose. In the field of ionizing
radiation the gray may also be used with other physical quantities also
expressed in joules per kilogram.'[8]

Greenwich Hour Angle [Greenwich, UK] *astronomy See* right ascension.

Greenwich Mean Time (GMT) *time* The clock scheme set to have 12:00 noon when the Sun is nominally overhead on the meridian through a defined point within the Observatory at Greenwich, but since 1972 effectively renamed †Universal Time. (The 'defined point' was the centre of a precisely positioned telescope assigned to the purpose.)

Established, along with the Greenwich meridian as the datum line for †longitude, by the International Prime Meridian Conference held in Washington in 1884, GMT formed the basis for the world-girdling scheme of time zones, generally set so that each spans 15° of longitude, with the time zone of Greenwich itself being from 0° to 15°W, and with an International Date Line following the 180° meridian except for minor deviations.

Hourly time zones, with GMT as the basis for them, and even Greenwich as the prime meridian, took years to be accepted. Germany dispensed with its multiple time zones only in the 1890s; France accepted GMT in 1911.

See equation of time for a discussion of the variation in the timing of noon.

Gregorian calendar [Pope Gregory XIII; Rome 1502–85] *See* calendar.

grex *textiles* $= \frac{1}{10}$ tex *See* yarn units. Decitex.

Gs *See* gauss.

G scale *geography* A proposed inverted geometric scale relating the area of a part of Earth's surface to the whole, the scale value being

$\log_{10} w/a$

where *a* is the area of the part, *w* that of the whole.[68] Since the total surface area of Earth is close to 500×10^6 km^2 (200×10^6 square miles), value 6 on the G scale is close to 500 km^2 (200 square miles).

g scale *acceleration* *See* g.

gtt. *medicine* *See* gutta.

Gunter's [A. Gunter; England 1581–1626] *See* chain; link.

gutt, gutta [Lat] *medicine*. Symbol gtt. *England archaic* Drop.

G value *radiation physics, chemistry* The number of moles of substance produced else consumed per joule of radiation; originally defined as the number of molecules per 100 eV of absorbed energy.

Gy *radiation physics* SI *See* gray, also prefixed variously, as in kGy = kilogray; *see also* SI alphabet.

H, h *Metric 1960* As a lower-case prefix, **h-** , *See* hecto-, e.g. ha = hectare. Until 1960, as an upper-case prefix, **H-**, *see* hecto-.
   electromagnetics As **H**, *see* henry; *see* SI alphabet for prefixes.
   fundamental constant As **h** and $\hbar$, *see* Planck constant.
   astronomy **H** *See* Hubble constant.

H-, h- *Metric* As symbol, *see* H, h.

Ha *area Metric* To 1960, hectare = 100 are. *See* hecto-; are.
   rheology See Hartman number.

ha *area SI 1960* Hectare = 100 are.

half *plane angle* Notably in a nautical context, a half of a compass point $= \frac{1}{2} \times \frac{1}{32}$ revolution $= \frac{\pi}{32}$ rad = 0.098 174 770~ rad (5° 37′30″).

half-life *time* The period over which an amount or other value of decaying substance is reduced to half. The term is used most notably for the spontaneous disintegration and transformation of one radionuclide into another. It is used similarly for the transformation of elementary particles and, by parallelism, to the elimination of drugs from the body and similar circumstances. The significance of half-life depends on the rate of transformation or elimination being proportional to the amount present, i.e. on the graph of substance remaining being an †exponential curve (called first-order decay in kinematics). Has also been called **half-value period**.

half-line *pressure* Applying line as a tenth of an inch to 30 inches of mercury in a barometer, $= \frac{1}{600}$ the standard atmospheric pressure; as **Russian half-line** applied at 62°F with the standard at 1 012.804 mbar, hence equal to 1.688~ mbar (168.8~ Pa).[11]

half-value An amount that absorbs one half of incident radiation, e.g. the grams of a substance per millilitre that must be added to a solvent to reduce the transmitted light through 1 cm by half.[74] *See also* half-life.

hand *length* Particularly for the height of horses, 4 in (101.6 mm).

H and D [F. Hurter; UK 1844–98 and V. C. Driffield; UK 1848–1915] *photography* See film speed.

hardness numbers *engineering* The pioneer scale for hardness of solids was the **Mohs** scale of 1824, using ten natural minerals as the comparators. Elaborated in 1933 to the fifteen-point **Ridgeway** or **modified Mohs** scale, it is shown, with values of the Knoop scale, in Table 22.[177]

For metals, scleroscopic tests using indentation with a controlled blow or pressure are normal.[75] They were pioneered by the **Brinell** scale of 1900; this uses a standardized indenter ball of 1 cm or other specified diameter pressed against the surface,[76] with the Brinell hardness number being the ratio of the force used to the curved surface area of the resulting indentation (preferably sufficient to make the indentation 25 to 50% the diameter of the ball, and generally expressed as kilogram-force per square millimetre). The **Meyer** scale is similar, giving slightly larger numbers, depending on ball size; the **Vickers** (called also **diamond hardness dynamic test**) of 1922 and the **Knoop** scale of 1939 use

Table 22

Mohs	Ridgeway	Material	Knoop
1	1	talc	
2	2	gypsum	32
3	3	calcite	135
4	4	fluorite	163
5	5	apatite	430
		glass	530
6	6	feldspar	560
	7	vitreous pure silica	
7	8	quartz	820
	9	garnet	1 360
8	10	topaz	1 340
9		corundum	1 800
	11	fused zirconia	2 100
	12	fused alumina	
	13	silicon carbide	2 480
	14	boron carbide	2 750
10	15	diamond	7 000

pyramidal diamond indenters. The former, regarded as very accurate for a wide range of thickness in the tested material, uses a square point, the latter a rectangular point, particularly useful for very thin or plated materials.

The **Rockwell** scales of 1922 use a ball or a pointed indenter, but with depth of indent being salient. It uses a range of set cones for different sectors of hardness, these being labelled with letters that must be appended to the penetration reading. A load of 1 kg is applied throughout the test, setting the zero datum before the test load is applied.

In the **Shore** (also **scleroscope hardness**) scale, a light-weight ball hammer is dropped from 25 cm and its proportion of bounce off the horizontal surface is measured on an equi-spaced scale of 140 units running from 0% to 100% of the drop.

Select approximate comparisons are shown in Table 23 (with WC = tungsten carbide).[77]

hartley *informatics* The unit of information corresponding to the storage of an arbitrary decimal digit, $= \log_2 10 = 3.321928\sim$ bits. (One cannot have a fraction of a bit, but the unit is relevant to the accommodation of multiple digits.)

Hartman number *rheology.* Symbol Ha. A dimensionless quantity characterizing flow of conducting fluid in a transverse magnetic field, being the product of the magnetic flux density, a representative length, and the square root of the ratio of electrical conductivity to viscosity.

Table 23

name:		Rockwell			Vickers		Brinell		Shore
scale:	C	A	D	superficial					
load:		60 kg-f	100 kg-f						
penetrator:							stnd	WC	
	68	85.6	76.9	93.2	75.4	940			97
	60	81.2	70.7	90.2	66.6	697		654	81
	52	76.8	64.6	86.4	57.4	544	500	512	69
	44	72.5	58.5	82.5	47.8	434	409	409	58
	36	68.4	52.3	78.3	38.4	354	336	336	49
	28	64.3	46.1	73.9	28.9	286	271	271	41
	20	60.5	40.1	69.4	19.6	238	226	226	34
		160	152	152					24

Hartree [D. R. Hartree; UK 1897–1958] *sub-atomic physics* Originally applied as a qualifier to any unit in a set introduced by Hartree;[78] now, as **Hartree energy** (symbol E_h), specifically attached to that for energy. Called by Hartree the **double Rydberg**, it is the energy level below which certain effects cease, $= 4.359\ 743\ 81(34) \times 10^{-18}$ J with †relative standard uncertainty 7.8×10^{-8}.[4] The †atomic unit of energy.

haze factor *meteorology* The ratio of the luminance of an object to the luminance of the engulfing mist or fog.

head of liquid *pressure* The result of the height differential in a liquid, either in the sense of the height to open surface from a submerged point (a reservoir outlet, for instance), or of the height of the surface facing a vacuum relative to a surface open to the atmosphere or other pressure, as in a barometer. The pressure of such a head depends on the liquid's inherent volumic mass, its temperature, and the gravitational acceleration that produces force from the head's mass; within such a context, the linear measure of the head (the height) provides the measure of its inherent pressure. For the reservoir, the head is controlled, the resulting force being available to drive turbines, etc. For the barometer, the head provides the balancing force to the extraneous pressure (of the atmosphere or such), hence a measure of it.

The traditional liquid in the barometer is mercury, for which the head is about 30 in or 760 mm for normal atmospheric pressure; standard conditions[79] are a volumic mass of $13.595\ 1 \times 10^3$ kg m^{-3} at 0°C and the standard gravity of $9.806\ 65$ m·s^{-2}. Under such conditions, a head of 760 mm of mercury (29.921 26~ in) is defined as the †standard atmosphere, identically a pressure of 101.325 kPa. The pressure is normally stated in terms of **millimetres of mercury** else **inches of mercury**, written also as **mm of Hg** or **mmHg** else **in of Hg** (Table 24). The mm of Hg, which is identically the torr, is the usual unit for blood-pressure readings. Water is an obvious alternative to mercury, but its markedly lesser volumic mass implies, for the same pressure range, an impractical column 13.6~ times as high as that for mercury. (The volumic mass of mercury is problematic to the level of precision shown here, probably because of variation in isotope proportions.[80])

Technically, one could have a head of gas, but this would lack a demarcatable surface and would have major variation of volumic mass. While a head of gas is of trivial importance as a tool, it is the essence of atmospheric pressure, which derives from the column of (progressively thinning) air above, reflected in actual pressure falling as elevation

Table 24

1 mm of Hg		1 in of Hg		1 ft of H_2O	
0.001 315 789 47$\sim$	atmos	0.033 421 052 6$\sim$	atmos	0.029 499 796 9$\sim$	atmos
1.333 223 68$\sim$	mbar	33.863 881 6$\sim$	mbar	29.890 669 3$\sim$	mbar
0.133 322 368$\sim$	kPa	3.386 388 16$\sim$	kPa	2.989 066 93$\sim$	kPa
1.0	torr	25.4	torr	22.419 845 69$\sim$	torr
0.039 370 087 0$\sim$	in of Hg	25.4	mm of Hg	22.419 845 7$\sim$	mm of Hg
0.044 603 340 0$\sim$	ft of H_2O	1.132 925	ft of H_2O	0.882 671 090$\sim$	in of Hg
0.019 336 774 7$\sim$	lb-f/in^2	0.491 154 077$\sim$	lb-f/in^2	0.433 527 504$\sim$	lb-f/in^2

increases (*see* atmosphere for details). Atmospheric statements addressing surface conditions are usually expressed relative to nominal sea level, and resident barometers are usually adjusted correspondingly for their altitude.

heat transfer factor *rheology* $St \times Pr^{2/3}$ where St is the †Stanton number for heat transport and Pr is the †Prandtl number.

hect- *Metric* Contracted form of †hecto- , as in hectare = ha.

hectare [hect- + are] *area*. Symbol now ha; until 1960 Ha. *Metric* 100 a = 10 000 m^2 (13 080.$\sim$ yd^2, 2.471 1$\sim$ ac).

Though a multiple of the more basic †are, the hectare is, at least in North America and for land generally, the standard metric areal unit (lesser areas being expressed in square metres). However, as with the base unit kilogram, the hectare itself is not subject to standard prefixing; a tenth of a hectare is a decare (a 'ten' are), not a decihectare. In practice no such other prefixing of the are is commonly used, leaving only are and hectare as areal units (other than m^2, etc., at the lesser end and km^2 at the greater). The 1978 decision of the †CIPM considering it acceptable to continue to use the hectare with the SI still stands; most official English-speaking national publications regard the hectare as the SI unit of area.

hectare·metre *volume Metric* Akin to †acre·foot, being the volume of 1 m × 1 ha = 10 000 m^3, used mostly in the measurement of irrigation water (353 146.7$\sim$ ft^3, 8.107 133$\sim$ acre·foot, 2 641 720.$\sim$ US gal).

hecto- [Gk: 'hundred'] Symbol h- since 1960, but H- prior to 1960. *Metric* The 100 multiplier, e.g. 1 hectogram = 1 hg = 100 g. Contractable to hect- before a vowel, e.g. hectare = 1 ha = 100 a.

Though used extensively in Europe, the hectometre, hectogram, and hectolitre are little recognized, so undesirable elsewhere. However, the hectare has become the effective base unit for land area in many jurisdictions while the hectopascal is common on meteorological maps. The long-established practice of expressing atmospheric pressure using the millibar, which is identical with hectopascal, has prompted the standard use of hPa as the SI unit.

The hectometre is not legal in the UK for trade.[28]

hectr. *area See* hectare.

hefner candle, hefner unit, Hefnerkerze [F. F. von Hefner-Alteneck; Germany 1845–1904] *flame luminosity*. Symbols HK, hefner unit. *Germany* One form of †candle, being the luminosity of a Hefner lamp burning amyl acetate with a flame 40 mm high ≈ 0.920 cd.

hehner number, hehner value *chemistry* Identically grams of water-insoluble fatty acids per 100 grams of a containing fat or oil.

helion mass [the nucleus of the helium atom] *sub-atomic physics*. Symbol m_h. 5.006 411 74(39) $\times 10^{-27}$ kg = 5 495.885 238(12) m_e with †relative standard uncertainties 7.9 $\times 10^{-8}$ and 2.1 $\times 10^{-9}$.[4]

helium scale *temperature* A practical scale for use in the ultra-low temperature range of 1 to 5 K, where the †international temperature scale is impractical. Agreed internationally in 1958, it uses the vapour pressure of He^4 as an indicator.

henry, henrys (pl.) [J. Henry; USA 1797–1878] *electromagnetic inductance, magnetic permeance*. Symbol H. The henrys of inductance of a circuit equal the ratio of the generated electromotive force in volts to the rate of change of the current in amperes per second.
$H = V \cdot (A \cdot s^{-1})^{-1} = V \cdot s \cdot A^{-1} = Wb \cdot A^{-1}$.
SI, Metric-m.k.s. (= $m^2 \cdot kg \cdot s^{-2} \cdot A^{-2}$ in base terms). The following are among the coherent derived units:
- H^{-1} for magnetic reluctance;
- $H \cdot m^{-1}$ for magnetic permeability.

This shows that the electrical and the magnetic are inseparable facets: inductance in an electric circuit is always associated with the generation of a magnetic field.
Metric-c.g.s. See abhenry; stathenry. *See also* practical unit.

History
The henry was agreed as a unit, but with the name **quadrant**, by the

International Electrical Conference in 1889, and renamed henry in 1893, joining the already established †practical units derived from the absolute units of the †e.m.u. system. This was the **practical henry**, $= 10^9$ abhenry. Like all the e.m. units, the abhenry was itself defined ultimately in terms of purely mechanical units.

Because the explicit laboratory specifications established for the ampere, ohm, and volt were subsequently found to be slightly discrepant from the intent, so was the henry, as a unit derived from them. The IEC of 1908 covered the discrepancy by adopting the distinct name **international henry**. Because of experimental vagaries, the value for conversions is normally referred to as the mean international henry, $= 1.000\,49\sim$ H. There is also the **US international henry**, $= 1.000\,495\sim$ H.

With the adoption of the †m.k.s.A. system in 1948, and its basing of electrical units on an ampere compatible with the original absolute units, the modern henry is essentially the old practical henry. Sometimes called the **absolute henry**, this became identically the henry of the SI.

herring unit *mass UK 1975* 100 kg $\{220.46\sim$ lb$\}$ = 4 box.

hertz [H. R. Hertz; Germany 1857–94] *frequency*. Symbol Hz. *Metric 1933* Applicable to any oscillating entity but usually applied only to electromagnetic and other waveforms, identically cycles per second ($= s^{-1}$ in base terms). Already established in use in Germany, it was approved by a majority vote at the International Electrotechnical Commission in 1933,[116] but is still commonly expressed as cycles per second (c.p.s.), or just 'cycles'.

hexa- [Gk: 'six'] Prefix for six, e.g. **hexagon** = six angles (hence six sides).

hexadecimal [a composite adjective from Greek and Latin for 16] With divisor/multiplier steps of 16 ($= 2^4$) in contrast with the steps of 2 for †binary, of 8 for †octal, of 10 for †decimal, etc.

hexadecimal notation, hexadecimal number A style of expressing numeric values similar to †decimal notation except for being based on 16 rather than 10, the graphic characters for which are now usually written

    0 1 2 3 4 5 6 7 8 9 A B C D E F
For example, the number 301 in the decimal system, being
$$= \mathbf{1} \times 256 + \mathbf{2} \times 16 + \mathbf{13} \times 1$$
$$= \mathbf{1} \times 16^2 + \mathbf{2} \times 16^1 + \mathbf{13} \times 16^0$$

is written in hexadecimal as 12D, the D standing for the 13. (All 16 graphic characters, including the six letters, are called 'digits' within this context.) Hexadecimal has the advantage with computers that 16 is a power of 2, and the lowest such power that can accommodate all 10 decimal digits. (*Compare* octal.) Since $16 = 2^4$, each hexadecimal digit equates with 4 bits, so fits conveniently precisely two digits to the now ubiquitous 8-bit byte. Some computers use hexadecimal as the base for †floating-point numbers, i.e. the binary number has its fractional point moved in steps of 4 bits, the exponent applicable to 16 rather than 10. Hexadecimal notation is of particular convenience for the human study of binary data, providing a compact form for expressing binary numbers that is segmented, unlike octal, consistently with the machine bytes.

HF [High Frequency] *physics* The waves of the †electromagnetic spectrum of about 10 MHz frequency, used, *inter alia*, for short-wave radio.

Hg *mass Metric* Until 1960, hectogram.

hg *mass SI 1960* Hectogram.

Hg ohm [Hg = symbol for the chemical element mercury] Mercury ohm.

hhd. *volume See* hogshead.

Hn *sub-atomic physics Metric* Sometimes used for henry, allowing its proper abbreviation H to be used for †Hartree.

hogshead Symbol hhd. A bulk-measure cask, with established volumes and quantities for various commodities in historic marketplaces, $= \frac{1}{4}$ tun.

Hoppus foot *volume* For lumber, 1 ft length of a log of 4 ft circumference, hence a volume of $1.273\,2\sim$ ft^3 ($0.036\,054\sim$ m^3), $\approx \frac{1}{100}$ cord.

horse (vernacular), **horse power** *power*. Symbol h.p. *BI 1809* 550 ft·lb·s^{-1} ($745.699\,872\sim$ W), originally the rounded estimated rate of working of a typical horse (when trudging a circle endlessly, dragging a boom to power machinery), and a unit notably not †coherent within any system. (The term appears to have been coined by Watt for promoting his steam engine.)

The h.p. of a petrol engine can be expressed on a theoretical basis of power at source (**indicated** and **nominal horse power**) or as available power external to the engine (**brake horse power** or **b.h.p.**). The last, generally reckoned as the power available at 4 000 revolutions per minute, is of the order of 80 to 200 for a typical car. In contrast, the horse power figure used for taxing cars in the UK up to 1947 was of the order of 8 to 20; this was a notional calculation, proportional to the aggregate cross-sectional area of the cylinders, i.e.

$$= 0.4 \times D^2 \cdot N$$

where D is the cylinder diameter in inches and N the number of cylinders. If P is the maximal pressure on the pistons in pounds per square inch and S is the stroke length in inches, then

indicated hp $\qquad = \frac{1}{4\pi} D^2 \cdot N \cdot P \cdot S/(33\,000)$
nominal hp $\qquad = (D^2 \cdot N \cdot P \cdot_3 \sqrt{S})/15.6$

Besides the original unit, there is the rounded

metric horse power = 75 kg-f·m·s^{-1} (0.986 320~ hp, 0.735 499~ kW

and several context-specific forms, namely:

electric horse power $\quad = 0.746\,\text{kW}$,
water horse power $\quad = 0.746\,043\text{~kW}$,
boiler horse power $\quad = 9.809\,50\text{~kW}$.

hounsfield unit [G. D. Hounsfield; UK 1919–] *radiation physics*. Symbol H. Actually a scale, providing a fine measure of the attenuation of x-rays by different tissues, the unit being equal to 0.1% of the attenuation coefficient of water and reflecting the ability of computer-assisted tomography to discriminate very small differences. If μ represents attenuation, then the number of Hounsfield units

$$= 10^{-3}\, \frac{\mu_{\text{tissue}} - \mu_{\text{water}}}{\mu_{\text{water}}}$$

Clearly H = 0 for water; the values for other tissues depend on the particular sample as well as its generic form. Representative approximate values include −1 000 for air and +1 000 for bone.

hour Symbol h in the SI. Traditionally $\frac{1}{24}$ of a †day (a unit varying in size depending on the qualifier) and sized by such fractioning. Since 1967, however, the hour of normal usage (derived from the mean solar day), routinely equalling 3 600 seconds, has been defined from the atomic. (*See* leap second for exceptions to the number 3 600.)

The otherwise unusual factor of 24 for hours in a day reflects an early history, when the day itself was divided into night-time and day-time, each of which was divided into 12 hours. Such hours were, of course, variable from day to day and, except near the equinoxes, from day-time to night, though the lower latitudes of the pertinent Middle Eastern civilizations kept such variation modest compared with Europe.

See also lunour for another form of hour.

astronomy See right ascension.

Hour Angle *astronomy, navigation See* right ascension.

hourly difference *navigation* The change in the elements of the nautical almanac caused by a 1-hour step of time.

h.p. *See* horse power.

HPa *pressure Metric* Until 1960, *see* hectopascal.

SI 1960 Hectopascal, except for hectare the only usage of the hecto-prefix in North America, where, because of being identical to the long-established millibar, it is common on meteorological charts.

hubble [E. P. Hubble; USA 1889–1953] *astronomy See* Hubble length.

Hubble constant *astronomy.* Symbol H. The ratio of receding speed to distance (usually as kilometres per second per megaparsec). Deriving from the phenomenon that extragalactic objects in general are receding from an observer on Earth at a speed linearly proportional to their distance away, the value is currently estimated, after various allowances for observational factors, at $71 \pm 7 \, \text{km} \cdot \text{s}^{-1} \cdot \text{Mpc}^{-1}$ (though originally valued much larger, and currently estimated by some at only $60 \, \text{km} \cdot \text{s}^{-1} \cdot \text{Mpc}^{-1}$). *See* Hubble length.

Hubble length *astronomy.* Symbol LH. The decimally rounded number of †light years approximating the radius of the visible (from Earth) cosmos, $= 10^{10}$ ly ($94.605\,36\sim \times 10^{21}$ km, $632.397\,8\sim \times 10^{12}$ AU, $58.785\,1\sim \times 10^{21}$ mi). The actual radius, which is about 40% greater, equals the ratio of the speed of light to the †Hubble constant. (The use of the name for 10^9 ly, the distance travelled in an eon, was proposed in 1968 by Gamow.[81])

Objects 1 LH distant are receding at 1 light year per year, hence at the limit to visibility. However, any objects still further away would be receding from us faster than light – an impossibility by current theory, save that the accepted view is that the fundamental structure of space-time is expanding, thereby carrying the galaxies further apart. The implication is that the sphere defined by the radius of visibility is identically the limit of the Universe which, despite all the conjectures about other planets having life like Earth's, would make Earth uniquely the centre of the Universe.

hundredweight (**quintal**) *mass* 112 lb (50.802~ kg) (**cwt, gross hundredweight, long hundredweight**) in Britain and elsewhere, but 100 lb (45.359~ kg) (also **net hundredweight, short hundredweight**) in North America. *See* pound for more precise values, and Table 25 for scales.

The cwt was removed from official UK measures in 1985.[28]

For **new hundredweight**, *see* cental (= 100 lb).

Table 25(a)

BI, Australia, New Zealand						SI
lb	...	...	...	...		454.~ g
14	stone		...	...	...	6.35~ kg
28	2	quarter		...	...	12.7~ kg
112	8	4	hundredweight		...	50.8~ kg
2 240				20	ton	1016.~ kg

For downward extension, *see* ounce.

Table 25(b)

US-C, Canada					SI
lb	...	...	...	...	454.~ g
25	quarter		...	...	11.3~ kg
100	4	hundredweight		...	45.4~ kg
2 000			20	ton	907.~ kg

For downward extension, *see* ounce.

weight, force See gravitational system, weight.

Hz [Hertz] *acoustics, electromagnetics Metric See* hertz. Also prefixed variously, as in kHz = kilohertz; *see* SI alphabet.

I, i As **I** else **i** the Roman numeral for 1, a representation of one finger.

mathematics (also **j**, particularly as used by engineers) As **i**, the imaginary value $\sqrt{-1}$, used very powerfully in applied mathematics in the form of complex †number.

Ilford [place and company name in UK] *photography* See film speed.

imaginary number, imaginary part *See* number.

Imperial system *See* British Imperial system. (It should be noted that many units labelled as imperial, e.g. the †cup, were never part of BI.)

improper fraction *mathematics* Having the numerator bigger than the denominator, e.g. $\frac{3}{2}$; *see* number, proper.

imputed A term in metrology relating to the ascribing of an amount measured in one quantity to a measurement in another. Such ascribing has been common in the grain trade, where the †bushel, a unit of volume equal to 64 pints and routinely used in that sense at farm level, has been ascribed corresponding values of 'weight'. The practice reflects the respective ease of measuring small amounts by volume but large amounts by weighing. It has been used similarly for other fluidic dry goods, and for such units as the bag and, for mercury, the flask.

in., inch [L: uncia, 'twelfth' – cognate with ounce] *length* Traditionally the width of the thumb nail. *British/American* Always $= \frac{1}{12}$ ft $= \frac{1}{36}$ yard, the inch of British provenance has for centuries been everywhere within 0.01% of its current international value.[73] However, the inch of modern times was technically different in each of the UK, the USA, and Canada prior to the agreed international standard,[142] though mutual differences were less than 0.001%. Introduced in Canada in 1951[82] then in the UK, the USA, Australia, and New Zealand in 1959 (though in the UK not completely until 1964), the international value has the inch an integer multiple of a tenth of the millimetre, hence the foot, yard, etc., likewise. However, through a succession of †prototypes it had probably been the same to at least six significant figures for centuries. See Table 26.

Internat, BI, US-C 1959 25.4 mm, so $1\,\text{m} = \frac{1}{0.0254}\,\text{in} = 39.370\,078\,7\sim$ in.
(However, because of their entrenched representation in so many land surveys and of the effect of even very minor differences in definition when magnified by geographical distances, the then extant US units displaced by adoption of the international inch in 1959 have been retained as the **survey** or **coast** units, based on the 1866 value.)

Canada 1951 25.4 mm.

BI 1898 A UK Order-in-Council declared that, on the basis of comparative measurement of the prototype yard and the international prototype metre, $1\,\text{in} = 25.399\,978\sim$ mm (making the prototype yard obsolescent, but not displacing it).

US-C 1893 It was declared that the 1866 equation $1\,\text{m} = 39.37\,\text{in}$ be based on the prototype metre, not the prototype yard, making the latter obsolete.

US-C 1866 For conversion from metric-expressed contracts to US-C, it was declared that $1\,\text{m} = 39.37\,\text{in}$, hence $1\,\text{in} = \frac{1}{39.37}\,\text{m} = 25.400\,050\,8\sim$ mm.

BI 1864 From comparative measurement of the prototypes it was declared that $1\,\text{in} = 25.399\,53\sim$ mm for conversion from metric-expressed contracts.

BI 1825, US-C 1838 $\frac{1}{36}$ the prototype yard.

History

The **international inch** of 25.4 mm represented virtually the mean of the differently defined UK and US sizes. The US definition of 1866, by its rounded basis of $1\,\text{m} = 39.37\,\text{in}$ aimed at simplification of conversion from metric, can be seen as defining an American metre from the yard prototype. This was reversed in 1893 through the †Mendenhall Order, defining the inch in terms of the international metre. The practice for BI

Table 26

Internat, BI, US-C, Australia, Canada, New Zealand, etc.						SI
inch	...	...	...	...	...	$254.\sim mm$
12	foot	...	...	...	...	$305.\sim mm$
36	3	yard	...	...	...	$914.\sim m$
198	33:2	11:2	rod	...	...	$5.03\sim m$
792	66	22	4	chain	...	$20.1\sim m$
7 920	660	220	40	10	furlong, eighth	$201.\sim m$
63 360	5 280	1 760	320	80	8	mile $1.61\sim km$

differed in that its figures were the best then ascertainable by relating the respective prototypes: the 1898 Order-in-Council figure 25.399 977 9~ mm was obtained by measuring the prototype metre bar at 39.370 113 in, the 1863 figure at 1 m = 39.370 81~ in. These British figures, since they related to a long-established prototype that was common to the UK and the USA, represented the size of the inch, to that level of precision, when BI came into effect in 1825, and the size of US-C inch prior to 1893. The rounding to have the inch equal to $\frac{1}{39.37}$ m made a shift of less than three parts per million, and was easily acceptable at the time. The much later change to exactly 25.4 mm, though of less than two parts per million, was not universally acceptable in the USA, and resulted in the old sizing, particularly for the consequent foot, being retained for survey purposes.

Careful measurement of the prototype BI yard still in practical use in the 1950s showed the British master copy as being 0.914 397 2~ m,[142] giving an inch of 25.399 92~ mm.

volume In a maritime context, *see* shipping cubic inch.

physics For pressure, e.g. as **inch of mercury**; *see* head of liquid.

Indo-Arabic numerals *See* numerals.

infinity *mathematics* Symbol ∞. A term meaning the boundlessly large, but not necessarily uncountable, actually inclusive of a range of distinct numbers, all being boundlessly large but not mutually equal. The smallest infinity corresponds to the number of integers. It is called **aleph-null** or **aleph-nought**, written $\aleph_0$, being the first letter of the Hebrew alphabet with a zero subscript, and, perhaps surprisingly, it can be shown to be identically the number of †rational numbers. Despite the general understanding that anything infinite is uncountable, to the mathematician this smallest infinity is regarded as **countable**; to count an unbounded group one just attaches 1 to the first, 2 to the second, et seq. Such a procedure will not only cover all the †natural numbers but also all the †integers, indeed all rational numbers. The integers can be so numbered by doubling each unsigned value and adding 1 if originally negative. The rationals m/n between 0 and 1 can be uniquely numbered as $m + \Sigma(n - 2)$, where Σ means the sum of all positive integers up to that value (numbering $\frac{1}{2}$ as 1, $\frac{1}{3}$ as 2, $\frac{2}{3}$ as 3, $\frac{1}{4}$ as 4, $\frac{3}{4}$ as 6, et seq). The full range of rationals can be shown to be similarly countable. However, rational numbers form only a small minority of numbers. There is no such method for attaching uniquely the integers to all numbers between 0 and 1; they cannot be uniquely numbered in the integer sense, since they are not countable. Their number is symbolized by the letter **C** or **c** (for continuum); C is not equal to $\aleph_0$.

Each of these two infinities, individually and separately, is such that most arithmetic on it leaves it unchanged. Multiplication, division, addition, or even subtraction by itself leaves it unchanged. (Hence, for instance, the number of rational numbers between 0 and 1 is the same as between 0 and 100, or any other value; likewise C is the number of numbers in any †proper interval.) However, exponentiation changes it; for any distinct infinity ∞, the exponentiated value ∞^∞ is not equal to the original ∞. The result of such exponentiation is a new infinite number; starting from aleph-null there is aleph-one; then, if this is self-exponentiated, aleph-two, and so on, i.e. $\aleph_0, \aleph_1, \aleph_2, \ldots$ Since there is no limit to this process, there is no limit to the number of distinct infinities. The number of such would appear to be countable in the mathematician's language, but uncountable in popular terminology. Since any number between 0 and 1 can be represented by a countable string of decimal (else binary else duodecimal else hexadecimal else ...) digits, C must equal any finite integer to the power $\aleph_0$. Since this is definitely greater than $\aleph_0$ but not greater than $\aleph_1$, it must equal the latter or some further distinct infinity between those two. So C is not by any means the largest infinity; what is remains a puzzle, as indeed does the meaning of all after $\aleph_1$.

infrared The waves of the †electromagnetic spectrum just outside the †visible light spectrum, ≈ 30 to 430 THz frequency (just below red light).

inhour [inverse hour] *sub-atomic physics USA* For the activity of a nuclear reactor, the inverse of its period in hours.[83] *See also* dollar; nile.

INM International Nautical Mile; *see* nautical mile.

integer, integral number *mathematics* Any whole †number.

intensity millicurie *radiation physics See* sievert.

intercalary [Lat: 'in between' + 'solemnly proclaim'] A promulgated insertion into a †calendar to make it correspond with the solar year.

international biological standards Standardized preparations of biologically active substances are held as samples of archive in Copenhagen (immunologicals) and London (pharmacologicals),[84] each a set mass of a specific form of the substance. The quantities are expressed as **international units (units of activity)** – the **IU** familiar on vitamins for instance. The pioneers are shown in Table 27.[66]

Since those years some have been redefined, while very many others have been established. Current definitions, using hash marks appended to indicate redefinitions (with one mark for a first redifinition, two marks for a second, etc.), include those set out in Table 28.

Table 27

diphtheria antitoxin	1922	62.8 µg
arsphenamine	1925	1. mg
neoarsphenamine	1925	1. mg
sulpharsphenamine	1925	1. mg
insulin (crude dry insulin hydrochloride)	1925	125. µg
pituitary (posterior lobe) powder	1925	500. µg
digitalis	1925	80. mg
ouabain	1925	80. mg
tetanus antitoxin	1928	154.7 µg
anti-dysentery serum (*Shiga*)	1928	500. µg
vitamin A* (mixed carotenes)	1931	1. µg
vitamin B1 (adsorption product of)	1931	10. mg
vitamin D (irradiated ergosterol solution)	1931	100. µg

Table 28

Antigens		
diphtheria toxoid (plain, alcohol purified)	1951	500. µg
diphtheria toxoid (adsorbed to Al(OH)$_2$, dried)	1955	750. µg
diphtheria – Shick test toxin (purified)	1954	4.2 µg
old tuberculin	1935″	10. µg
pertussis vaccine (dried)	1957	1 500. µg
swine erysipelas vaccine (adsorbed to Al(OH)$_2$, dried)	1959	750. µg
tetanus toxoid (alcohol purified)	1951	30. µg
Antibodies (hyperimmune horse serum except as shown)		
gas-gangrene serum (dried)		
perfingens Chlostridium welchii Type A	1953″″	113.2 µg
perfingens Chlostridium welchii Type B	1954	13.7 µg
perfingens Chlostridium welchii Type D	1954	65.7 µg
vibrion septique	1954‴	11.8 µg

* but *see* RE.

oedematiens	1952″	113.5	μg
hystoliyticus	1951″	200.	μg
Sordelli	1938	133.4	μg
staphylococcus α antitoxin (dried)	1938″	237.6	μg
scarlet fever staphylococcus antitoxin (dried)	1952	49.	μg
swine erysipelas serum (anti-N)	1954	140.	μg
antistreptolysinanO (human, dried)	1959	21.3	μg
antipneumococcus serum type 1 (dried)	1934	88.6	μg
antipneumococcus serum type 2 (dried)	1934	89.4	μg
anti-*Brucella abortus* serum (bovine, dried)	1952	91.	μg
anti-Q-fever serum (bovine, dried)	1953	101.7	μg
anti-rabies serum (dried)	1955	1 000.	μg
anti-A blood-typing serum (human, dried)	1950	346.5	μg
anti-B blood-typing serum (human, dried)	1950	352.	μg
syphilitic human serum (dried)	1958	3 617.	μg
Antibiotics			
penicillin (sodium benzyl …)	1944	0.6	μg
	1952″	0.5988	μg
phenoxymethylpenicillin (sodium benzyl …)	1957	0.59	μg
streptomycin (… sulphate)	1958	1.282	μg
dihydrotreptomycin (… sulphate)	1953	1.316	μg
bacitracin	1953	18.2	μg
tetracycline (… hydrochloride)	1957	1.01	μg
chlortetracycline (… hydrochloride)	1953	1.	μg
oxytetracycline (… dihydrate)	1955	1.11	μg
oeryhromycin (… dihydrate)	1957	1.053	μg
polymyxin B (… sulphate, purified)	1955	0.127	μg
Hormones			
insulin (pure crystalline insulin)	1935	45.5	μg
poxytoxic, vasosuppressor and anti-diuretic substances			
(posterior ox pituitary, acetone-dried, powdered)	1957‴	500.	μg
prolactin (active principle anterior ox pituitary, dried)	1939	100.	μg
corticotropin ACTH (pig pituitary, crude)	1955″	880.	μg
thyrotropin (anterior ox pituitary, crude)	1954	13 500.	μg
Vitamins and Miscellaneous			
digitalis (dry powdered leaf of *D purpurea*)	1949″	76.0	μg
hyaluronidase (bovine testes, dried)	1955	100.	ng
vitamin D$_3$	1949″	0.025	μg

(contd.)

Table 28 (*contd.*)

vitamin B$_1$ (pure synthetic form)	1938	3.0 μg
vitamin C (*t*-ascorbic acid)	1934	50.0 μg
vitamin E (α-tocopheryl acetate)	1941	1.0 mg
heparin	1942	7.7 μg

The 'marks indicate successor versions.

international gravimetric system *See* gravitational system.

International Nautical Mile *See* nautical mile.

international (practical) temperature scale (IPTS, ITS)
temperature Metric Introduced in 1927 to provide a common
reproducible basis for scientific and industrial measurement of
temperatures over a large range of values, this scale has been
progressively refined and redefined as technologies have improved.[85]
Whereas the familiar everyday scale uses just two points (originally the
freezing and boiling points of water, more recently 'absolute zero' and
the triple point of water, i.e. the point of equilibrium between solid,
liquid, and vapour), the compatible ITS's, to meet practical necessity, use
multiple points between which it interpolates (all within defined
experimental conditions).[86]

The **ITS-27** was succeeded in 1948 by **ITS-48**, and re-labelled in 1960 as
IPTS-48 when amended but without numeric change. A greatly
augmented scale, with some changes to common points too, was
introduced in 1968 as **IPTS-68**.[87] In 1975 it was also amended without
numeric change. An adjustment to the measured value of the triple point
of water, and hence to the size of the kelvin, was agreed in 1988, to take
effect in 1990, with **ITS-90**.[88] The chosen points and their cited values
for the three later versions are shown in Table 29.

Above the uppermost points, temperatures are measured by means
of the Planck radiation formula.[89] Below 13.81 K the scale was
undefined prior to ITS-90; the †helium scale and †Curie temperature
scale providing in part for this ultra-low zone. Using helium, ITS-90
extends the scale to 0.65 K; for 0.65 to 3.2 K this uses the relation of
vapour pressure to temperature, from 3.2 to 24.556 1 K it uses the
constant-volume gas thermometer. Work is active in extending the
range down to 0.001 K, based on the melting pressure of
helium.[92]

International Union of Biochemistry unit *See* enzyme unit.

Table 29

IPTS-48	IPTS-68		ITS-90	
°C	°C	K	K	
	−259.34	13.81	13.803 3	triple point of hydrogen
	−256.108	17.042		intermediate equilibrium point of hydrogen
	−252.87	20.28		liquid hydrogen and its vapour
			24.556 1	triple point of neon
	−218.789	54.361	54.358 4	triple point of oxygen
			83.805 8	triple point of argon
−182.970	−182.962	90.188		liquid oxygen and its vapour
			234.315 6	triple point of mercury
0			'ice point'	ice and air-saturated water
	0.01	273.16	273.16	triple point of water
			302.914 6	freezing point of gallium
100	100	373.15	'steam point'	liquid water and its vapour
			429.748 5	melting point of indium
444.600				liquid sulphur and its vapour
			505.078	melting point of tin
	419.58	692.73	692.677	melting point of zinc
			933.473	melting point of aluminium
960.8	961.93	1 235.08	1 234.93	melting point of silver
1 063.0	1 064.43	1 337.58	1 337.33	melting point of gold
			1 357.77	melting point of copper

The element-specific points are often called the 'sulphur point', etc.

international unit *electromagnetics* Metric The units developed
under the title †electromagnetic units, forming the †e.m.u. system, were
adjusted by decimal multiplication to give the †practical units, adopted
at the first International Electrical Conference in 1881 as suitable for
everyday electrical work. Accompanying explicit laboratory definitions
of the ohm, ampere, and volt were established, making them base
practical units instead of derived units as in the e.m.u., where these
electrical units were defined in terms of mechanical equivalents. These
new definitions were soon found to be discrepant with the intended
and, despite some adjustments, are continuingly so (albeit by less than
0.05%). They were not even mutually self-consistent until the volt was

made a derived unit, at the IEC of 1908, when the residual discrepancies were accepted via the use of the distinctive label 'international ohm', etc.

By Act of Congress in 1984, the USA created its own definitions of the 'international units', slightly different from the others, hence creating a set of **US international units**.

The true international units were in turn discarded at the end of 1947 in favour of new definitions within the metric †m.k.s.A. system that became the †SI system, these latest units often being called 'absolute', and generally agreeing with the original practical units rather than the 'international' ones. Typical values in SI terms are as follows:

1 international ampere = $0.99985\sim$ A
1 international coulomb = $0.99985\sim$ C
1 international farad   = $0.999510\sim$ F
1 international henry   = $1.00049\sim$ H
1 international mho   = $0.999510\sim$ S
1 international ohm   = $1.00049\sim$ Ω
1 international volt   = $1.00034\sim$ V

food and drugs See international biological standards.

interval *music* In British/American practice the familiar diatonic scale is represented by the first seven letters of the alphabet, expressed repeatedly to represent the further notes (*see* pitch). Intervals on this scale are usually expressed relative to the number of named notes they span, using the ordinal rather than the cardinal number: for example, the interval A to C is termed a 'third' because it spans the three note letters A, B, and C, except that what would be called 'first' is called unison. Since the spacing of the named notes is not uniform, and any can be modified sharp or flat, the size of such an interval is ambiguous unless qualified. Assuming the scale to be of equal temperament, the interval between consecutive natural notes (i.e. the notes of those letters without modification by a sharp or a flat accidental) is either a **tone** or, half that, a **semitone**. (Also called **step** and **half-step** respectively, and, as qualified, a †second.) Using a 'bullet' to mark the whole tones (hence the black keys of the piano, which are designed to split each full tone into its halves), that scale is of the form

A • B C • D • E F • G • A • B C • D • E F • G

As discussed under pitch, each note corresponds to a frequency, the interval of a whole †octave being a doubling of the frequency, a step up of a tone being an increase of just over 12%, and that of a semitone of just under 6%. The intervals of this scale that give the frequency ratios of 1:1, 1:2, and (approximately) 2:3 and 3:4, because of their superior harmony, are called **perfect**; these are respectively unison, the octave, the fifth of 3 tones plus 1 semitone, and the fourth of 2 tones plus 1 semitone. A fourth, fifth, or octave a semitone less than perfect is **diminished**; any a semitone greater than perfect is **augmented**. The other intervals, less harmonious and each of two sizes in the above scale, are qualified as **minor** and **major**; if a semitone is yet smaller or greater by accidental, it too is **diminished** or **augmented**. In summary, Table 30 shows the interval size on the left in semitones, on the right in frequency of the higher note relative to the frequency of the lower (assuming the scale to be of equal temperament).

Ninths, tenths, et seq., are also used in some circumstances, representing the octave plus a second, third, et seq.

Various numerical schemes have been introduced for expressing interval size. These include the †cent with 100 to a semitone, the †centi-octave with 100 to an octave, and the †savart with (approximately else precisely) 300 to an octave.

Table 30

	unison or first 1st	second 2nd	third 3rd	fourth 4th	fifth 5th	sixth 6th	seventh 7th	eighth 8th	Ratio
0	perf	dim	…	…	…	…	…	…	1.000
1	aug	min	…	…	…	…	…	…	1.059
2	…	maj	dim	…	…	…	…	…	1.122
3	…	aug	min	…	…	…	…	…	1.189
4	…	…	maj	dim	…	…	…	…	1.260
5	…	…	aug	perf	…	…	…	…	1.335
6	…	…	…	aug	dim	…	…	…	1.414
7	…	…	…	…	perf	dim	…	…	1.498
8	…	…	…	…	aug	min	…	…	1.587
9	…	…	…	…	…	maj	dim	…	1.663
10	…	…	…	…	…	aug	min	…	1.782
11	…	…	…	…	…	…	maj	dim	1.888
12	…	…	…	…	…	…	aug	perf	2.000
13	…	…	…	…	…	…	…	aug	2.119

iodine number *biochemistry* The grams of iodine absorbed per

100 grams for a containing fat or oil, which indicates the proportion of unsaturated linkages.

IPTS *See* international temperature scale.

I.Q. [Intelligence Quotient] The usual measure of intelligence, now assessed by standardized tests that supposedly measure the innate, untrainable intelligence of an individual in a fully objective way. Originally created to measure the mental 'age' of children as a percentage of real age, a value of 100 is thus inherently normal. Values obtained for adults range fairly widely, from about 20 to 200. There is continuing scepticism about IQ tests, and doubts about their objectivity, the innateness of what they measure, and their correlation with meaningful mental ability.

irrational number Any †number that cannot be expressed as the ratio of two †natural numbers, π and $\sqrt{2}$ being familiar examples.

ISCC-NBS system *colorimetry* [Inter-Society Color Council – National Bureau of Standards] *USA* A formal system for expressing colour, with standard symbols for hue, qualified for lightness and saturation. (See also †CIE system.) Derived from the †Munsell system, its basic codes for hue are:

B = blue, bG = bluish green.

Br = brown, brO = brownish orange, brPk = brownish pink.

G = green, gB = greenish blue, gY = greenish yellow

O = orange, OY = orange yellow.

Ol = olive, OlBr = olive brown, OlG = olive green.

P = purple, pB = purplish blue, pPk = purplish pink, pR = purplish red.

Pk = pink.

R = red, rBr = reddish brown, rO = reddish orange, rP = reddish purple.

V = violet.

Y = yellow, yBr = yellowish brown, yG = yellowish green, yPk = yellowish pink

YG = yellow green.

ITS, ITS-48, ITS-68, ITS-90 *See* international temperature scale.

IU, i.u. *See* international unit; international biological standards.

J

J, j Particularly as **j**, sometimes the †Roman numeral for 1 when occurring as a terminal symbol following the one, two, or even three i symbols, e.g. as iij for the value 3.

mathematics As used by engineers in place of the more usual i, as **j**, the imaginary value $\sqrt{-1}$, used very powerfully in applied mathematics in the form of complex †number.

mechanics, electromagnetics, etc. As **J**, *see* joule; *see* SI alphabet for prefixes.

jansky [K. G. Jansky; USA 1905–50] *astrophysics See* flux unit.

jar *electric capacitance* That of a typical Leiden or Leyden storage jar, designed to equal 1 000 electrostatic †centimetres, $= 10^3$ statF $= 1.1126$ $\sim$ nF, but often accepted as being $\frac{10}{9}$ nF. The storage jar was the first practical reference unit for electricity.[90]

JD *time See* Julian date.

jerk *engineering* An informal unit of acceleration, $= 1$ ft s^{-2}.

Josephson constant [B. D. Josephson; UK 1940–] *fundamental constant*. Symbol K_J. The quotient of frequency divided by the potential difference corresponding to the $n = 1$ step in the Josephson effect, used in the determination of the †volt etc.[6] The conventional value, agreed in 1988 for implementation in 1990, is $K_{J-90} = 483\,597.9$ GHz·V^{-1}. The true value, denoted K_J, is evaluated as $483\,597.898(19) \times 10^9$ Hz·V^{-1} with †relative standard uncertainty 3.9 $\times 10^{-8}$.[4]

joule [J. P. Joule; UK 1818–89] *energy, work, quantity of heat*. Symbol J. The joules of work done by a †steady current equal the product of the number of coulombs of charge and the electromotive force in volts. $J = C \cdot V$.

SI, Metric-m.k.s. A. Also $= N \cdot m$ ($= m^2 \cdot kg \cdot s^{-2}$ in base terms). The following are among the coherent derived units:

- J·m^{-3} for energy density;

- $J \cdot kg^{-1}$ for specific energy;
- $J \cdot kg^{-1}$ = gray for absorbed radiation;
- $J \cdot kg^{-1}$ = sievert for dose equivalent;
- $J \cdot s^{-1}$ = watt for power, radiant flux;
- $J \cdot K^{-1}$ for entropy, heat capacity;
- $J \cdot mol^{-1}$ for molar energy;
- $J \cdot mol^{-1} \cdot K^{-1}$ for molar entropy or molar heat capacity;
- $J \cdot V^{-1}$ = coulomb for quantity of electricity;
- $J \cdot C^{-1}$ = volt for electromotive force, potential difference;
- $J \cdot kg^{-1} \cdot K^{-1}$ for specific entropy or specific heat capacity.

See also practical unit.

History

The joule was recognized internationally in 1889, at the second International Electrical Conference, as a derived addition to the †practical units of the †c.g.s. system; it was defined as the energy dissipated in 1 second by current of 1 ampere flowing through a resistance of 1 ohm. Hence the **practical joule**. While discrepancies, for the underlying ampere and ohm, between measured absolute values (in centimetre-gram-second terms) and their laboratory specifications led the IEC in 1908 to rename units based on the latter as unadorned international units, differing by $0.05 \sim$ % from the practical, the joule as a mechanical unit was unequivocally a force of 10^7 dyne acting over 1 centimetre (= 10^7 erg).

With the implementation of the †m.k.s.A. system in 1948, this became 1 newton acting over 1 metre (leaving its electric definition unchanged). The same year, the joule was accepted as a unit of heat; prior to this, and despite the initial electric definition of 1889 using the expression 'the energy disengaged as heat in one second', the joule was regarded as a mechanical equivalent of, but not a measure of heat.

1946 CIPM '*Joule* (unit of work or energy) The joule is the work done when the point of application of 1 MKS unit of force [newton] moves a distance of 1 metre in the direction of the force.'[8]

Julian date, Julian day number [G. Julius Caesar; Rome 102–44 BCE]
astronomy See astronomical day system.

informatics A vernacular term meaning a date expressed purely as a year, then number of day within year, e.g. 1997/99 for 9 April in that year, but 1996/100 for 9 April in the preceding year, being a leap year.

Julian year *time* The average †year under the scheme, introduced by

Julius Caesar, that has every fourth year as a leap year (*see* calendar) = 365.25 mean solar †days. Still used in astronomy, also called the †Besselian year.

K, k *Metric* As a lower-case prefix, **k-**, *see* kilo-, e.g. kg = kilogram.
Metric Until 1960, as an upper-case prefix, **K**, *see* kilo-.

   temperature Metric As **K**, *see* kelvin, a reference to thermodynamic
temperature. Note that, though previously written as °K (K as a symbol
for scale rather than unit), the prefixed symbol for and the term degree
were abolished under the SI in 1968. Can be prefixed variously, when
representing a difference between two points on the scale, as in kK =
kilokelvin; *see* SI alphabet.

   electromagnetics Metric Sometimes used without the alien Ω to
represent kΩ, i.e. kilohm. A deprecated practice.

   thermodynamics As **k**, *see* Boltzmann constant; more distinctively **k$_B$**.

   astronomy As **k**, *see* Gaussian gravitational constant.

   informatics See kilo-.

K-, k- *Metric* As symbol, *see* K, k.

Kalantaroff system [P. Kalantaroff; USSR] *See* base units.

kapp, kapp line, kappline [G. Kapp; UK 1852–1922] *magnetic flux*
Metric-c.g.s.-e.m.u. and *-c.g.s.-Gaussian* 600 maxwell. Technically defined in
a three-dimensional system, it corresponds in the four-dimensional
electromagnetic sector of the †SI system to 6 µWb.

karat *See* carat.

kat *See* katal.

katal *catalytic activity.* Symbol kat. *SI 2000* mol·s^{-1}. (Previously
sometimes **catal**.[91])

1999 13th CGPM: 'considering
    • the importance for human health and safety of facilitating the use of SI
units in the fields of medicine and biology,
    • that a non-SI unit called "unit", symbol U, equal to 1 µmol·min^{-1}, which
is not coherent with the SI, has been in widespread use in medicine and
biochemistry since 1964 for expressing catalytic activity,

- that the absence of a special name for the SI coherent derived unit of mole per second has led to results of clinical measurements being given in various local units, that the use of SI units in medicine and clinical chemistry is strongly recommended by the international unions in these fields,
- that the International Federation of Clinical Chemistry and Laboratory Medicine has asked the Consultative Committee for Units to recommend the special name katal, symbol kat, for the SI unit mole per second,
- that while the proliferation of special names represents a danger for the SI, exceptions are made related to human health and safety (15th General Conference, 1975, Resolutions 8 and 9, 16th General Conference, 1979, Resolution 5)'.

noting that the name katal, symbol kat, has been used for the SI unit mole per second for over thirty years to express catalytic activity,

decides to adopt the special name katal, symbol kat, for the SI unit mole per second to express catalytic activity, especially in the fields of medicine and biochemistry,

and **recommends** that when the katal is used, the measurand be specified by reference to the measurement procedure; the measurement procedure must identify the indicator reaction.[92]

kayser [J. H. G. Kayser; Germany 1853–1940] Symbol k, previously also rydberg.

physics Particularly for spectroscopy, a version of †wave number, being the number of waves per 1 cm.[93]

k_B *thermodynamics* *See* Boltzmann constant.

kelvin [W. Thompson, 1st Baron Kelvin, a river in Glasgow; UK 1824–1907] *temperature*. Symbol K. *SI* The base unit for (thermodynamic) temperature, being the fraction $\frac{1}{273.16}$ of the thermodynamic temperature of the triple point of water (a point minutely above freezing point – see below), and the name of a temperature scale using that unit.

Note that the standard expression of a kelvin temperature does not, since 1968, use the familiar (symbol for or verbalized) expression for degree; whereas one writes 10°C or 50°F, for kelvin one writes 283 K. This last expression on its own is ambiguous in that it can mean a single point on the scale else a difference between two points, the context invariably resolving the matter (whereas for Celsius the recommendation is to use 'deg' or 'degree' rather than the symbol when meaning difference). As shown below, prior to 1968, the term kelvin was prefixed by 'degree' in some sense, and, from 1948, the symbol ° was for scale position only.

The **kelvin scale** was derived directly from the †Celsius scale, its unit being identical to the Celsius degree, and its scale differing by having its zero at the †thermodynamic null (so-called 'absolute zero'), evaluated as −273.15°C, hence the synonyms **absolute temperature scale** and (**absolute**) **thermodynamic temperature scale** for the kelvin scale. In 1954 the roles were reversed, the kelvin becoming defined as stated above, the Celsius derived from it. The new definition uses the 'triple point of water', which had been determined to be very close to 0.01°C and is much more precisely definable than freezing point. The triple point is specified to be 273.16 K which, with 0 K continuing to be the null point, defines the size of the unit. Zero on the Celsius scale is then defined to be 273.15 K rather than the freezing point as such. As with other measurements, marking the triple point is subject to increasing accuracy as technology progresses. Following laboratory evaluations that showed the triple point slightly discrepant from 273.16 K as then established, a minute adjustment to the size of the kelvin was agreed in 1988, to take effect in 1990. (The adjustment applied identically to the degree Celsius, and technically repositioned the Celsius scale.)

See temperature for other scales and conversions between scales.

While most people think of graduated thermometers as the means for measuring temperature, such instruments are severely limited in their usefulness. Calibrating for and measuring temperature precisely over its huge practical range requires a matching range of well-defined points in addition to the triple point of water. *See* international temperature scale.

1948 9th CGPM: 'degree absolute °K
To indicate a temperature interval or difference, rather than a temperature, the word "degree" in full, or the abbreviation "deg", must be used.'

1954 10th CGPM: 'The 10th CGPM decides to define the thermodynamic temperature scale by choosing the triple point of water as the fundamental fixed point, and assigning to it the temperature 273.16 degrees Kelvin exactly.'

1967–68 13th CGPM: '*decides*
1. the unit of thermodynamic temperature is denoted by the name "kelvin" and its symbol is "K";
2. the same name and the same symbol are used to express a temperature interval;
3. a temperature interval may also be expressed in degrees Celsius;

4. the decisions of 1948 re the use of degree and deg in association with K or kelvin are abrogated ... and *decides* to express "the" definition as follows:

"The kelvin, unit of thermodynamic temperature, is the fraction $\frac{1}{273.16}$ of the thermodynamic temperature of the triple point of water".'[8]

electromagnetics Has been used to mean †kilowatt·hour.

kerma *radiation physics* The kinetic energy released in material by uncharged particles.

K factor *radiation physics* The dose rate in roentgens per hour of gamma rays 1 cm from a 1-millicurie point source, being an indicator of the gamma-ray-emitting strength of the emitting substance.

Kgf, Kg-f, kgf, kg-f *engineering* See kilogram-force.

kibi- [kilo- binary] *informatics*. Symbol Ki-. The 1 024 = 2^{10} multiplier, as **KiB** = kibibytes and **Kib** = kibibits. This arises as a matter of convenient terminology for memories and similar entities organized on a binary basis. (But see below re communication channels.)

Concerned at the widespread use of SI decimal prefixes for nearby but distinct binary multipliers, the International Electrotechnical Committee in 1999 adopted as an international standard (outside of the SI) a range of revised multiplier prefixes, derived from those of the SI, implying the true binary values. Specifically and uniformly, the familiar decimal multiplier prefixes were modified by the appendage of the letters 'bi' to the initial two letters of the SI prefix (the appendage being pronounced 'bee', the beginning as with the SI). The symbols are similarly those of the SI with an appendage, namely the single letter i, except that, for consistency within the new set, the capital K was retained. These prefixes are shown in Table 31.

Table 31

Symbol	Prefix	Value	Etymology
Ki	kibi-	$(2^{10})^1 = (1\,000)^1 \times 1.024$	< kilo-
Mi	mebi-	$(2^{10})^2 = (1\,000)^2 \times 1.048\,576$	< mega-
Gi	gibi-	$(2^{10})^3 = (1\,000)^3 \times 1.073\,741\,824$	< giga-
Ti	tebi-	$(2^{10})^4 = (1\,000)^4 \times 1.099\,511\,627\,776$	< tera-
Pi	pebi-	$(2^{10})^5 = (1\,000)^5 \times 1.125\,899\,906\,842\,624$	< peta-
Ei	exbi-	$(2^{10})^6 = (1\,000)^6 \times 1.152\,921\,504\,606\,846\,976$	< exa-

It should be realized that many memories and other devices have only nominally the stated size if interpreted as precisely the binary power. Further, with the '56k' modem the symbol k is neither the decimal 1 000 nor the binary 1 024. The nominal speed is 57 600 bps, making k = 1 028.6~ so close to the binary kibi-. (The figure derives from multiplying 300, the first †baud rate above telegraph speed. This rose binarily to 4 800 before eight-layer signalling made a tripling to 14 400 bps. Successive doubling of this figure brought the 57 600.)

kil- *Metric* contracted form of †kilo-, as in kilohm = kΩ.

kilare [kilo- + are] 1 000 a = 10 ha. A rare term.

kilderkin A small cask, usually = $\frac{1}{2}$ barrel = $\frac{1}{16}$ tun.

kilerg [kilo- + erg] 1 000 erg.

kilo- [Gk: 'thousand'] Symbol now k-, until 1960 K-. *Metric* The 1 000 = 10^3 multiplier, e.g. 1 kilogram = 1 kg = 10^3 g; contractable to kil- before a vowel, e.g. 1 kilohm = 1 kΩ = 10^3 Ω.
  A legal prefix in the UK for trade only as kilometre and kilogram.[28]
  informatics. Symbol often K-. Often the value 1 024 = 2^{10}, being the binary power close to 10^3, particularly as kilobytes (KB) and kilobits (Kb). This arises as a matter of convenient terminology for memories and similar entities organized on a binary basis. But see kibi- .

kilocalorie The Calorie; *see* calorie.

kilogram [kilo- + gram] *mass*. Symbol now kg, until 1960 Kg. *Metric* 1 000 grams (= $\frac{1}{0.45359237}$ lb = 2.204 622 62~ lb), its value is that of the International Prototype Kilogramme created for the archives in 1889.
  The base unit for mass in all †m.k.s. systems, including the SI. The following are among the coherent derived units:
- kg·m^{-3} for mass density;
- kg·m^{-1}·s^{-2} = J·m^{-3} for energy density;
- kg·m^{-1}·s^{-2} = pascal for pressure, stress;
- kg·m·s^{-2} = newton for force;
- kg·m^{-1}·s^{-1} = N·s·m^{-2} for dynamic viscosity;
- kg·s^{-2} = N·m^{-1} for surface tension.

See SI unit for full repertoire.
  Note that, although the kilogram is the base unit for mass in the SI, the standard decimal prefixes do not apply to it; they apply in the original way, to the gram (= $\frac{1}{1000}$ kg) from which the kilogram is derived in the

standard metric manner. However, it is appropriate to express mass within the SI using the kilogram multiplied by powers of ten, in parallel with like multiplication of metre and other base, supplementary, and derived units. In such a context it is quite appropriate to indicate 1 gram as 10^{-3} kg, 1 milligram as 10^{-6} kg and so on. (But note that the use of powers that are not multiples of 3 is deprecated under the SI; 20 g can be written as 20×10^{-3} kg, but is not acceptable as 2×10^{-2} kg.)

History

The kilogram began, with the name **grave**, as merely the 1 000 multiple of the gram, that unit being originally defined as the mass of 1 cm³ of pure water at its maximal volumic mass. However, the kilogram soon became the reference unit of mass for the †metric system, in 1799 through the creation of a metallic †prototype **Kilogramme des Archives**. It was subsequently shown that the volume of a kilogram of water (at its specified maximum volumic mass and under normal atmospheric pressure) as measured by the prototype was discrepant from the intended cubic decimetre.[94] In 1872 the definition was changed to be direct from the prototype. Measurements in 1901 indicate that this had meant an increase of 0.0028~% for the mass units. *See* litre for ramifications.

The international prototype kilogram has been the standard for the pound of US-C since the †Mendenhall Order of 1893 declared it equal to 2.204 622 34 lb, and internationally since 1959 when the international pound was adopted as 453.592 37 g (i.e. 7 000 times the standard international grain of 64.798 91 mg) making 1 kg = 2.204 622 6~ lb. However, it was only in 1964 that the UK formally discarded its prototype pound in favour of that for the kilogram.[95]

In due course the †electron mass (m_e), †neutron mass (m_n) else †proton mass (m_p) may replace the prototype as the standard for the kilogram, as light wave lengths have replaced the prototype metre:

$$1 \text{ kg} = 1.097\,78\sim \times 10^{30}\, m_e = 0.597\,041\sim \times 10^{27} m_n = 0.597\,864\sim \times 10^{27}\, m_p.$$

Alternatively, these could be used collectively; *see* mole. Currently, however, only fundamental physical constants that involve mass, notably the †Newtonian constant of gravitation, have been seen as feasible alternatives to the prototype as standards for such a †base unit.[92]

1889 1st CGPM: referring to an established prototype block made of 90% platinum and 10% iridium, and to the kilogram as the unit of mass 'This prototype shall henceforth be considered to be the unit of mass.' The accuracy mentioned is just to 1 mg.

1901 3rd CGPM: in pursuit of clarity for mass vis-à-vis weight 'The Conference
declares:

1. The kilogram is the unit of mass: it is equal to the mass of the
international prototype of the kilogram

2. The word weight denotes a quantity of the same nature as force: the weight
of a body is the product of its mass and the acceleration due to gravity; in
particular, the standard weight of a body is the product of its mass and the
standard acceleration due to gravity

3. The value adopted in the International Service of Weights and Measures
for the standard acceleration due to gravity is 980.665 cm/s², a value already
stated in the laws of some countries.'

1967 CIPM 'declares that the rules … apply to the kilogram in the following
manner: the names of decimal multiples and sub-multiples of the unit of
mass are formed by attaching the prefixes to the word "gram".'[8]

kilogram-calorie See calorie.

kilogram-equivalent See equivalent weight.

kilogram-force force. Symbols kgf, kg-f, kilopond. Metric A unit of
weight, i.e. force, represented by 1 kilogram of mass (kg or kg-m) subject
to standard gravity; hence

$1 \, \text{kg-f} = 1 \, \text{kg-m} \times 9.806\,65 \, \text{m} \cdot \text{s}^{-2}, = 9.806\,65 \, \text{N} \, (2.204\,622\,62 \sim \text{lb or lb-f}).$
This is a base unit of the †gravitational system created from the SI units,
but is SI-deprecated. Hence

• kg-f·second per square metre for dynamic viscosity = 9.806 65 Pa·s.

kilogramme, kilogramme- See kilogram, kilogram-.

kilogram-mole See mole.

kilolitre, kiloliter [kilo- + litre] volume. Symbol now kL else kl, until
1960 Kl. Metric 1 000 litres (264.17~ US gal). A proper term for large
volumes, particularly applicable to liquids (large volumes of dry goods
usually being weighed), but rarely used. Effectively the cubic metre.

kilometre, kilometer [kilo- + metre] length. Symbol now km, until
1960 Km. Metric 1 000 metres (1 093.6~ yd). Used even astronomically.

kilopond force Metric 1 000 pond = 1 kilogram-force.

kilopound See klb.

kiloton mechanics 1 000 ton. Relative to explosions, particularly of
nuclear bombs, the energy of one thousand (short, long, or metric) tons of

the explosive trinitrotoluene (TNT) $\approx 4.2\sim$ GJ (coincidentally $\approx$ 1 Gcal). The atomic bombs dropped in 1945 on Japan had the energy of about 20 kilotons. Modern 'tactical' atomic weapons, relying primarily on fission, have a power in the range 0.1 to 20 kilotons from less than a kilogram of plutonium; 'strategic' weapons may be of several hundred kilotons, using fusion triggered by a small fission bomb. *See also* megaton.

kilovolt·ampere *power*. Symbol kVA, properly kV·A. For a †steady current, identically the product of electromotive force in kilovolts and current strength in amperes, used in rating transformers and such. *See* r.m.s. re alternating current.

kilowatt·hour *energy*. Symbol kWh, properly kW·h. For a †steady current, identically the product of power in kilowatts and time in hours, used particularly as a measure of electric energy consumed, = 3.6 MJ. *See* r.m.s. re alternating current.

UK 1882 The 'energy contained in a current of one thousand amperes flowing under an electromotive force of one volt during one hour', called also **Board of Trade unit**, and at one time the **kelvin**.

kine [Gk: 'move'] *speed* Metric-c.g.s. cm·s^{-1}, but used only briefly.[96]

kintal *mass BI See* Cental.

kip [kilo imperial pounds] *mass US* 1 000 lb.

K$_{J, J-90}$ *electromagnetics See* Josephson constant.

Kl *volume Metric* Until 1960, *see* kilolitre.

kL, kl *volume SI 1960 See* kilolitre.

klb [kilo + lb] *mass* Kilopound, i.e. 1 000 lb, the typical precision for many specifications in USA (e.g. truck weights), and used in this publication, minimally, for convenience.

Klitzing constant *electromagnetics See* von Klitzing constant.

Km *length Metric* Until 1960, *see* kilometre.

K$_m$ *biochemistry See* Michaelis constant.

km *length SI 1960 See* kilometre.

Kn *rheology See* Knudsen number.

Knoop hardness number *See* hardness numbers.

knot *length* See below.

   speed 1 †nautical mile per hour.

Internat 1955 1.852 km·h^{-1} (0.514 4$\sim$ m·s^{-1}, 1.687 8$\sim$ ft s^{-1},
   1.1508$\sim$ m.p.h.). The 1978 decision of the †CIPM considering it
   acceptable to continue to use the knot with the SI still stands.

UK To 1975, 6 080 ft·h^{-1} (1.853 2 km·h^{-1}, 0.514 77$\sim$ m·s^{-1},
   1.151 5$\sim$ m.p.h.).

USA To 1954, 6 080.2 ft·h^{-1} (1.853 2 km·h^{-1}, 0.514 79$\sim$ m·s^{-1},
   1.151 6$\sim$ m.p.h.).

   The term relates to knotted markers at regular intervals on the log line, this being drawn out from a ship's stern by a float that would effectively stay fixed (relative to the sea rather than earthly position, hence measuring speed relative to the surface currents rather than the map). It was common in earlier times to have a sand-glass of appropriate duration against which to count the knots being drawn out, e.g. a 28-second timer and knots tied at nearly 4 fathoms spacing (correctly 3.985 2$\sim$ fathom, 14.41$\sim$ m, 47.29$\sim$ ft) would give a direct result in knots. As a unit of length, the term knot applies to this distance. (However, confusion sometimes results in the term being used to mean the nautical mile, with the speed expressed in knots per hour; such usage is grossly erroneous.)

Knudsen number [M. H. C. Knudsen; Denmark 1871–1949] *rheology.* Symbol *Kn.* Relating to momentum transport in rarefied gases, the dimensionless ratio of the mean free path of the molecules to a representative length.[61]

kunitz unit [M. Kunitz; Russia, USA] *biochemistry* A unit specific to the enzyme ribonuclease, being the amount required to cause a decrease of 50% per minute in the ultraviolet light of wavelength 300 nm absorbed at 25°C by a 0.05% solution of yeast nucleic acid in a 0.05 molar acetate buffer solution of pH 5.0.

kV · A *electrics See* kilovolt·ampere.

kW · h *electrics See* kilowatt·hour = 3.6 MJ.

kW · yr *electrics See* Kilowatt·year = 8 766 kW·h $\approx$ 32 GJ.

L

L, l Generally as **L**, the †Roman numeral for 50, notably in post-Roman use.

   volume *Metric* As **l** else, since 1979 and now preferably, as **L**, *See* litre. Also prefixed variously, as in mL = millilitre. Note that, unique among SI symbols, the initial letter of litre can be expressed officially in upper or lower case, the standard use of lower case for the original metric units being varied to avoid the confusion of lower-case letter l with the digit 1 that has occurred so widely on printed matter. (Since the litre is a derived unit, and derived units generally have upper-case symbols, the use of upper-case L is far from inconsistent.) The alternative technique of adopting italics or some other graphic variation is deprecated.

   fundamental constant As *L*, *see* Avogadro constant.

   luminance *Metric-c.g.s. 1920* As *L*, *See* lambert.

lambda [name of the Greek letter 'l'] *length*. Symbol λ. A superfine measure of equipment and machining tolerance, based now usually on the wavelength of light from a neon laser ($0.632\,8\sim\mu$m, $0.0249\,1\sim \times 10^{-3}$ in), formerly usually on the mercury green line ($0.546\sim \mu m$, $0.021\,5\sim \times 10^{-3}$ in). Precision in grinding astronomical mirrors, for instance, is often expressed as a fraction of lambda, e.g. lambda-by-forty means the said value divided by 40.

   volume An old name for the microlitre, adopted by the †BIPM of 1880, but now an SI-deprecated unit.

   chemistry Former name for the microlitre.

lambert [J. H. Lambert; Germany 1728–77] *luminance, irradiance*. Symbol L. *Metric-c.g.s. 1920* For a point on a diffuser, identically candela·steradian per square centimetre = $cd\cdot sr\cdot cm^{-2}$, = $10^4 cd\cdot sr\cdot m^{-2} = 10^4 lm\cdot m^{-2}$. (The corresponding m.k.s. unit was the apostilb, = 10^{-4} L. There is no corresponding specially named unit in the SI, but the lux or illuminence has the same basic dimensions as the apostilb.) For a perfect diffuser, 1 L corresponds to the emission of $\frac{1}{\pi}$ $cd\cdot cm^{-2}$, = $0.318\,3\sim cd\cdot cm^{-2}$, = $0.318\,3\sim$ stilb. The millilambert (mL) was usual.

langley [S. P. Langley; USA 1834–1906] *astrophysics* Metric-c.g.s. Of solar
radiation, the energy-flux density of one 15°-calorie per square
centimetre, else, as used originally, the rate of the said amount per
minute.[97] The amount = 41.840 kJ·m^{-2} (41.840 Mg·s^{-2}), the rate
= 697.33~ W·m^{-2}, which is about half the average solar radiation rate
for Earth. Effectively re-named in 1942 as the pyron.

latitude *geography, astronomy* Applying to a point on a spherical or
similar surface (the heavens being effectively a celestial sphere
surrounding Earth), the distance of the point from a datum bisecting
section, usually in terms of the angle subtended at the centre of the
sphere by the direct line along the spherical surface from the point to the
datum plane. For a rotating sphere that plane is routinely the bisecting
surface transverse to the axis of rotation (notably the equatorial plane for
Earth, the ecliptic for the heavens). Points of equal latitude within one
hemisphere form a ring that is the intersection with the spherical
surface of a plane parallel to the datum plane, so such a ring is routinely
called a **parallel**, accorded the definite article if qualified by a specific
latitudinal angle, e.g. the 60th parallel for the ring of points subtending
60°. For everyday use, latitude is expressed in (normal geometric) degrees
north else south of the datum plane (i.e. the equatorial plane for Earth),
hence ranges 0° to 90° arithmetically, 90°S to 90°N geographically.
    Latitude is normally paired with **longitude** to provide unique
coordinates for a point. While expressed also in terms of a central angle,
longitude differs in its method by relating to the angle subtended at the
centre of the relevant parallel plane by the surface line along the parallel
from the point to a datum plane transverse to the parallel plane and
passing through its centre. Further, there is rarely an inherent datum
position for longitude in any circumstances. The worldwide standard
datum reference for locations on Earth has, since 1884, been the
Greenwich meridian, i.e. the planar semicircle passing through the
North and South Poles (hence including the axis) and a specific point
within the national observatory at Greenwich, England (*see* Greenwich
Mean Time for particulars and additional discussion); previously each
country of significance had its own datum line. Longitude is normally
expressed in degrees east else west of that meridian, and hence ranges 0°
to 180°W, 0° to 180°E geographically. Points of equal longitude within
one hemisphere form a semicircle that is the intersection with the
spherical surface of a plane containing the axis, called a **meridian** for a
hemisphere stretching Pole to Pole (and accorded the definite article if

qualified by a specific longitudinal angle). The distance between two points of different longitude but identical latitude depends on the value of the common latitude as well as their distinct longitudes, while the distance between two points of different latitude but identical longitude depends only on their latitudes.

Because Earth is not a true sphere, the angle of latitude differs as to whether it relates to the line through the centre of Earth or the line †orthogonal to the local tangent plane (i.e. the line of a plumb-line). Called respectively **geocentric** and **geodesic** latitude, the latter is the only practical one. (The two are identical at the Equator and at the Poles, otherwise the geodesic is greater. The difference increases progressively with increasing latitude until 45°, at which it is about a fifth of a degree; then it decreases progressively. An interesting consequence is that the natural †nautical mile increases progressively from Equator to Poles, despite the radius decreasing.)

Terrestrial latitude and **terrestrial longitude** refer to the familiar latitude and longitude for Earth's surface, measured as indicated relative to the Earth's equatorial plane and the Greenwich meridian. Duly qualified and defined, latitude and longitude are applied similarly to the Sun, its other planets, and other objects in the solar system.

The **celestial latitude** of a star or such is measured relative to the ecliptic, with Earth as the centre; **celestial longitude**, which ranges 0 to 360°, has the †First Point of Aries as its zero value, a moving reference reset each year. *Compare* right ascension for a form of longitude for celestial objects that uses Earth's equatorial plane as a datum, the related latitude being called declination.

The **galactic latitude** of a star within our galaxy (the Milky Way) is measured relative to the plane of symmetry of the galaxy (the galactic equatorial plane, conventionally accepted as being at angle 62° to the celestially extended equatorial plane), with the nearest point in the galactic plane to Earth as its centre. The centre of the Galaxy (in constellation Saggitarius, at sidereal hour angle 94.4° along the galactic Equator) is given as 0° **galactic longitude**, which is routinely positive.[22]

lb [Lat: libra] *mass See* pound.

lb-f *force See* pound-force.

lb·ft *torque See* pound·foot.

Le, Le *rheology See* Lewis number.

league *length* On land, usually 3 (statute) miles; in a maritime context three †nautical miles.

leap second *time* A second of time added to the duration of a day, as measured by the atomic clock, to maintain consistency with the astronomically measured diurnal pattern; *see* Universal Time. A true parallel to the leap year, after which it is named, would call the relevant day a leap day rather than the added second a leap second. It should be noted, however, that there may be more than 1 second added on any such day and, alternatively, second(s) could be subtracted – a leap backwards.

The leaps are applied at the close of the day, meaning that the final minute and hour, along with the clock day, are discrepant from their standard definitions.

leap year *time* The year with an added day to correct for Earth's time for circling the Sun being a fraction more than 365 days. *See* calendar.

legal size *paper and printing* North America A sheet of 8.5 in × 13.5 else 13 in (215.9 mm × 342.9 else 330.2 mm). *See* foolscap.

Leiden jar *electric capacitance* *See* jar.

lentor *rheology* Metric-c.g.s. An early name for stokes, the unit of kinematic viscosity.

Leo *astronomy* *See* zodiac; right ascension.

leo [Galileo Galilei; Pisa, Florence 1564–1642] *acceleration* Metric dam·s^{-2}, = 10m·s^{-2} (32.808 4~ft·s^{-2}),[11] very close to normal gravitational acceleration at Earth's surface, and to †standard gravity.

letter size *paper and printing* North America A sheet of 8.5 in × 11 in (215.9 mm × 279.4 mm). *See* quarto.

Lewis number [G. W. Lewis] *rheology*. Symbols Le, *Le*. A dimensionless quantity characterizing a particular substance, the ratio of the thermal diffusivity to the diffusion coefficient, identically the ratio of the †Schmidt number to the †Prandtl number.

Leyden jar *electric capacitance* *See* jar.

LF [Low Frequency] *physics* The waves of the †electromagnetic spectrum of about 100 kHz frequency, used, *inter alia*, for long-wave radio.

Libra *astronomy* *See* zodiac; right ascension.

lifetime *time* For the human being, 70 years is the typical inference, though not only is reality very variable, but the average has reached that figure only recently for any significant proportion of the world's people.

light watt *photics* The nominal unit in the reciprocal ratio of the efficacy of a source of light to the maximum efficacy of 680 lm·W^{-1}, which occurs at a wavelength of 555 nm (for the light-adapted eye). (The reciprocal of this maximum, i.e. 1.47$\sim$ mW·lm^{-1}, is the **mechanical equivalent of light**.)

light-microsecond *astronomy* The distance travelled by light in 1 μs, = 299.792 458 mm (11.803$\sim$ in). Not much used, but the natural unit that would have better served as the metric unit of length (*see* metre).

light-nanosecond *astronomy*. Symbol lns. The distance travelled by light in 1 ns, = 0.299 792 458 mm (0.011 803$\sim$ in).

light-second *astronomy* Symbol ls. The distance travelled by light in 1 s, = 299.792 458 $\times$ 10^3 km (1.862 8$\sim$ $\times$ 10^5 mi, 0.002 003 960$\sim$ AU).

light-year *length*. Symbol ly. The distance travelled by light in one mean year, = 9.460 536$\sim$ $\times$ 10^{12} km (63 239.8$\sim$ AU, 5.878 505$\sim$ $\times$10^{12} mi) $\approx$ 10^{-10} hubble.

line *length* UK trad $\frac{1}{12}$ in = 2.116 7$\sim$ mm; elsewhere sometimes $\frac{1}{10}$ inch.

line of electrostatic induction *electromagnetic induction* Metric-c.g.s. The equivalent of the maxwell.

line of induction *magnetic flux* Metric-c.g.s. Identically the maxwell.

link *length* $\frac{1}{100}$ chain, hence:
    Gunter's link, in USA = **surveyors' link**
    Ramsden's link = 12 in (304.8 mm).
    Rathborn's link = 3.96 in (100.58 mm).
    surveyors' link = 7.92 in (201.16$\sim$ mm).

liq. oz., liquid ounce *See* US fluid ounce.

litre, liter *volume*. Symbol l, since 1960 also L. *Metric* Perhaps the most familiar and most used but the most problematical of the metric units, the litre is simultaneously very convenient yet utterly superfluous, since volume can be expressed in terms of the cube of the linear unit. In fact it was not in the original †metric system of 1791, being introduced two years later, as a sop to public demand and (with the †are for area) the first example of official concession in the fight to establish what we now call metrication. Initially it was named the pinte, the cadil, and the litron (oddities in the context of post-revolutionary France, being relics from monarchist days, while the other units had been accorded original names), before acquiring its now-familiar name two years later. Of more concern to measurement, the litre has been ambiguous in its size, albeit only very modestly so.

The litre was conceived as 1 dm^3 and identically the volume of 1 kg of water, since the gram was defined as the mass of 1 cm^3 of water. However, following the creation in 1799 of the physical metallic-block †prototype kilogram to avoid the use of the water standard, and as accurate measuring became more precise, it was discovered that the volume of a kilogram of water (at its specified maximum volumic mass and under normal atmospheric pressure) as measured by the prototype was discrepant from the intended 1 dm^3. Construction of a revised prototype kilogram to fit the litre (hence the extant definition of the gram) was considered but rejected in 1872. Instead, the international committee asserted the 1799 prototype as defining the kilogram.[94] This re-sized the kilogram, but no change was made to the litre, hence entrenching the discrepancy. Careful measurement in 1901 gave the mass of 1 dm^3 as only $0.999\,970\,7\sim$ kg.[98] The first official reaction to this discrepancy was to continue to leave the litre unchanged; thus, effective from 1901, it was accepted as $1.000\,029\sim \text{ dm}^3$, making 1 mL not equal to 1 cm^3 or '1 cc' (and the litre not †coherent as it had been ostensibly in the rarely used †d.k.s. system). (Later measurements gave marginally smaller figures;[99] see also below.)

The 1901 definition was formally abrogated in 1964, with the world congress asserting that the litre be regarded as a special name for, and hence identically the intended 1 dm^3; simultaneously, though, it recommended the unit not be used in high-accuracy contexts. The litre is not part of the SI, though it is admitted for use with it. It is now normally regarded as precisely 1 dm^3 for most scientific use, but presents problems relative to extant standard volumetric vessels. Working to five significant figures, the discrepancy affects only a minority of values, and them only to one unit in the last place; hence it can be effectively ignored in such a context. Where greater precision is intended, use of the unambiguous term dm^3 (or similar) is appropriate and recommended, though scientific practice probably points to a complete equating of the two.

The final problem with litre concerns its abbreviation; officially it has long been the lower-case initial. However, since this character is often not clearly distinguishable from the unit digit, that official form is unsatisfactory for many documents. The upper-case initial is the simplest and commonest solution, but a switch to italics or a different (e.g. cursive) face is an alternative that can meet the lower-case policy. International standards allowed the capital as of 1979, but deprecate the use of italics or alternative faces. The capital form must be expected to prevail. (The litre not being a base unit, there is nothing inconsistent with the use of a capital rather than a lower-case letter.)

1901 3rd CGPM: '*The Conference declares:*
   1. The unit of volume, for high accuracy determinations, is the volume occupied by a mass of 1 kilogram of pure water, at its maximum density and at standard atmospheric pressure: this volume is called the "litre".'

1960 11th CGPM: '*considering* that the cubic decimetre and the litre are unequal and differ by about 28 parts in 10^6, that the determinations of physical quantities which involve measurements of volume are being made more and more accurately, thus increasing the risk of confusion between the cubic decimetre and the litre, *requests* the CIPM to study the problem and submit its conclusions to the 12th CGPM.'

1961 CIPM: 'The CIPM requests that the results of accurate measurements of volume be expressed in units of the International System and not in litres.'

1964 12th CGPM: '*considering* Resolution 13 adopted by the 11th CGPM in 1960 and the Recommendation adopted by the CIPM in 1961
   1. *abrogates* the definition of the litre given in 1901 by the 3rd CGPM
   2. *declares* that the word "litre" may be employed as a special name for the cubic decimetre
   3. *recommends* that the name litre should not be used to give the results of high-accuracy volume measurements.'

1979 16th CGPM: '*recognizing* the general principles adopted for writing the unit symbols in Resolution 7 of the 9th CGPM (1948), *considering* that the symbol l for the unit litre was adopted by the CIPM in 1879 and confirmed in the same Resolution of 1948, *considering* also that, in order to avoid the risk of confusion between the letter l and the number 1, several countries have adopted the symbol L instead of l for the unit litre, *considering* also that the name litre, although not included in the International System of Units, must be admitted for use with the System, *decides*, as an exception, to adopt the two symbols l and L as symbols for the unit litre *considering* further that in the future only one of these two symbols should be retained, *invites* that the CIPM to follow the development of the use of these two symbols and to give the 18th CGPM its opinion as to the possibility of suppressing one of them.'[8]

livermore loops [a laboratory in California] *informatics* A specific computer program package representative of intensive calculation typical of nuclear physics in the 1980s, used as a benchmark to measure the power of computers, usually as †MFLOPS, by standardized conversions.

livestock unit *agriculture* A measure of equivalent grazing or other load for the various types of livestock, in such scale as:
0.7 camels = 0.8 water buffalo or horses = 1 cow = 4 pigs = 8 sheep/goats. However, the relative loads vary considerably even within a species, between breeds, by sex, and with maternal status; clearly a Clydesdale can far out-eat a small horse, and a suckling sow needs much more than an average pig. In North America the **cow-calf unit** is a common livestock unit, being equated to an average beef cow with calf at foot.

lm *photics See* lumen; *see also* SI alphabet for prefixes.

ln *mathematics* Abbreviation for natural †logarithm.

lns. *astronomy See* light-nanosecond.

log *See* logarithm.

logarithm [Gk: λογοσ 'ratio'] *mathematics*. Symbol log. For any positive number N, and some reference base number b, also positive, the power to which b must be raised to equal N. That is, if $b^n = N$, then n is said to be the logarithm of N to the base b, usually written
$n = \log_b N$
or, where b, commonly 10, is implicitly known, just
$n = \log N$.
Inversely, N is said to be the **antilogarithm** of n (briefly **antilog** n).
The advantage of logarithms is the facilitation of multiplication and exponentiation, including the derivation of square and other roots, particularly non-integral roots. The first advantage comes from the fact that multiplication of numbers equates with addition of indices. That is
if $M = b^m$ and $N = b^n$, then $M \cdot N = b^{m+n}$ and $M \cdot N = $ antilog $(m + n)$.
The advantage rises with the digit-length of the numbers. Until the

emergence late in the 20th century of the electronic calculator, logarithms were the essential means for any extensive calculation with many-digit numbers. Even with the powerful assistance of the computer, logarithms remain the means for elaborate root taking, using the fact that

if $N = b^n$ and r is positive, then $N^r = b^{n \cdot r}$ and $r\sqrt{N} = N^{\frac{1}{r}} = b^{n/r}$.

With electronic calculators, the machine calculates any required logarithm, as it does for cosines and other trigonometric values, using the various mathematical formulae. However, prior to the calculator, all such values were obtained by table look-up and interpolation, a process laborious in itself but far less so than raw calculation. To be useful, such tables had to be consistently on one base, and have a matching reciprocal set, i.e. a table of antilogarithms showing the value of b^n for all values of n as well as the table of log N for all N.

John †Napier is credited with inventing this technique for converting multiplicative problems into additive ones, etc., but he achieved this by a graphical technique with one axis the normal uniformly spaced †arithmetic scale, the other one with progressively shortening steps that we would now call †logarithmic. His base was apparently very large, though his name is often attached, in the form **napierian logarithm**, to logarithms with their base the special number †e ($= 2.718\,281\,8\sim$); usually such logarithms are termed **natural logarithms**. Besides having the general style $\log_e N$, natural logarithms are often written as ln N (allowing the common log to the base 10, discussed next, to be written without suffix).

Because of our decimal number system, manual calculation is far more convenient with 10 as the base; it particularly restricts the extent of the tables needed, since, using the unqualified log to mean its usual $\log_{10}$,

$\log(10^c \cdot N) = \log\ 10^c + \log\ N = c + \log\ N$.

Using the appropriate integer c (called the **characteristic**), this means that all needs can be met by just the logarithms for numbers between 1 and 10 (with log values between 0 and 1, this fractional part being the **mantissa**); antilogs can be likewise contained, to values between 0 and 1. The advantages of base 10 were first noted by H. Briggs, who constructed the first pertinent tables, so this style is sometimes called †**briggsian logarithm**; but more usually the **common logarithm**, or just logarithm, unqualified.

See †negative number for a note on the unusual notation for negative-valued logarithms.

logarithmic [from logarithm] *mathematics* A logarithmic scale has member values corresponding with the logarithm of the values of the measured entity; it is synonymously a †geometric scale.

The term logarithmic scale applied to the axis of a graph indicates that equal-length steps along it represent recurrent multiplication of the actual variable by a constant amount (of value depending on the size of the steps). If that axis is for observed values and the other axis is linear, then it represents a geometric measurement scale for the observed variable. If both axes are logarithmic (as on 'log-log' paper), then, if the fundamental variables are x and y, the graphical relationship is between $\log x$ and $\log y$, e.g. if linear

$a \cdot \log\ x = c + b \cdot \log\ y$, i.e. $\log\ x^a = c + \log\ y^b$

for some values a, b, and c; hence, assuming common logs,

$x^a = 10^c \cdot y^b$.

logit [from logarithm + digit] *See* decibel.

long., longitude *See* latitude.

Lorentz unit [H. A. Lorentz; Netherlands 1853–1928] *photics See* Zeeman splitting constant.

Loschmidt constant [J. Loschmidt; Austria 1821–95] *fundamental constant*. Symbol n_0. From Loschmidt's number $= 2.686\ 777\ 5(23)$ $\times 10^{25}$ m^{-3} with †relative standard uncertainty 1.7×10^{-6}.[4]

Loschmidt's number *fundamental constant* The number of molecules per cubic metre in an ideal gas at †s.t.p., $= 2.686\ 8\sim \times 10^{25}$, but *see* Loschmidt constant. (Originally defined within metric-c.g.s. as per cubic centimetre, $= 2.686\ 8\sim \times 10^{19}$.) The value equals †Avogrado's number divided by †mole volume in relevant volumetric unit. It has itself been called Avogadro's number.

loudness unit *acoustics*. Symbol LU. *USA 1942* For magnitude of auditory sensation,[100] now superseded by the sone, with 1 sone $=$ 1 000 LU.

***l*P** *sub-atomic physics See* Planck length.

ls *astronomy See* light-second.

L.U. *acoustics See* loudness unit.

lumberg *quantity of light See* lumerg.

lumen [Lat: 'light'] *luminous flux*. Symbol lm. *SI, Metric-m.k.s. 1948*
Having been proposed as lumen in 1896 and long accepted[101], then
accepted into the SI in 1948 as **new lumen**, identically
candela·steradian (cd·sr). The following are among the coherent derived
units:

- $lm·m^{-2}$ = lux for luminous exitance, illuminance;
- $lm·W^{-1}$ for luminous efficacy;
- lm·s for quantity of light;
- lm·h for quantity of light, used, similarly to kW·h, in illumination.

The lumen is the luminous flux emitted from a point-source of uniform
intensity of 1 candela into unit solid angle; the total spherical intensity of
such a source is therefore 4π lm = 12.566 37$\sim$ lm. The power of such a
flux depends on the wavelength of the light, 540 THz being the standard
for the candela.

History
The term appears to have been introduced before 1900, and adopted
officially in France in 1919, applying to the candle in its successive forms,
before being adopted internationally (as new lumen) in 1946 with the
'new candle'; *see* candela for the history of the candle.

1946   CIPM: '4. *New lumen* (unit of luminous flux). – The new lumen is the
       luminous flux emitted in unit solid angle (steradian) by a uniform point
       source having a luminous intensity of 1 new candle.'
1948   9th CGPM 'lumen' adopted in place of new lumen.[8]

lumerg [luminous erg] *quantity of light* Metric-c.g.s. cd·sr·s, = lm·s.

lump *engineering* For ore processing, material sieving 35 to 88 mm.

lunation Synodic month.

Lundquist number, Lundqvist number *rheology* A dimensionless
quantity characterizing the unidirectional Alfven waves occurring in a
conducting fluid (such as a physical plasma) subject to a magnetic field.
The number is the ratio of the product of the magnetic flux density, the
electrical conductivity and a representative length to the square root of the
ratio of permeability to density.[102]

lune *time* The 'day' of which precisely 30 span the mean interval
between consecutive new moons, = $\frac{1}{30}$ †synodic month,
= 85 048.1$\sim$s (23.624 4$\sim$ h, 0.984 353$\sim$ d) currently, but is slowly
increasing.

lunour	...	...	59.1~ min
24	lune	...	23.6~ h
	30	synodic month	29.6~ d

lunour *time* The 'hour' of the 'day' of which precisely 30 span the mean interval between consecutive new moons, $= \frac{1}{24}$ lune, $= 3\,543.67\sim$ s (59.061 2~ min, 0.984 353~ h) currently, but is slowly increasing.

lusec [litre per sec] *high-vacuum technology* An evacuation rate of 1 litre per second when at a pressure of only 1 μm of mercury (i.e. at 0.136 Pa). *See also* clusec.

lux [Lat: 'light'] *illuminance*. Symbol lx. *SI 1948* Having been in use for some years,[101] identically lumen per square metre, $= \mathrm{lm \cdot m^{-2}}$ $= m^{-2} \cdot cd \cdot sr$. Hence $\mathrm{lx \cdot s^{-1}}$, $\mathrm{lx \cdot h^{-1}}$ for light exposure.

luxon *optometry See* troland.

lx *photics SI See* lux. Can be prefixed variously, as in klx = kilolux; *see also* SI alphabet.

ly *astronomy See* light-year.

M, m [notably from Lat: mille, = 1 000] The †Roman numeral for 1 000.
Metric As an upper-case prefix, **M-**, *see* mega-, e.g. MHz = megahertz.
Hence more generally as **M**, *see* million.
Metric As a lower-case prefix, **m-**, *see* milli-, e.g. mg = milligram.
North America As **m**, thousand, then mm = million and so on, particularly
in business literature (and in contrast with the metric meaning of a
thousandth), e.g. as in mFBM for 1 000 FBM of sawn wood, mm.c.f. for
1 000 000 cubic feet of natural gas.

length *Metric* As **m**, *see* metre. Also prefixed variously, as in mm =
millimetre(s); *see also* SI alphabet. Strictly lower case.
UK On highway signs, as **M**, metre; as **m**, mile: an unfortunate practice
when the lower-case letter is the only proper symbol for metre, but
understandable given the widespread investment in and established
use of m for mile.

electromagnetics *Metric* **M** is sometimes used without the alien Ω to
represent MΩ, i.e. megohm.

electrics and mechanics [magnification] *See* Q factor.

astrophysics For †stellar magnitude, **M** is absolute magnitude, **m** is
apparent magnitude.

paper and printing Often used Roman-style for 1 000 sheets, etc.

m^3gas Cubic metre of natural gas, nominally of energy content
37.26 MJ; *see* b.o.e.

M-, m- *Metric* As symbol, *see* M, m.

Ma *time* With geochronologic scale, a †mega-†annum, i.e. 1 000 000
years.

rheology *See* mach.

mach (number) [E. Mach; Austria 1838–1916] *rheology*. Symbol *Ma*.
Relating to momentum transport, the dimensionless ratio of the speed of
the object to the speed of sound, in the same situation and conditions.[61]
Values less than 1 are termed 'subsonic' and those greater than 1
'supersonic' or, if significantly higher, 'hypersonic'.

Since the speed of sound varies considerably with both pressure and volumic mass of a medium, translation between Mach number and more familiar expressions of speed is local in context. In the atmosphere the speed of sound, and hence the meaning of 'Mach 1', varies from about $340 \, \text{m·s}^{-1}$ ($1\,115 \, \text{ft s}^{-1}$, 760 m.p.h.) in air at 20°C near sea level to about $290 \, \text{m·s}^{-1}$ ($950 \, \text{ft s}^{-1}$, 650 m.p.h.) at an altitude of 11 000 m (35 000 ft).

mache [E. Mache; Germany 1876–1954] *radiation physics* Of radioactive emanation, that which sets up a saturation electric current of 10^{-3} statampere, $= 3.6 \times 10^{-10}$ curie $= 1.368{\sim}$ Bq.

MacMichael degree *rheology UK* A scheme for measuring the viscosity of a fluid as indicated by its transference of torque, being the twist of a cylinder immersed in the rotating fluid. The reading depends on the stiffness of the suspending wire as well as the diameter of the cylinder; typically 1 to 3° MacMichael is 1 poise (10^{-1} Pa·s).

Madelung constant *physical chemistry* A value of electric charge characterizing ionic crystals of a specific chemical compound, being, for a reference ion, the product of the nearest-neighbour distance and the cumulative sum for all local ions of the ratio of charge to distance.[103]

magic number *sub-atomic physics* The number of electrons else of protons/neutrons characterizing very stable conditions, corresponding to filled shells in the relevant structure. Magic numbers include the atomic numbers of the noble gases (helium = 2, neon = 10, argon = 18, krypton = 36, xenon = 54, radon = 86) and, for nuclei, the values 2, 8, 20, 28, 50, 82, and 126, plus possibly higher values for artificial elements.[104]

magnetic constant *electromagnetics.* Symbol μ_0. *SI* The adopted value for the magnetic permeability (*see* permittivity) of a vacuum $= 4\pi \times 10^{-7} \, \text{H·m}^{-1} = 1.256\,6{\sim} \times 10^{-6} \, \text{N·A}^{-2}$.

magnetic flux quantum *electromagnetics.* Symbol Φ_0. The ratio of the †Planck constant to twice the †elementary charge $= 2.067\,833\,636(81) \times 10^{-15}$ Wb with †relative standard uncertainty 3.9×10^{-8}.[4]

magnetic Reynolds number Symbol Re_m, Rm. *See* Reynolds number.

magnetic temperature *mechanics See* Curie temperature scale.

magneton *sub-atomic physics* For magnetic intensity, a unit of magnetic moment proportional to the ratio of charge to rest mass, being qualified as the **Bohr magneton** (symbol μ_B) for an electron, the **nuclear magneton** (symbol μ_N) for a proton. The value of each magneton is $e \cdot h/4\pi \cdot m$ where m is the rest mass of the respective particle. The values,[4] each of †relative standard uncertainty 4.0×10^{-8}, are

for the electron $\mu_B = e \cdot h/4\pi \cdot m_e$ $9.274\,008\,99(37) \times 10^{-24}\,\text{J} \cdot \text{T}^{-1}$
for the proton $\mu_N = e \cdot h/4\pi \cdot m_p$ $5.050\,783\,17(20) \times 10^{-27}\,\text{J} \cdot \text{T}^{-1}$

These are used as units of magnetic dipole in the fields of electron-spin resonance and nuclear magnetic resonance spectroscopy, respectively.

magnification *electromagnetics*. Symbol m. *See* Q factor.

magnitude *astronomy* For stars, *see* stellar magnitude.
  geophysics For earthquakes, *see* Mercalli scale; Richter scale.

mantissa *mathematics See* logarithm.

Margoulis number [M. Margoulis; France 1856–1920] *rheology*. Symbol *Ms*. *See* Stanton number.

mass energy conversion factor *physics* The energy equivalent of the dalton, i.e. the mass of 1 proton, neutron, or such, $= 931.501\,6\sim$ MeV.

mass transfer factor *rheology* $St^* \times Sc^{2/3}$ where St^* is the †Stanton number for mass transfer and Sc is the †Schmidt number.

maxwell [J. C. Maxwell; UK 1831–79] *magnetic flux*. Symbol *Mx*. *Metric-c.g.s.-e.m.u.* and *-c.g.s.-Gaussian 1930* Identically abvolt·second, i.e. the magnetic flux whose expenditure in 1 second produces 1 abvolt per turn of a linked circuit. Technically defined in a three-dimensional system, it corresponds in the four-dimensional electromagnetic sector of the †SI system to 10 nWb, and is an impractically small unit. In use for some years, the name was agreed by the International Electrotechnical Committee in 1930, along with a corresponding †practical unit, the pramaxwell (or pro-maxwell) $= 10^8$ maxwell.[71] Equivalently a line of induction; *see also* kapp line, $= 600$ Mx.

mayer [J. R. von Mayer; Germany 1814–78] *heat capacity Metric* A scarcely used term $= \text{J} \cdot (\text{g} \cdot \text{K})^{-1}$ ($= 1\,000\,\text{m}^2 \cdot \text{s}^{-2} \cdot \text{K}^{-1}$ in base terms).[105]

mb *pressure Metric* An improper representation for millibar; *see* bar.

mc.d. *radiation physics France See* millicuries-destroyed.

Table 32

cu. ft					
1 000	m.c.f.	...	...		thousand cubic feet
1 000 000	1 000	mm.c.f.	...		million cubic feet
1 000 000 000		1 000	b.c.f.	...	billion cubic feet
1 000 000 000 000			1 000	t.c.f.	trillion cubic feet

m.c.f. [Roman numeral m, then cubic feet] *petroleum* North America Particularly for natural gas, $1\,000\,\text{ft}^3$ ($28.316\,85\sim \text{m}^3$). The standard range of units for natural gas is as shown in Table 32.

mcg [microgram] *mass* North America Particularly for ingredients of consumer pharmaceutical products, alongside the international mg for milligram, a representation of microgram that avoids use of the alien graphic μ; *see* micro-. (Though improper within the rules of the SI, the prefix for mc could well join the prefix da that replaced the distinct capitalized D for deca- as a practical solution to an inherent problem. The letter u is an alternative.)

m_d *sub-atomic physics* See deuteron mass.

m_e *sub-atomic physics* See electron mass.

mean See average.

measurement ton As maritime unit, *see* shipping ton.

mebi- [mega- binary] *informatics*. Symbol Mi. The $1\,048\,576 = 2^{20}$ multiplier, as mebibytes (MiB) and mebibits (Mib). *See* kibi-.

mechanic *acoustics* See mechanical ohm.

mechanical equivalent For heat, *see* calorie; for light, *see* light watt.

mechanical ohm *acoustics* Metric Originally coined in a general mechanical context as a parallel of the ohm used for electric impedance,[96] a unit for the mechanical impedance to sound waves. Quantitatively, at a given surface, it is the ratio of average effective pressure (i.e. force) to the volume velocity (i.e. area times speed †orthogonally thereto) of the resulting waves through it (but *compare* acoustic ohm[3]). E.g.
Metric-m.k.s. $\text{N}\cdot(\text{m}^{-1}\cdot\text{s}^1) = \text{N}\cdot\text{s}\cdot\text{m}^{-1} = \text{m}\cdot\text{kg}\cdot\text{s}^{-1}$.

Metric-c.g.s. $\text{dyn} \cdot (\text{cm}^2 \cdot \text{cm} \cdot \text{s}^{-1})^{-1} = \text{dyn} \cdot \text{s} \cdot \text{cm}^{-1} = \text{cm} \cdot \text{g} \cdot \text{s}^{-1}$.
Called also **mechanic ohm** and, for the latter, **mechanic abohm**.

median *See* average.

meg- *Metric* Contracted form of †mega- , as in megohm = MΩ.

mega- [Gk: 'great'] Symbol M-. *Metric* The 10^6 multiplier, e.g. 1 megajoule = 1MJ = 10^6 J; contractable to meg- before a vowel, e.g. 1 megohm = 1 MΩ = 10^6 Ω.
   informatics Sometimes $1\,048\,576 = 2^{20}$, but *see* mebi- then kibi- .

megaline *magnetic flux* 10^6 line of electrostatic induction.

megaton *mechanics* Relative to explosions, particularly of hydrogen bombs, the energy of 1 million (short, long, or metric) tons of the explosive trinitrotoluene (TNT) $\approx 4.2\sim$ TJ (coincidentally, $\approx$ 1 Tcal).

mel [melody] *acoustics, music* USA 1951 A linear scale for subjective pitch, referenced to 1 000 mel for a pure tone of 1 kHz.[106]

Mendenhall Order *USA* Issued 5 April 1893 by Thomas C. Mendenhall, this order is usually seen as defining the †US Customary units in terms of metric units instead of their own †prototypes.
   Mendenhall, long an advocate of the †metric system, was at the time Superintendent of the Coast Survey, and thereby responsible for the modest national Office of Weights and Measures. His action, concurred with by the Secretary of the Treasury, effectively implemented a statute of 1866 that legalized the use of metric units. The values in the Order were
   1 m = 39.37 in,   giving 1 in = 25.400 051∼ mm
   1 kg = 2.204 622 34 lb, giving 1 lb = 453.592 43∼ g
The former was precisely that of the 1866 statute; the latter differed from the statute slightly, because of the more recent appraisal of the reference standard. The exact expression of metric units in US units, as distinct from the exact expression of US (and British) in metric as applies now, pointed to the facilitating of conversion from metric to local customary units.
   The 1866 statute, in authorizing the use of metric in contracts, etc., and specifying their values in normal US units, did not invalidate the basing of the national units on the prototype yard and pound. The Mendenhall Order, in contrast, is invariably interpreted as redefining those national units in metric, though argument has been made that, since it lacked legislated authority, it could not effect the consequent change, albeit minor, in those units. Indeed, it has been said that the

order merely defined the metre and the kilogram for American consumption, changing them slightly in size! Whatever the legality, the reality is that it changed the basing of the US units. Mendenhall's definitions lasted until 1959 when slightly different international definitions became the law; they are:

$$1\,in = 25.4\,mm, \quad giving\ 1\,m\ = 39.370\,078\,7{\sim}\,in$$
$$1\,lb = 453.592\,37\,g, giving\ 1\,kg = 2.204\,622\,62{\sim}\,lb$$

mensem Month, hence **per mensem** is per month.

mEq Milliequivalent; *see* equivalent weight.

Mercalli scale [G. Mercalli; Italy 1850–1914] *geophysics* A qualitative scale for the intensity of an earthquake based on observed effects (hence its values are local, and various for any one earthquake), ranging from 1 upwards, often written using †Roman numerals. Created as a 10-level scale in 1902[107] from the widely used Rossi Forel scale of 1883, the Mercalli scale was expanded to 12 levels and progressively adjusted to give the **MM Scale** or **modified Mercalli Scale** of 1931.[108] A single number quoted for a particular earthquake represents the maximum, but, because of different distances from the epicentre and the further complication of the depth below the epicentre of the true centre, this number does not characterize the earthquake as a geophysical event, but as a geographic if not purely socio-economic event. (The †Richter scale, in contrast, is computed singly for any one earthquake, and represents the geophysical dimension.) The following brief interpretation of the MM Scale is based on that of Richter.[109]

I   Registered by seismographs, but felt by very few people and those only in especially sensitive conditions.

II Felt by a few people at rest, especially on upper floors of buildings. Delicately suspended objects may swing.

III Felt noticeably by many people indoors, especially on upper floors of buildings, but not necessarily recognized as an earthquake, vibration being similar to that of any passing truck. Standing vehicles may rock. Has discernible duration.

IV Dishes, windows, doors and hung items disturbed; walls make cracking sound. Some sleepers wakened. Sensation as from heavy truck striking building. Outdoors felt by some, standing vehicles rock noticeably.

V   Felt generally indoors and out. Doors and lighter loose objects displaced.

VI Felt by all, frightening for most, causing them to seek the outdoors. Fragile items break and many items fall from shelves and walls. Furniture displaced. Some cracking of weak masonry.

VII Some cracking of ordinary masonry and plasterboard; displacement of exposed and weaker masonry, falling bricks, etc. Waves on water bodies, earth-slides on unstable slopes, people falling over.

VIII Some disturbing of goods including reinforced masonry, with general collapsing of chimneys, towers, monuments. Frame houses shifted. Trees broken, moving cars displaced. More slipping and cracking of the ground.

IX Widespread collapsing of all but specially resistant masonry, conspicuous cracking of ground with damage to reservoirs and pipelines. General panic.

X General destruction of all but specially resistant buildings, bridges, etc. Major landslides and horizontal shifting of sand, mud and water bodies; serious damage to associated earthworks. Some bending of steel rails.

XI General destruction of buildings and widespread damage to underground works. Major damage to steel rails.

XII Extensive displacement of ground, virtually complete destruction of structures.

mercury *See* head of liquid.

meridian mile [mile along a meridian] *See* geographic mile.

Mersenne prime [M. Mersenne; France 1588–1648] *See* prime number.

mesh number *See* sieve number.

met [metabolism] A casual unit equated with the metabolic rate of a seated person at rest, equated with $58.15 \, \text{W} \cdot \text{m}^{-2}$.[110]

meter The term 'meter' has a wide generic meaning within mensuration, not just for a device that measures (e.g. water meter) but also for units of measurement already mentioned or inferable. Thus to measure by the 'common meter' can mean the standard measuring stick, rod, or rope of the particular industry or trade good. The use in North America of the spelling 'meter' for the international metre is irreversible, and official in the USA, though the latter is the official standard in Canada and through most of the English-speaking world. *See* metre.

Metonic cycle [Meton; Ancient Greece *c*.450–*c*.400 BCE] *astronomy* The period of whole days over which the visible lunar and solar periods almost re-synchronize, namely 6 939.65 days, = 234.998 7~ synodic months $= 19.000\ 1\sim$ mean tropical years. The difference between 235 synodic months and 19 mean tropical years is barely 2 hours. *See* calendar; saros; Dionysian period.

metre, meter *length* Symbol m and strictly the lower-case form, though used in uppercase conspicuously, in contrast to the traditional m for mile, on British road-signs. *Metric* The base unit for length, defined (1983) as equal to the length of the path travelled by light in a vacuum during a time interval of $\frac{1}{299792458}$ of a second. The following are among the coherent derived units:

- $m \cdot s^{-1}$ for speed;
- $m \cdot s^{-2}$ for acceleration;
- $m^2 \cdot s^{-1}$ for kinematic viscosity.

See SI unit for full repertoire.

Using the 1959 definition of the foot, $1\ m = \frac{1}{0.3048}$ ft $= 3.280\ 839\ 895\sim$ ft $(39.370\ 078\ 740\sim$ in).

History

The metre was created in 1793 as the base unit for the †metric system, its value being defined on a †natural basis as $\frac{1}{10000000}$ of the length of the quadrant from the North Pole to the Equator along the meridian through Paris (at nominal sea level). The kilometre of 1 000 metres thus became a hundredth of a hundredth of that quadrant, hence, with the concomitant centesimal division of the right angle (*see* grade), effectively a †geographic mile. Its size was set by a physical survey from Dunkirk in northern France to Barcelona in Spain, and recorded in metal standards. A repeat survey gave a discrepant result, prompting definition via a solid physical prototype rather than the natural reference, using the standards already created. Enshrined as a platinum bar, the **Mètre des Archives**, and with copies provided to standards authorities around the globe, the metre has become the standard of length in virtually all countries. Both the UK and the USA incorporated it in legislation in 1866, but only within a common (not quite identical) non-metric system. The USA adopted it as the basis for defining the foot, etc. (with $1\ m = 39.37$ in); the UK, though allowing it in contracts, did not make it the fundamental length until 1959, when it set $1\ yd = 0.914\ 4\ m$ exactly, this being a rationalized value adopted by Canada in 1951,[82] and by the USA and internationally in 1959.

Following the international conference on the metre in 1875, an International Prototype Metre was created in 1879 as a (more elaborately specified) successor to the original prototypes. This survived until 1960, when a natural reference, using light, was established (see below). As developing technology allowed and demanded greater accuracy, older definitions were demonstrated as inadequate, so the definition repeatedly improved. Definition of the angstrom, with an intention of it being exactly 10^{-10} m, in 1907 in terms of light (in the form of the red line of cadmium) was a precursor of the modern definition of the metre, and simultaneously an example of the accuracy problem, for it proved to be 1.000 000 2$\sim$ times that intention. The definition of the metre in 1960, based on krypton,[111] was expressly made consistent with the definition of the angstrom, hence deliberately varied the metre by a like proportion.

The current definition, that of 1983, relates the metre to the speed of light, rounding the metre so that the distance travelled by light in 1 second is an integral number of metres, i.e. 299 792 458. Technically this changed the size of the metre again, now at the tenth decimal place. Though far in format from its original geographic definition and not practical for everyday use, this newest definition puts the metre onto a natural base again and into a context that is much more practical to the scientist. As laboratory instruments allow ever more precise measurement of the speed of light, so the metre will change, but only ever more finely.

The speed of light is of great fundamental significance in physics, especially in electromagnetics (*see* e.m.u. system). It is ironic that, had such a natural base been adopted at the outset, and the metre been defined on a natural base like the †light-nanosecond, not only would the pertinent electromagnetic calculations and relationships have involved the simple numerical factor of 10^9, but also the vital new unit of length would have been effectively a foot, specifically 299.79$\sim$ mm (0.983 6$\sim$ ft), just over 11 Paris inches of the time, and almost exactly the Roman foot.

The international prototype metre, then its successor definitions, became the standard for the inch, etc., of US-C by the †Mendenhall Order of 1893, which declared it equal to 39.37 in, and internationally since 1959 when the international inch was adopted as 25.4 mm, so 1 m $= \frac{1}{0.0254}$ in $= 39.370\,078\,7\sim$ in. However, it was only in 1964 that the UK formally discarded its prototype yard in favour of that for the metre.[95]

1889   1st CGPM: referring to an established prototype block made of 90% platinum and 10% iridium, and to the metre as the unit of length 'This prototype, at the temperature of melting ice, shall henceforth be considered to be the unit of length.' The accuracy mentioned is just to 0.01 mm.

1927 7th CGPM: 'The unit of length is the metre, defined by the distance, at 0°, between the axes of the two central lines marked on the bar of platinum-iridium kept at the BIPM and declared Prototype of the metre by the 1st CGPM, this bar being subject to standard atmospheric pressure and supported on two cylinders of at least one centimetre diameter, symmetrically placed in the same horizontal plane at a distance of 571 mm from each other.'

1960 11th CGPM: '*considering* that the international Prototype does not define the metre with an accuracy adequate for the present needs of metrology, that it is desirable to adopt a natural and indestructible standard, *decides*

1. The metre is the length equal to 1 650 763.73 wavelengths in vacuum of the radiation corresponding to the transition between the levels $2p_{10}$ and $5d_5$ of the krypton 86 atom.'

1983 17th CGPM: 'The metre is the length of the path travelled by light in a vacuum during a time interval of $\frac{1}{299792458}$ of a second.'[8]

metre-atmosphere *See* atmo-metre.

metre-kilogram-second system *See* m.k.s. system.

metre-tonne-second system *See* m.t.s. system.

metric slug [slug] *engineering Metric See* engineering mass unit.

metric system The comprehensive decimal system created as a national standard in France in the 1790s, following the revolution, subsequently adopted on an international basis as the primary standard in many countries, and then the standard for science. It is now the primary system for measurement in virtually all developed countries other than the USA; even there it provides the official basis for the measurement standards as well as the scheme for most science. Today it has become the †SI system, with various technical differences from its predecessor. Other manifestations include the †c.g.s. system, the †m.k.s. system, and an †m.t.s. system. (Within general entries in this book, these are labelled Metric-c.g.s., Metric-m.k.s., and so on, sub-qualified where appropriate.)

History
The metric system grew out of proposals made at least as early as 1670, considered officially then enacted in post-revolutionary France in the 1790s, though not widely accepted until 1845. The initial fundamentals were a metre (symbolically m), equal in length to $\frac{1}{10000000}$ of a meridional

quadrant of Earth and a gramme (gram, symbolically g), equal to the mass of $\frac{1}{1000000}$ cubic metre of water (at normal atmospheric pressure and the temperature of its maximum volumic mass, i.e. close to 4°C). Such definitions were seen as 'natural', and certainly untainted by reference to any despised royalty, but problems (*see* metre; kilogram) led by 1799 to the creation of †prototypes. Called the Mètre des Archives and the Kilogramme des Archives, the former became the defining entity, the latter merely a practical reference until 1872, but with what proved to be a small discrepancy of lasting impact; *see* litre; kilogram. (The 'gramme' has become the 'gram' very widely, the 'metre' and the 'litre' the 'meter' and 'liter' in US usage. The first of these changes is adopted in this publication but the others not, in deference to the more usual international practices.)

Rather than have independent names for the successive terms in any range of units, a range of standard decimal multiplier prefixes was introduced, with Greek terms for the integral multipliers and Latin for the fractional ones. By name and value significance, with etymological derivation and abbreviation, these were as shown in Table 33. This has since been extended and varied, as shown under the †SI system.

It should be realized that this scheme of steps of ten, the literally decimal approach, was based on the idea that people did not use big integer values, as is common in North America, but always moved to larger units, as was characteristic in Europe. Thus, just as one would say 2 feet 8 inches in England and the equivalent elsewhere in Europe for what would usually be called 32 inches in America, the original metric terminology would have said, for a similar length, 8 decimetres 1 centimetre 2 millimetres rather than 812 mm or such. The decimal scheme, of course, meant no awkward conversions during additions,

Table 33

Symbol	Prefix	Value	Etymology
m	milli-	10^{-3}	'thousand' in Latin
c	centi-	10^{-2}	'hundred' in Latin
d	deci-	10^{-1}	'ten' in Latin
D	deca-	10^{1}	'ten' in Greek
H	hecto-	10^{2}	'hundred' in Greek
K	kilo-	10^{3}	'thousand' in Greek
My	myria-	10^{4}	'ten thousand' in Greek

etc., as applied, for instance, when feet and inches were employed. But it facilitated just as readily, and more importantly, the discarding of the multi-unit expression in favour of the single-unit practice, particularly as the use of decimal fractions became more familiar. The current practice within metric of using a single unit has evolved over time as this fact climbed above ingrained habits, and as the populace became more arithmetically literate. (The easy North American practice of using single units with non-decimal British units, which largely obviates the consequences of them being non-decimal, is illustrated by measuring tapes of many feet in length but marked predominantly in inches, e.g. to 120 inches on a 10-foot tape. It is very conspicuous on freight vehicles, where the load limit is expressed in such style as 75 000 lb, when the British equivalent said 33 ton 9 cwt 2 qtr.)

Though the decimal steps within metric persist in everyday use in Europe and other traditional metric areas (even to such expressions as 'a third of a decimetre'), North American metric practice largely abjures units other than the thousand multipliers, and indeed should be described as a †millesimal scheme rather than a decimal one. The 13th †CGPM in 1968 recommended that this practice should be general, at least in the sense that expressed powers of ten should have indices that are multiples of 3. While the SI has retained all the decimal prefixes from milli- to kilo-, it dropped the one decimal step beyond (the myria-) and, in expanding the range of prefixes, restricted them to millesimal steps. (Using only millesimal steps minimizes but does not remove the risk of misreading measurements on plans, for instance.)

The metre was a relatively large unit for common use, the gram a relatively small one; most market transactions would involve fractions of a metre but rarely be more precise than 10 grams, suggesting that the new scheme was rather 'academic', little orientated to the people at large, despite them being the essence of the recent revolution. The square metre was a reasonable unit of area, but the cubic metre extremely large for volume. Each could serve its respective dimension, along with the squares and cubes of the decimal multiples of the metre, but this would mean steps of 100 and 1 000 respectively in these compound dimensions for each decimal step in the linear. (For example, while $1 \text{ cm} = 10 \text{ mm}$, $1 \text{ cm}^2 = 100 \text{ mm}^2$ and $1 \text{ cm}^3 = 1 000 \text{ mm}^3$.) To obviate this problem and make the system more acceptable to the public, the †are for measuring area and the litre for volume were early additions to the system. Rather than being the obvious (and †coherent) square and cubic metre, these introductions were set at one square decametre and one cubic decimetre, sizes more appropriate to their everyday jobs of measuring land and goods in the marketplace. (As discussed under litre, that unit, because of the manner of its definition, proved to be discrepant from the intended.)

A law of 1812 renewed in France the use of several old names but with metricized values, e.g. the toise of 2 metres, the pied of a third of a metre, the livre for half a kilogram. This accommodation of traditional terms was discontinued officially in 1840, but established a practice that continued in France and extended elsewhere.

Decimalization was also applied to time and to angles; the latter persisted as the †grade, the former persisted only briefly, relative to the calendar and to the fractioning of the day.

Various derived terms combining the basic pair of metre and gram, plus the well-established second as the unit of time, were progressively introduced later, for scientific usage particularly. These included the dyne for a force of 1 centimetre·gram per second squared ($= 1 \, cm \cdot g \cdot s^{-2}$) and the erg for the energy of 1 dyne operating along 1 centimetre ($1 \, cm^2 \cdot g \cdot s^{-2}$). These, having their unit values involving the centimetre rather than the metre but otherwise the base units gram and second, were termed a †c.g.s. system or scheme, which was termed 'coherent' because with such units all relationships involve only singular units. The British Association for the Advancement of Science extended this development to the electrical domain in the 1860s, based on defining a unit of electrical resistance, called the BA unit, in mechanical terms of the metre, gram, and second.[112] These, because of their reference back to non-electrical units, were called 'absolute' units. They were revised in the 1870s to have their unit coherence based on the centimetre rather than the metre, thus joining the dyne and other derived units within the c.g.s. scheme, forming the †electromagnetic units. As discussed under that heading, these units were mostly far from the amounts to be measured, so decimally derived †practical units, and subsequently slightly different †international units, were adopted, before being replaced in 1948 with the current units which, in 1960, became part of the newly labelled †SI system.

The use of the centimetre, along with the gram, in the various manifestations of the c.g.s. system illustrated the relative incompatibility of the metre and the gram. After early efforts with the natural combination of metre-gram-second, various other combinations were used to produce a convenient coherent system. Late in the 19th century Gauss used a millimetre-milligram-second system. At the other end of the scale, in 1919 France adopted an m.t.s. system based on the metre, tonne, and second. The modern vogue, and the structure of the SI, is the metre-kilogram-second combination. This m.k.s. system was adopted in 1946 as the primary international standard, effective for the beginning of 1948. The formal title Le Système International (the SI) was adopted in 1960. As this includes one electrical unit among its base units, chosen to be the ampere (the unit of current strength), the SI is classed as an †m.k.s.A. system. (It has both the kelvin and the mole as physical base units too, plus the somewhat subjective candela.)

Metric-c.g.s. *See* c.g.s. system.

Metric-c.g.s.-e.m.u. *See* e.m.u. system.

Metric-c.g.s.-e.s.u. *See* e.s.u. system.

Metric-c.g.s.-Gaussian *See* Gaussian system.

Metric-prac *See* practical unit.

MeV *sub-atomic physics* USA Mega-electron-volts (originally a million electron volts).

Meyer hardness number *See* hardness numbers.

MF [Medium Frequency] *physics* The waves of the †electromagnetic spectrum of about 1 MHz frequency, used, *inter alia*, for AM-style radio. *North America* Range 0.535 to 1.605 MHz for AM radio.

The letters MF have also been used to symbolize microfarad, properly μF (MF = megafarad).

mFBM *volume* For sawn wood, 1 000 FBM.

MFLOPS ['megaflops', = million floating-point operations per second] *informatics* A measure of the numeric calculating power of a computer, relevant particularly to numerically intensive computing. Technically the number of such operations that can be executed in a second, but this, for any one machine, varies considerably with the mixture of instructions – add, divide, etc. – that a particular program has. The practical reality was usually measured by standardized programs like †dhrystone, †livermore loops and linpack. *Compare* MIPS.

m.g.s. system (Metric-m.g.s.) *Metric* The inherent †coherent †metric system, deriving its units coherently from the metre, the gram, and the second (in contrast with the metre, the kilogram, and the second of an

†m.k.s. system and its contemporary form the SI). In an m.g.s. system, for example, the unit of force gives an acceleration of $1\,\text{m}\cdot\text{s}^{-2}$ to a body of $1\,\text{g}$, a thousandth of the newton of the SI, which gives an acceleration of $1\,\text{m}\cdot\text{s}^{-2}$ to a body of $1\,\text{kg}$, but a hundred-fold the dyne of c.g.s., which gives an acceleration of $1\,\text{cm}\cdot\text{s}^{-2}$ to a body of $1\,\text{g}$. The m.g.s. approach was pursued quite naturally in the 19th century, but the major size disparity between the metre and the gram in most circumstances of joint applicability (e.g. the mass of 1 cubic metre of water is 10^6 grams) soon prompted the consideration of schemes that used a multiple of the gram, or a fraction of the metre, notably as pioneer the †c.g.s. system.

m_h *sub-atomic physics See* helion mass.

mho [reverse spelling of ohm] *electric conductance* Metric The conductance of a circuit in mhos is the reciprocal of its resistance in ohms, i.e. it equals the ratio of the current produced in amperes between two points of a conductor to the †r.m.s. potential difference in volts across these points (the conductor not being the seat of any electromotive force).

Metric-m.k.s.A. 1948 Ω^{-1} ($= \text{m}^{-2}\cdot\text{kg}^{-1}\cdot\text{s}^3\cdot\text{A}^2$ in base terms). The mho is identically the SI unit siemens in non-inductive circumstances; the name mho has not been retained within the SI, although it is †coherent with the contemporary system.

Metric-c.g.s. See abmho; statmho. *See also* practical unit.

History
The mho entered use in 1883, joining the already established practical units derived from the †e.m.u. system. This was the practical mho, $= 10^{-9}$ abmho. Like all the e.m. units, the abmho was itself defined ultimately in terms of purely mechanical units.

Because the explicit laboratory specifications established for the ampere, ohm, and volt were subsequently found to be slightly discrepant from the intended, so was the mho, as a unit derived from them. The International Electrical Conference of 1908 covered the discrepancy by adopting the distinct name **international mho**. Because of experimental vagaries, the value for conversions is normally referred to as the mean international mho, $= 0.999\,510\sim$ S. There is also the **US international mho**, $= 0.999\,505\sim$ S.

At the IEC in 1933, the distinct name siemens was adopted in place of the contrived name mho, within the m.k.s. system.

With the implementation of the Metric-m.k.s.A. system in 1948, and its basing of electrical units on an ampere compatible with the original absolute units, the modern unit became essentially the old practical mho.

mi. *length See* mile, expressly here the BI and US mile.

mic *electric inductance UK* 1 μH, used routinely by the British Admiralty prior to 1940.

Michaelis constant [L. Michaelis; Germany, USA 1875–1949] *biochemistry.* Symbol K_m. If an enzyme reacts at rate V at saturation in a substrate but at rate v at sub-saturation concentration c, then

$$K_m = c(V - v)/v$$

which is identically the concentration when $v = \frac{1}{2}V$.

micri-erg *energy Metric* A scarcely used unit, $= 10^{-14}$ erg $(10^{-21}$ J).[113]

micro- Symbol μ-, but *see also* mcg. *Metric* The 10^{-6} multiplier, e.g. 1 microgram $= 1$ μg $= 10^{-6}$ g.

microequivalent *See* equivalent weight.

micro-inch *length.* Symbol μin. A vernacular term for 10^{-6} inch $= 10^{-3}$ thou.

micron *length.* Symbol μ. *Metric* An SI-deprecated name for what is the micrometre; 10^{-6} m $= 1$ μm.

1879	CIPM: introduced
1890s	symbol μ adopted
1948	9th CGPM: confirmed
1967–68	13th CGPM: '*decides* … to remove … the unit name "micron" and the symbol "μ" which had been given to that unit but which has now become a prefix.'[8]

pressure From the above, equals 1 μm of mercury (0.133 322 4~ Pa).

midrange *See* average.

mil A 1 000th part, such that 1 000 of them make a whole.
length (**thou**) $\frac{1}{1000}$ in (0.0254 mm, 0.001 in).
volume UK 1907 For pharmaceuticals, the millilitre (16.231~ minims).
plane angle A variety of units having 1 000 to a convenient large unit, i.e. to a right angle, to a radian, or to a revolution or turn. Since, for very small angles, the tangent of an angle is equal to the size of the angle expressed in radians, a convenient angular scale has the radian divided

by the range for expressing the height (else width) of the observed object. For example, having thousandths of the radian allows simple progression from the height of the target object in metres to the range in kilometres; it is often called the **infantry mil**, being 0.001 rad, and gives 6 283.~ mil per turn. An adjusted version has a rounded 6 400 per turn, giving five layers of binary division to get 100 mil; this is the **artillery mil**, $= 0.000\,982\sim$ rad $(0.056\,25°)$.[114] An older US artillery mil had 1 000 to the right angle, hence 1 mil $= 0.09°$ $(0.001\,571\sim$ rad).

mile *length Internat, BI, US-C* Pedantically as **statute mile** (*compare* geographic mile or nautical mile), 1 760 yd $= 5\,280$ ft $= 63\,360$ in $= 1.609\,0\sim$ km; *see* inch for greater details, including reference to **coast** or **survey mile** of $1\,760 \times \frac{36}{39.37}$ m $= 1.609\,347\,22\sim$ km.

millennium *time* 1 000 years. Because the familiar calendar started with year 1, 'the millennium' is pedantically the span, for instance, from 1001 to 2000 rather than 1000 to 1999. Modern practice increasingly ignores the pedantic and favours the obvious; the millennium, along with the twentieth †century and its last †decade, ended with 1999 (though the Christian second millennium surely included the year 2000).

   More generally and without the definite article, any identified span of 1 000 years can be called a millennium. Note that 'the millennium' is also used to mean the end-point of such a time period.

millesimal With divisor/multiplier steps of 1 000, in contrast with the steps of 10 for †decimal, of 100 for †centesimal.

   The †SI system, although basically decimal through its metric heritage, is largely millesimal. Official policy is for scaling beyond a factor of 100 to be effected in steps of 10^3 and, while the prefixes for 10 and 100 multiplication and division are retained, they are not universally legal for commerce.

   Steps of 1 000 have the advantage of minimizing the scope for confusion of the unit involved, e.g. on architectural drawings.

milli- [Lat: 'thousand'] Symbol m- . *Metric* The $\frac{1}{1000} = 10^{-3}$ multiplier, e.g. 1 milligram $= 1$ mg $= 10^{-3}$ g.

   The only prefix legal in the UK for all of metre, gram, and litre.[28]

milliard *Europe, UK* 1 000 million $= 10^9$. *See* thousand.
   volume 10^9 m^3 $= 1$ km^3 $(810\,713.\sim$ ac·ft, $264.2\sim \times 10^{-9}$ US gal).

millicuries-destroyed *radiation physics.* Symbol mc.d. *France* A peculiar expression for x-ray dosage, equivalent to a fall in the radioactive source of 1 millicurie. *See* curie; becquerel.

milliequivalent Symbol mEq. *See* equivalent weight.

millier *mass* BI 1878 1 000 000 g.

millihg Millimetre of mercury (element Hg); *see* head of liquid.[115]

millik *sub-atomic physics Canada See* nile.

millimetre of mercury (mm of Hg) *See* head of liquid.

millimicro- $10^{-3} \times 10^{-6}$-, i.e. 10^{-9}- = nano-. This prefix was coined before the nano- prefix was introduced in 1960; it is no longer acceptable.

millimicron 1 nm, i.e. 10^{-3} micron; the term is no longer acceptable.

million (mn.) 1 000 thousand = 1 000 000 = 10^6. *See* thousand for scales. Unlike the related terms of higher value (billion et seq.), this term has an unambiguous meaning, so can be used readily, even in a scientific context. Million also has the advantage of having the same initial letter as its SI equivalent prefix, namely mega- (M-) as in 'megabytes' and 'megabucks'. However, there is the confusion that, from Latin, M also means thousand, and has often been used so in various business and other writing.

Besides the abbreviation mn., and the metric M, the construct mm is often used in technical business writing in North America to mean a million, e.g. mm.c.f. = 10^6 ft^{-3}; *see* mm-.

million acre *volume* Particularly for capacity or contents of irrigation reservoirs, 1 000 000 acre·foot (1 233 482~ m^3, 326.~ × 10^9 US gal).

minim [Lat: 'least'] *volume* BI, USA For apothecaries, $\frac{1}{60}$ fluid drachm. BI 59.193 9~ μL. Removed from official UK measures in 1970.[28] US-C Solely for liquids, 61.611 5~ μL.

minute [Lat: 'small'] Symbol ′. Usually the first subdivisional unit of some more established unit, a term adopted in medieval times when the Babylonian sexagesimal scheme of division was common, hence rarely other than a sixtieth of the larger unit. Often represented by a single prime or hash mark, e.g. 3′ for 3 minutes of time, angle, etc.; its next layer of subdivision, being the second layer, is invariably into units called seconds, represented by double hash marks.

length $\frac{1}{60}$ of 1 degree of a great circle of Earth, being 1 minute of longitude along the Equator and equivalent elsewhere, particularly meridionally, approximately 1.85 km or 1.11 mi (essentially the †geographic mile and †nautical mile). Also a minute of longitude along an identified parallel of latitude, which means a lesser and progressively declining length as latitude increases. Specific values include:

longitude	along Equator	1.854 93~ km (1.152 60~ mi)
latitude	near Equator	1.842 52~ km (1.144 89~ mi)
latitude	near Pole	1.861 15~ km (1.156 46~ mi)

The minute of latitude probably set the size of the early fathom, foot, etc.

time. Symbol min in the SI. Traditionally $\frac{1}{60}$ of an hour, thereby $\frac{1}{1440}$ of a †day (a unit varying in size depending on qualifier) and sized by such fractioning. Since 1967, however, the minute of normal usage (derived from the mean solar day), routinely equalling 60 seconds, has been defined from the atomic †second. (*See* leap second for exceptions to the number 60.)

plane angle (also **arcmin**, **minute of arc**) The first layer of fractioning the degree, the traditional measure of plane angle, by definition $= \frac{1°}{60} = 0.016\,667\sim°$, $= \pi/(60 \times 180)$ rad $= 0.000\,290\,89\sim$ rad.

astronomy Note that minute for †right ascension is the minute of the sidereal clock, while for declination and most other purposes it is the minute of plane angle; the two differ by a ratio of 1:15.

See also centesimal minute.

minute difference *astronomy* The change in the elements of the nautical almanac caused by a 1-minute step of time.

minute of arc *geometry See* minute.

MIPS [million instructions per second] *informatics* A measure of the overall processing power of a computer including integer arithmetic, logic, and other operations. Technically the number of such operations that can be executed in a second, but this, for any one machine, varies considerably with the mixture of instructions (add, divide, branch, etc.) that a particular program has. The practical speed is usually measured by standardized programs like †whetstone, the value 1 MIPS having been arbitrarily set, in the 1970s, to be the power of a $\frac{VAX11}{780}$ computer. *See also* MFLOPS.

mired [micro reciprocal degrees] *temperature* A scheme for expressing very high temperatures, the reciprocal of that in kelvins, expressed at the 'micro' level. E.g. 25 000 K, having reciprocal 40×10^{-6}, is 40 mired.

MJD *time* Modified †Julian date.

MKS system, m.k.s system (Giorgi system, Metric-m.k.s.) *Metric*
The version of the †metric system, notably now in the form of the †SI
system, that has its derived constants relating †coherently to the metre,
the kilogram, and the second, in contrast with the centimetre, gram, and
second of the long-used †c.g.s system, for instance. In the m.k.s. system
the unit of force, for example, is the newton, which gives an acceleration
of $1\,\text{m}\cdot\text{s}^{-2}$ to a body of 1 kg, rather than the dyne of the c.g.s. system,
which gives an acceleration of $1\,\text{cm}\cdot\text{s}^{-2}$ to a body of 1 g, hence of only
$10^{-5}\,\text{m}\cdot\text{s}^{-2}$ to a body of 1 kg.

The m.k.s. basis for coherency began by accident, in the form of the
†practical units of electromagnetics in the 19th century, derived by
relatively arbitrary scaling of the †electromagnetic units of the c.g.s.
system. Their coincident fit to m.k.s. was realized and publicized in 1901
by Giorgi, who stimulated its adoption generally to the mechanical and
other fields. What was originally known as the **Giorgi system** was
accepted by the International Electrotechnical Commission in 1933 as
appropriate for electro-technicians without affecting the use of the
classical c.g.s. system by physicists.[116] In 1946 the †CIPM decided that,
effective from the beginning of 1948, an m.k.s. system would displace
c.g.s. as the primary system for science, etc. In 1960, the adoption of the
SI system brought a new title; 'm.k.s.' survived as the formula for
coherency, but the label became historic.

When the label first came into use, Kg stood for kilogram. Now, under
the SI, the symbol for kilo- is k. So m.k.s. is correct today, but m.K.s. was
correct yesterday.

The system adopted in 1948 incorporated the ampere as a base unit; it
could, for instance, have been the volt instead, or, as with the †e.m.u.
system, could have related all electrical units to their mechanical
equivalents, obviating the definition of any base electric unit. In
consequence of this decision, the SI is called an †m.k.s.A. system (q.v. for
further notes on the use of lower- versus upper-case letters in the system
title).

Most of the older derived units of m.k.s. survive in the SI (the mho as
the siemens), exceptions among coherent units being, for:
- acoustics – specific impedance: rayl $= \text{N}\cdot\text{s}\cdot\text{m}^{-3} = \text{Pa}\cdot\text{s}\cdot\text{m}^{-1}$;
- acoustics – mechanical impedance: mechanical ohm $= \text{N}\cdot\text{s}\cdot\text{m}^{-1}$;
- acoustics – mechanical mobility: mohm $= \text{m}\cdot\text{N}^{-1}\cdot\text{s}^{-1}$;
- dynamic viscosity: Pl $=$ poiseuille $= \text{N}\cdot\text{s}\cdot\text{m}^{-2} = \text{Pa}\cdot\text{s}$;

- luminance: nit = $cd \cdot m^{-2}$;
- luminous energy: talbot = $s \cdot cd \cdot sr$.

Non-coherent units of the old m.k.s. system included apostilb, bar, darcy, langley, mayer, poncelet, and rad.

m.k.s.A. system, m.k.s.a. system (Metric- **m.k.s.A.**) *Metric* The formal recognition of the m.k.s. system augmented for electric quantities by use of the ampere rather than the volt (or other) as the relevant base unit. (The original †c.g.s. system had all electrical units defined in mechanical terms, while subsequently, for a time, the ampere and the volt had independent definitions. With the SI, the ampere alone is defined, making our contemporary form of metric an m.k.s.A. system.)

The use of lower- versus upper-case letters in the system titles is very varied. If one regards the title in itself as deserving of capitalization, then upper-case letters are appropriate. Otherwise, if the title is composed from the initial letters of the names of the units, lower-case letters are appropriate. However, since the symbol for ampere is distinctively the capital A, many people find the use of small a unacceptable. But, while lower-case m for metre and s for second are proper symbols, k was not always the symbol for kilo-. This system first came into use when K rather than k stood for kilo-, and neither alone stands for kilogram. So m.k.s.A. is perhaps best today, but m.k.s.A., or m.Kg.s.A., was correct yesterday and M.K.s.A. and m.k.s.A. are defensible; m.k.s. and m.k.s.A. are used in this publication, the latter, as Metric-m.k.s.A., to mean the international standard system adopted in 1946, effective for 1948, which relegated the volt to derived status. (Interestingly, it is MKS in the official 1946 definition of newton.)

ML, Ml *volume* *Metric* Megalitre; *see* mega-; litre.

mL, ml *volume* *Metric* Millilitre.
   optics (mL) Millilambert.

MM Scale Modified Mercalli Scale; *see* Mercalli scale.

mm *length* *Metric* Millimetre.
   See also mm-.

mm- *North America* Particularly in business literature, by compounding the single letter m representing thousand, $(1\,000)^2 = 10^6$: e.g. as in **mm.c.f.** for natural gas, $= 1\,000$ m.c.f. $= 10^6\,ft^3$ ($28\,317.\sim m^3$); **mmFBM** for sawn wood, $= 1\,000\,mFBM = 10^6\,FBM$. Sometimes written isolated from the relevant unit, notably with the local abbreviation mt for metric ton, e.g. as **5mm mt** for 5 000 000 t; it also occurs fully isolated with currency statements, e.g. $500mm to mean $500 000 000 (also written $500m and $500mn, and, being half a billion, alternatively as $0.5bn).

mmHg, mm of Hg Millimetre of mercury (element Hg); *see* head of liquid.

m_n *sub-atomic physics See* neutron mass.

mn. *See* million.

module [Lat: modus, 'measure'] A standard length or other measurement set within an industry, for instance 8 inches has been used widely in the building industry, for bricks, door-frames, panelling, etc., these having integer multiples of the said module for all their gross dimensions. In grander architecture, the module might be the arbitrary, even indefinite, size of a major component, for instance the base diameter of the characterizing columns. An example of a less direct use is with metric-sized gear wheels, where the module is the effective diameter divided by the number of teeth.

modulus *mathematics* Expressed as $|z|$ for the †number $z = a + b\,\mathrm{i}$ and defined as $|z| = +\sqrt{a^2 + b^2}$, hence it is always a positive value. If z is a real number, then $b = 0$ and $|z| = -z$ for $z < 0$, $|z| = z$ for $z \geq 0$; called also the †arithmetic value of z. Used likewise for functions, but also more elaborately in certain fields.

mohm [mobile ohm] *acoustics Metric-m.k.s.* and *-c.g.s.* A unit of mechanical mobility for sound waves, being the reciprocal of the mechanical ohm unit of impedance, i.e., for an acoustic medium, the ratio of the flux or volumic speed (area times particle speed) of the resulting waves through it to the effective sound pressure (i.e. force) causing them, the unit being qualified, according to the units used, as m.k.s. or c.g.s. Thus

$1\,\mathrm{mohm_{m.k.s.}} = 1\,\mathrm{m\cdot N^{-1}\cdot s^{-1}}$ $(\mathrm{s\cdot kg^{-1}}$ in base m.k.s. terms),

$1\,\mathrm{mohm_{c.g.s.}} = 1\mathrm{cm\cdot dyn^{-1}\cdot s^{-1}} = 10^3\,\mathrm{km\cdot N^{-1}\cdot s^{-1}}$ $(\mathrm{s\cdot g^{-1}}$ in base c.g.s. terms).

Mohr cubic centimetre [C. F. Mohr; Germany 1806–79] *saccharimetry* The volume occupied by 1 gram of water at 17.5°C $(1.002\,38\sim \mathrm{cm}^3)$ or other indicated temperature.[117]

Mohs hardness number [F Mohs; Germany, Austria 1773–1839] *See* hardness numbers.

mol *chemistry* SI *See* mole.

molal *chemistry* Indicating that an amount of a solute substance is relative to its habitat, specifically to the amount of solvent; *compare* molar.

molality *chemistry* SI The amount of a substance in moles per kilogram of solvent. Molality corresponds with but varies from the milliequivalent. For univalent ions (e.g. Na^-, K^-, Cl^+, CO_3^+) the two are identical; however, for others they differ by the valency, e.g. for divalent ions (e.g. Ca^{--}, SO_4^{++}) the molality is half the mEq/L. *Compare* molarity.

molar *physics* Implies quantity per amount of substance, typically per mole, the unit involved usually being expressly included in the expression.

  chemistry Indicating that an amount of a solute substance is relative to its habitat, specifically to the amount of solution; *compare* molal.

molar gas constant *fundamental constant.* Symbol R. $8.314\,472(15)$ $J \cdot mol^{-1} \cdot K^{-1}$ with †relative standard uncertainty 1.7×10^{-6}.[4]

molar mass constant *physical chemistry.* Symbol $M_{\underline{\underline{u}}}$. 1×10^{-3} $kg \cdot mol^{-1}$.

molarity *chemistry* SI The amount of a substance in moles per litre of solution. *Compare* molality.

molar volume of ideal gas *fundamental constant* At 273.15 K and 101.325 kPa, i.e. at †s.t.p., $22.413\,996(39) \times 10^{-3}$ $m^3 \cdot mol^{-1}$ with †relative standard uncertainty 1.7×10^{-6}.[4] *See* standard volume; amagat unit.

mole [molecule] *chemistry.* Symbol mol. SI The base unit for an amount of a substance: the amount of a substance that contains as many of its basic entities (e.g. atoms for chemical elements, molecules for compounds) as there are atoms in 12 g of ^{12}C.[118]

  The relevant basic entity should be defined or clearly implied in any expression of mole measurement; besides the common atom and molecule, it can be ions, electrons, or other 'elementary' particles, or groups of such particles. For any pure substance, one mole has a mass in grams numerically equal to its molecular weight in daltons. (Despite the change in †metric systems from gram to kilogram as the base unit for mass, the mole remains tied in this manner to the gram. This is sometimes emphasized by calling it the **gram-mole**, allowing the **kilogram-mole** to mean its SI-†coherent thousand-fold multiple.)

The number of molecules of any substance, else, for an element, of atoms, contained in 1 mole of it is †Avogadro's number, = 602.213 67(36) $\times 10^{21}$. This could provide a †natural definition of the mole independent of the kilogram and, using the masses of electron, neutron, and proton, a natural definition of that unit of mass too.

See also dalton.

History

Originally introduced early in the 1900s as the **gram-molecular weight**, i.e. the number of grams of a substance that equalled its molecular weight, it was then defined as the amount of substance of a system which contains as many atoms as 16 g of oxygen; called also **gram-molecule**. However, this proved ambiguous because physicists interpreted it as referring to the common isotope of oxygen, while chemists used the heavier naturally occurring and somewhat variable mixture of isotopes 16, 17, and 18. In 1940 the International Commission on Atomic Weights standardized the ratio as being 1:1.000 275, then brought in the unifying carbon-based scheme in 1961.[119] This gave values of 0.0043% less than the chemists', and 0.0315% less than the physicists'. The carbon base results in more elements having integer values than applied with oxygen, whose scale had similar merit over hydrogen.

See weight re usage of that word.

1950/60 'relative amount of substance' based on carbon 12.

  1967 CIPM; 1969 CIPM; 1971 14th CGPM:

    '1. The mole is the amount of substance of a system which contains as many atoms as 0.012 kilogram of carbon 12: its symbol is "mol".

    2. When the mole is used, the elementary entities must be specified and may be atoms, molecules, ions, electrons, other particles, or specified groups of such particles.

    3. The mole is a base unit of the International System of Units.'

  1980 CIPM: 'in this definition, it is understood that unbound atoms of carbon, at rest and in their ground state, are referred to.'[8]

molecular weight Now correctly †relative molecular mass.

mole volume *See* molar volume of ideal gas.

month [Moon] *time* Any of the various periods associated with the passage of the Moon about Earth, and hence of roughly 30 days. As discussed under †time, there are various lengths of month. The obvious natural month is the time between successive full moons or other lunar phase-point. Because of the ellipticity of the orbits of Earth and of the Moon, the length of such a period varies slightly. The mean value, approximately 29.5 (mean solar) days, is termed the †synodic month (also **lunation**). Since Earth is simultaneously travelling about the Sun, with the same anti-clockwise turning viewed from the (North) Pole Star as the Moon's orbiting and thereby changing the relative position of the illuminating Sun, the synodic month is greater than the †sidereal month, the time for one orbit relative to the stars, by about one twelfth of that value. Other close but distinct periods defined are the †tropical month, being the time for the mean position to increase 360° (effectively one lap around Earth); the draconic or †nodical month, being the period between northward crossings of the ecliptic, which is shortened by the marked regressive gyration of the lunar orbital plane; and the †anomalistic month, the period between successive perigees (closest approaches). The values of these various months (in mean solar days) at the opening of the year 2002 were[10]

1 synodic month	= 29.530 589∼ days = 29 days 12 h 44 min 2.9∼ s
1 sidereal month	= 27.321 662∼ days = 27 days 7 h 43 min 11.6∼ s
1 tropical month	= 27.321 582∼ days = 27 days 7 h 43 min 4.7∼ s
1 anomalistic month	= 27.554 550∼ days = 27 days 13 h 18 min 33.4∼ s
1 draconic month	= 27.212 221∼ days = 27 days 5 h 5 min 35.9∼ s

Despite our visual perception being close to 30 days, the common understanding and hence the vernacular **lunar month** has 28 days; this gives a month of 4 whole weeks, but misleads people widely concerning the incidence of the full moon, etc. The average **calendar month** of the familiar Gregorian calendar, which has about 30.4 days, is closer to the visible month than the vernacular lunar one. The variation of the Gregorian calendar month, from 28 to 31 days, is well known; being greater on average than the synodic month means that one or two months each year contain two new moons, while February can have no new moon. Similarly, the seasons contain three but can contain four full moons. The term 'blue moon' applies to the penultimate full moon in a season with four (though it is often applied to the more frequent second in a month).[120] The human menstrual [Lat: 'month'] period, sometimes believed to be in phase with the Moon, ranges roughly from 20 to 40 days.

(It appears that the lunar month was about 30.5 days, with 13.1 such months in a solar year of about 400 days, a billion years ago.)

See calendar for a discussion of months relative to years.

See lune for a special division of the synodic month.

Mooney unit [M. Mooney; USA] *engineering* For rubber, a measure of plasticity, being the torque registered by a disc in raw rubber at 100°C within a cylinder rotating at 2 r.p.m., the result being read on an arbitrary scale, and expressed relative to the number of minutes of operation.

morgan [T. H. Morgan; USA 1866–1945] *chemistry* The unit of gene separation, being the distance along a chromosome for which the recombination probability is 1%. It is estimated to be about 300 μm.

mouse unit *biochemistry* A measure for the toxicity of a substance, being the amount lethal to a mouse.

m_P *sub-atomic physics* See Planck mass.

m_p *sub-atomic physics* See proton mass.

Mpc *astronomy* Megaparsec. *See* parsec.

mph *optics* Metric-c.g.s. Milliphot.

m.p.h. (usually **mph**) Miles per hour. 1 m.p.h. = 1.609 3∼ km/h.

Ms *rheology* See Margoulis number.

m.t. *North America* Particularly in business literature, metric tonne.

m.t.c.e. *engineering* Metric ton of coal equivalent, being for fuels other than coal the amount having an energy content equal to that of a tonne of coal, which is accepted as 29.29 GJ. *See also* b.o.e.

m.t.o.e. *engineering* Metric ton of oil equivalent, being for fuels other than oil the amount having an energy content equal to that of a tonne of oil, which is accepted as 44.76 GJ. *See also* b.o.e.

m.t.s. system (**Metric-m.t.s.**, also **MTS system**) *Metric* A version of the general system that has its derived constants relating †coherently to the metre, the tonne, and the second, in contrast with the metre, the kilogram, and the second of the †m.k.s. system and its contemporary form, the SI. In the m.t.s. system the unit of force, for example, is the sthene, which gives an acceleration of $1 \, \text{m} \cdot \text{s}^{-2}$ to a body of 1 t, a thousand-fold the newton of the SI, which gives an acceleration of $1 \, \text{m} \cdot \text{s}^{-2}$ to a body of only 1 kg. The system was promulgated officially in

France in 1919, but with little consequence; it is long obsolete. The distinctive units of m.t.s. included:
- force: sn = sthene = $m \cdot t \cdot s^{-2}$ (= 1 kN);
- pressure: pz = pieze = $1 \, sn \cdot m^{-2} = m^{-1} \cdot t \cdot s^{-2}$ (= 1 kPa);
- heat energy: thermie = raise 1 t of water 1 deg Celsius (= $4.185 \sim$ MJ).

mu [Anglicized name of Gk letter μ, pronounced 'mew' in English] *See* μ.

M$_u$ *physics See* molar mass constant.

***m*$_u$** *physics See* unified atomic mass constant.

Munsell system [A. H. Munsell; USA 1858–1918] *colorimetry* A scheme for specifying the three components hue, chroma (saturation), and value of a †colour, related to published samples but alphanumerically coded on a measured basis. Hue is coded from the capitalized initial letter of the labels red, yellow, green, blue, and purple arranged around a disc, singly interleaved with adjacent pairings. The numbers are prefixed with a number from 1 to 10 to show degree of gradation toward the next code, then onwards, e.g. 9 R, 10 R, 1 YR, 2 YR, ... , 10 YR, 1 Y. This is followed by a number from 1 to 10 representing value then by a like number for chroma. The value is gauged from a series of equi-graded grey samples ranging from white to black by matching to the one equivalent in lightness. The chroma (saturation) represents the difference between these two matched items. The composite expression is punctuated as, e.g., 2 YR 4/6. The chroma is often seen as a radius, giving the Munsell wheel, with value expressed axially.[121]

   Though introduced more for the arts, the Munsell system has been widely applied to science.[122] Adjusted over time, mainly through re-scaling, it gave the **Munsell renotation system**,[123] and was a foundation for the †ISCC-NBS system.

muon mass *sub-atomic physics*. Symbol m_μ. The theoretical rest mass of the never-resting particle, = 1.883 531 09(16) $\times 10^{-28}$ kg = 206.768 265 7(63) m_e with †relative standard uncertainties 8.4×10^{-8} and 3.0×10^{-8}.[4]

Mx *See* maxwell.

My- *Metric See* myria-; long defunct and never part of the SI.

myria- [Gk: 'countless', also 'ten thousand'] Symbol My-. *Metric* The old $10\,000 = 10^4$ multiplier, e.g. 1 **myriagram** = 1 Myg = 10^4 g. Long defunct and never part of the SI.

m_α *sub-atomic physics See* alpha particle mass.

m_μ *Metric* Old symbol for millimicron.

m_μ *sub-atomic physics See* muon mass.

m_τ *sub-atomic physics See* tau mass.

N, n *Metric* As a lower-case prefix, **n-**, *see* nano-, e.g. ng = nanogram.
mechanics As **N**, *see* newton; *see also* SI alphabet for prefixes.
chemistry See normal solution.
radiation physics See n unit.

n- *Metric* As symbol, *see* N, n.

n_0 *sub-atomic physics* See Loschmidt constant.

N_A [Number Avogadro] *fundamental constant* See Avogadro constant.

NACA [National Advisory Committee on Aeronautics] *USA* A body
advising on such matters as †standard atmosphere.

nano- [Gk: 'dwarf'] Symbol n-. *Metric* The 10^{-9} multiplier, e.g. 1
nanogram = 1 ng = 10^{-9} g.

nanon *length* In parallel with the equally undesirable term 'micron',
the nanometre.

Napier, napier [J. Napier; Scotland 1550–1617] *mechanics See* neper.

napierian logarithm *mathematics* A synonym for natural
†logarithm, i.e. to the base †e, on the false belief that †Napier used that
number as the base in his work pioneering logarithms.

natural Describing a unit nominally reproducible anywhere (rather
than an †arbitrary unit, defined by a physical †prototype). In reality, of
course, until very recent times, the natural references for units have
been much harder to reproduce consistently than have those of a
prototype.[41] The idealistic geodesic definition for the metre was
discarded after just two years, in favour of a prototype, to be replaced by
light-based natural definitions nearly 200 years later. Such fundamental
constants as the †speed of light in a vacuum, the †Newtonian constant of
gravitation, the †elementary charge, the †Planck constant, the †Compton
wavelength, and the †Bohr radius are now seen as preferable. Given an
adopted time element, the first can fix a reference length, the next from

that a reference mass. The Bohr radius and the Compton wavelength provide alternative fundamental units of length; the elementary charge of the proton provides a unit more fundamental for electricity than ampere or volt. The Planck constant, interrelating mass, length, and time, could replace any one of its components.

The †atomic units of physics form one such set, using the Bohr radius for length; the explicitly named **natural units**, using the Compton wavelength plus other natural sub-atomic units, form another. Named generically **natural units of action**, etc., the members of this †coherent set, and their current values (with 10^8 times the †relative standard uncertainty appended to each)[4] are shown below.

Base units: natural unit of

action	$\hbar$	Planck constant over 2π
length	$\lambda = \hbar/m_e \cdot e^2$	Compton wavelength over 2π
mass	m_e	electron mass
velocity	e	speed of light

Derived units: natural unit of

energy	$m_e \cdot e^2$	$= 8.187\,104\,14(64) \times 10^{14}$ J	7.9
momentum	$m_e \cdot e$	$= 2.730\,923\,98(21) \times 10^{-22}$ kg·m·s^{-1}	7.9
time	$\hbar/m_e \cdot e$	$= 1.288\,088\,655\,5 \times 10^{-21}$ s	0.73

natural logarithm (ln) A †logarithm to the base e, $= \log_e x$ or $\ln x$.

natural number Any positive integer, i.e. 1, 2, 3, et seq. *See* number.

natural unit *sub-atomic physics See* natural.

naut *See* nautical mile.

nautical mile (**admiralty mile**, **knot**, **meridian mile**) *length*
Generally, the surface distance along any great circle that subtends 1 minute of angle at Earth's centre, represented by 1 minute of latitude, hence a particular form of †geographic mile. Because of Earth's oblateness, the general unit varies from 1 842.~ m (6 045.~ ft) at the Equator to 1 862.~ m (6 106.~ ft) near the Poles. In an English-speaking maritime context the locally appropriate unit tends to be called sea mile, whereas the nautical mile has a standardized value. Thus the UK long had a standard of 6 080 ft while the USA had a standard of 6 080.2 ft, but since 1954 there has been an international standard. (The UK and the USA based their settings on a spherical Earth of surface area equal to that of the 1866 Clarke ellipsoid in use geodesically to define mean sea level.[11]) The usual scale on the nautical mile is

cable		
10	nautical mile	
	3	nautical league

but the †cable is often adjusted to be a round number of fathoms, etc.

Internat 1954 (**International (Standard) Nautical Mile, INM**) 1 852 m = $\frac{1852}{0.9044}$ yd = 6 076.115∼ ft (6 076.103∼ US survey ft). The 1978 decision of the †CIPM considering it acceptable to continue to use the nautical mile with the SI still stands.

UK To 1975 (also **admiralty mile**, **geographic mile**), 6 080 ft (1 853.184 m), but often taken for shorter distances as 6 000 ft (1 828.8 m). Note distinct and variable †sea mile.

USA To 1954, 6 080.2 ft (1 853.244 96 m, 6 080.188∼ US survey ft).

Internat 1929 6 076.1 ft (1.851 995∼ m, 6 076.088∼ US survey ft).

Ne$_c$ *textiles See* yarn units.

negative number In contrast to the positive number, typically referred to without qualification, the negative number is a notional entity. Written characteristically with a dash or 'minus sign' preceding the digit(s), a negative number conceptually represents the amount that would have to be added before there would be nothing, i.e. zero. The amount needing to be added is the said number without the preceding minus sign. In mathematics that positive amount would be termed the **arithmetic value** or **absolute value** of the original number, identically the value of the corresponding positive number; the term **algebraic value** pertains to the distinct negative/positive signed value of any number.

For integers, the expression of negative numbers is simple and unambiguous. The same might be said of fractions, etc., but the traditional practice with †logarithms presents a contrary picture; there the negative sign applies only to the integer part, the fractional being always positive. Thus what would normally be written −1.234 5 is, being identically −2+ 0.765 5, written for logarithms as −2.765 5 (or, to be distinctive, with the minus sign above the integer). The reason for this peculiarity can be seen by attempting to add, without machine assistance, a mixture of negative and positive numbers of several digits each; the usual resolution is to add the positives, then add the negatives and subtract their total. Logarithms are necessarily of several digits in the fractional part, but rarely of more than one digit in the integer part. The dichotomous practice with the sign allows straight addition of the fractional, constrains the awkwardness of mixed addition and subtraction to a simpler domain, and simplifies the task of looking up logarithms and their functional reciprocals (antilogarithms) in tables.

In financial statements, enclosure by curved brackets is typically used for negative values, rather than a preceding minus sign. The labels DB for debit and CR for credit are used more often in basic accounting to express opposite signs, the pertinent mathematical signs depending on context.

Ne$_L$ *textiles See* yarn units.

neper (napier) [I. Nepero = J. Napier] *engineering.* Symbol Ne. A †geometric scale for power ratios, akin to, and close to, the bel but using the natural logarithm and addressing primarily the amplitudes, defined quantitatively for amplitudes A_1 and A_2 as

$$\ln \tfrac{A_1}{A_2} = \tfrac{1}{2}\ln(A_1/A_2)^2 = \tfrac{1}{2}\ln\tfrac{P_1}{P_2}$$

where P_1 and P_2 are the respective powers. Each unit increase thus represents an increase in the amplitude ratios of †e = 2.718 3$\sim$, in the power ratio of e^2 = 7.389 1$\sim$. The bel, in contrast, addressing primarily the power ratios, is calculated as

$$\log_{10}\tfrac{P_a}{P_b}$$

each unit increase representing a ten-fold increase in the power.

nepit [neper bit] *informatics.* Symbol nit. A unit of information, the amount in nepits of information received being $\ln P_r/P_t$ where P_t is the probability at transmission and P_r the probability after receival; the latter may well be 1 but the former, unless transmitting information already received, is typically very small, essentially 2^{-n} for a random message of n bits.

neutron *sub-atomic physics* Values[4] of associated fundamental constants, with †relative standard uncertainties, are:

neutron gyromagnetic ratio (γ_n)	1.832 471 88(44) $\times 10^8$ s$^{-1}\cdot$T^{-1}	2.4 $\times 10^{-7}$,
neutron magnetic moment (μ_n)	$-$ 0.966 236 40(23) $\times 10^{-26}$ J$\cdot$T^{-1}	2.4 $\times 10^{-7}$,
neutron mass (m_n)	1.674 927 16(13) $\times 10^{-27}$ kg	7.9 $\times 10^{-8}$,
	= 1 838.683 655 0(40) m_e	2.2 $\times 10^{-9}$.

For the charge of the neutron, *see* elementary charge.

New *textiles See* yarn units.

new candle *See* candle; candela.

newton [I. Newton; England 1642–1727] *force.* Symbol N. *SI, Metric-*

m.k.s. Identically $kg \cdot m \cdot s^{-2}$, i.e. the force that gives to a mass of 1 kilogram an acceleration of 1 metre per second, per second ($= m \cdot kg \cdot s^{-2}$ in base terms). The following are among the coherent derived units:

- $N \cdot m$ for moment of force;
- $N \cdot m$ = joule for energy, work, quantity of heat;
- $N \cdot m \cdot s^{-1}$ = watt for power, radiant flux;
- $N \cdot s \cdot m^{-2}$ for dynamic viscosity;
- $N \cdot m^{-1}$ for surface tension;
- $N \cdot m^{-2}$ = pascal for pressure.

The name was advocated in 1935[125] and soon became accepted generally.

1946 CIPM '*Unit of force.* – The unit of force [in the MKS (Metre, Kilogram, Second) system] is the force which gives to a mass of 1 kilogram an acceleration of 1 metre per second, per second.'[8]

Newtonian constant of gravitation (constant of gravitation, gravitational constant) *fundamental constant.* Symbol G, γ. The

coefficient of proportionality in Newton's law of gravitation has probably been measured more often but with less precision than any other constant.[126] Officially evaluated as $6.673(10) \times 10^{-11}$ $m^3 \cdot kg^{-1} \cdot s^{-2}$ with †relative standard uncertainty 1.5×10^{-3},[4] but an unconfirmed evaluation[127] has refined this to $6.674\,215(92) \times 10^{-11}$ $m^3 \cdot kg^{-1} \cdot s^{-2}$ (close to that of Cavendish in 1798[128]).

For bodies that are spherically symmetric or closely so relative to their distance apart, the force exerted on a body of mass M_1 kilograms by a body of mass M_2 kilograms at a centre-to-centre spacing of D metres is G $M_1 \cdot M_2 \cdot D^{-2}$ newtons, i.e. $G M_2 \cdot D^{-2}$ N per kilogram. Where one body orbits another, this force defines the periodicity of the orbiting. Since the period is easy to measure, the formula provides a ratio between the masses of the two bodies, and hence the mass of one if the other is known.

Although here expressed in terms of mass, etc., γ has been seen as the unchangeable unit embracing mass; hence it is far preferable to the †prototype kilogram as the underlying fundamental for this base unit.[19]

See also Gaussian gravitational constant.

newtonian time *See* Ephemeris Time.

Newtonian universal constant *See* Newtonian constant of gravitation.

nibble [by parallelism of bite with byte] *informatics* A half-byte, equalling 4 bits.

nile *sub-atomic physics* UK A nuclear reactivity of 10^{-2}; hence a millinile $= 10^{-5}$, i.e. the order of magnitude of the ratio of excess production constant to the effective production constant for a live nuclear reactor.

The millik is identically the millinile. *See also* dollar; inhour.

nine *Metric* For use applied as a suffix to units, *see* tenth gram.

ninth gram, ninth metre *Metric See* tenth gram.

nit *luminance Metric-m.k.s. 1948.* Symbol nt. Identically $cd \cdot m^{-2}$.
informatics See nepit.

NM. *length* In a maritime context, *see* nautical mile.

Nm *textiles See* yarn units.

nm *length* SI Nanometre, i.e. 10^{-9} m, but note also next entry.

nm. *length* In a maritime context, *see* nautical mile.

nodical [node] *astronomy* Referring to the intersections of a celestial orbit with a plane, notably those of the Moon's orbit with Earth's orbital plane (the ecliptic). For **nodical month**, *see* month.

noise figure *electronics* For an amplifier, the amount by which the signal-to-noise ratio is degraded.

noise rating number *acoustics* Particularly for measuring aircraft and similar noise, the product of noise level in decibels and a factor dependent on frequency (factor $= 1$ at 1 kHz), the level being measured relative to a minimally perceptive increment of 20 μPa in †r.m.s. pressure.[129]

non- [Lat: 'nine'] Prefix for nine, e.g. **nonagon** $=$ nine angles (hence nine sides).

nonillion *See* Table 55 under thousand.

normal atmosphere *See* standard atmosphere.

normal solution (N solution} *chemistry* A solution that contains one †gram-equivalent of replaceable hydrogen per litre of solution, multiples of this concentration being expressed as like multiples of N. Thus one litre of solution containing 1 gram †relative molecular mass of HCl would be an N solution, but the like of H_2SO_4, because of the two hydrogen atoms per molecule, would be a 2N solution. *See also* demal.

normalized *mathematics* Applied to numbers expressed as a mantissa multiplied by an exponential expression, e.g. $8.97\sim \times 10^{20}$, the term applies to having the mantissa within some defined range. The above is identically $0.897\sim \times 10^{21}$, $89.7\sim \times 10^{19}$, $897.\sim \times 10^{18}$, $0.0897\sim \times 10^{22}$, and any other of the unlimited equivalent variants. Clearly any non-zero number can be written such that the mantissa is within some pre-set range that spans, multiplicatively, the size of the base. For base 10, as in the above example, we can fix the range to be from 1 (inclusive) to 10 (exclusive), i.e. 1 to $9.999\sim$; just as easily it can be 0.1 to $0.999\sim$, but it could be 0.5 to $4.999\sim$, for example. In computers, where a number in such two-part form is called a †floating-point number, a common choice for normalization is to have the mantissa of maximal size less than 1, e.g. from 0.1 to $0.999\sim$ for a decimal-based scheme, from 0.0625 to $0.999\sim$ for a hexadecimal-based scheme. This results in any multiplication of two mantissas being also less than 1, any addition being less than 2, so accommodated by just one bit of integer place. For desk-top use it is preferable to use 1 as the minimum, giving range 1 to $9.999\sim$ in the familiar decimal context, but the recommendation of the †SI system to use only multiples of 3 for an exponent, which effectively makes the SI a millesimal scheme, requires mantissas in the range 1 to $999.999\sim$ (else 0.001 to $0.999\sim$).

physics As applied to measure-unit systems, it refers to a system with †natural units in which the †base units are chosen to make selected physical constants equal to unity. The pioneer example had the †speed of light, the †Newtonian constant of gravitation and the †elementary charge all equal to 1.[165] The †Planck constant over 2π was substituted for elementary charge by Planck; the Planck length, etc., was the result.

nox *illuminance Germany* Identically the millilux (mlx).

noy [annoyance] *acoustics USA* 1 noy = the loudness of a random noise in the band 910 to 1090 Hz at 40 db above the standard listener's threshold of 20 μPa. (*Compare* sone.) Higher values correspond to the increase in physical intensity in the same proportion increase in loudness relative to intensity.[160]

N solution *chemistry See* normal solution.

Nt *mechanics* *Metric-m.k.s.* An obsolete representation of newton, also prefixed, as in kNt = kilonewton. The SI symbol for newton is N.
  textiles See yarn units.

Nu, **Nu*** *rheology See* Nusselt number.

Nu value *optics* An indicator of the optical non-dispersiveness of glass and similar transparent materials, quantitatively defined:
  $$= n_D - 1/n_F - n_C$$
where the variables are the refractive indices for, respectively, the mean of the sodium D lines (589.3 nm), the hydrogen red line (656.3 nm), and the hydrogen blue line (486.1 nm).[110] For typical window glass Nu $\approx$ 60.

nuclear magneton Symbol μ_N. *See* magneton.

number The numbers used for counting, namely 1, 2, 3, and so on, are called the **natural numbers**; they exclude zero. If a and b are two such numbers and a is bigger than b then there must be a natural number c such that $a = b + c$. Adding or multiplying two natural numbers always produces another natural number, but subtracting them does not necessarily produce a natural number, and dividing them rarely does. There are unboundedly many natural numbers, but they are self-evidently countable; one can put a unique number to each.

  If we omit the assumption that a is bigger than b, then c may have to be equal to or less than zero to have $a = b + c$. The term **integer** includes zero and negative whole numbers as well as the natural numbers. Adding, subtracting, or multiplying two integers always produces another integer but dividing them rarely produces an integer. There are unboundedly many integers, but they are obviously countable in the sense that one can put a unique natural number to each, e.g. double the unsigned part, then add one if not a natural number, two otherwise.

  Dividing two integers produces a **rational number**, an entity often called a **fraction** or, in North America, a **ratio**. Integers represent a special case, but are rational numbers too. If a is divided by b, we get the fraction represented by a/b, such styling being termed a **vulgar fraction**. If a is less than b this would be called a †proper, otherwise it is an †improper fraction, expressible as an integer plus a proper fraction. Decimal notation (common practice only in recent centuries) provides for the fractional part to be expressed as a string of one or more decimal digits, following the integer part and separated from it by a 'decimal point'. It is an essential feature of any rational number that such a string be either finite in length or settle to being a repeating pattern of at most b

uncountably
∧

* *algebraics are countable*

digits: thus $\frac{1}{2} = 0.5$; $\frac{1}{3} = 0.333\,3\sim$ with the single digit 3 repeating endlessly; $\frac{1}{7} = 0.142\,857\sim$ with this six-digit string repeating endlessly. The rational numbers are unlimitedly numerous, with the integers forming an infinitesimal subset; however, as discussed under infinity, the totality of rational numbers can be equated with the totality of integers by an unambiguous mapping formula, so the rational numbers are countable too.

Not all numbers can be expressed as the quotient of one integer over another, i.e. not all are rational; the well-known entity $\pi = 3.141\,6\sim$ is such an exception, as is $\sqrt{2} = 1.414\,2\sim$ (though not its approximants, like $\frac{22}{7}$). Any such number is called an **irrational number**; the vast majority of numbers are such. If we think of all possible numbers between 0 and 1, for instance, and visualize them as strings of digits, it is fairly obvious that only an infinitesimal proportion would have a finite number of or a repeating pattern of digits. The irrational numbers are not countable; together with the rational numbers they form the totality of †real numbers (q.v. for special discussion re use of this term in informatics).

The number $\sqrt{2}$ solves the equation $x^2 = 2$, identically $x^2 - 2 = 0$. Any irrational number that is a solution of any such polynomial equation, provided it involves only integers for powers of the variable and for coefficients, and is finitely long, is called an **algebraic number**. All *⤫* square roots, cube roots, etc., of natural numbers are obviously algebraic (some being natural numbers themselves). The number π (†pi) and the number †e are among the exceptions; no such polynomial exists of which either is a solution. Such numbers, infinite in number, are called *//* **transcendental**. *∧*

Nor do all polynomials have solutions within the realm of numbers in the ordinary sense, the simple polynomial $x^2 + 1 = 0$ being an example. The problem here is that x^2 is positive for any real number, positive or negative. To evade this restriction, mathematicians created the special entity labelled i, such that $i^2 = -1$, that is $i = \sqrt{-1}$. Written also as j, particularly by engineers, this forms the basis of **complex numbers**, these being of the form $a + b\,i$, where a and b are real numbers (termed the **real part** and the **imaginary part** of the composite number, a complex number with zero real part being called a pure **imaginary number**). Any polynomial, even one with complex numbers among its coefficients and exponents, can be solved completely within the realm of complex numbers.

In reference to everyday counting, for example as six apples in a bowl or eleven players on a soccer team, a number is a **cardinal number**.

Expressed in an ordering context, such as the sixth apple put in the bowl or the sixth largest of the apples there, the corresponding term is **ordinal number**.

The term 'digit' should be used for the individual numeric characters forming a number; e.g. 3 is the first digit of the value for π shown above.

See also negative number; prime number; infinity; numeral.

number density *chemistry*. Symbol N. The number of moles per unit volume.

numeral Any graphic character having an assigned numeric value. The best-known scheme of numerals throughout the world is the European adaptation of a scheme used in India, originating several millennia ago. Enshrined in the Arab world by the great Al-Kwharazimi in 820 CE then brought to Spain by the Moorish Arabs, it was translated by 1100 into Latin, and so into European culture. The writings of Leonardo Fibonacci of Pisa around 1200 made it popular. Inevitably called **Arabic numerals**, these characters are more appropriately called at least **Indo-Arabic numerals**. The two critical assets of this scheme over its dominant predecessor in Europe (†Roman numerals) was the inclusion of a character for zero and the limitation of other characters to the range 1 to 9.

See also hexadecimal; Roman numerals; sexagesimal.

n unit *radiation physics* Various units labelled with a single letter have been used for the biological effects of radiation. All obsolete now, they included **e unit** equal to the energy 'liberated in 1 cc by 1 roentgen, which is 82.7 ergs' and the n unit of twice that. [124] The **E unit** appears to have been effectively the e unit per second.[44]

Nusselt number [W. Nusselt; Germany 1882–1957] *rheology*
As **Nu**, relating to heat transport, the dimensionless ratio of the product of the coefficient of heat transfer and a representative length to thermal conductivity.[61] The name **Biot number** has been used when Nusselt number is used for purely convective heat transport.

As **Nu***, the **Nusselt number for mass transfer** (also **Sherwood number**), relating to transport of matter in a binary mixture, the dimensionless ratio of the product of speed and a representative length to the product of volumic mass and the diffusion coefficient.[61]

oct. *music See* octave.

octa- [Gk: 'eight'] Prefix for eight, e.g. **octagon** = eight angles (hence eight sides).

octal With divisor/multiplier steps of $8 = 2^3$, in contrast with the steps of 2 for †binary, 10 for †decimal, 16 for †hexadecimal, etc.

octal notation, octal number A style of expressing numeric values similar to †decimal notation except for being based on 8 rather than 10, the graphic characters for which are the obvious 0, 1, 2, ... , 7. For example, the number 301 in the decimal system, being

$$= 4 \times 64 \quad + 5 \times 8 \quad + 5 \times 1$$
$$= 4 \times 8^2 \quad + 5 \times 8^1 \quad + 5 \times 8^0$$

is written in octal as 455. Octal notation has the advantage relative to computers that 8 is the highest power of 2 able to be accommodated by the ten decimal graphic characters. Since $8 = 2^3$, each octal digit equates to 3 bits. Once used routinely for humanly readable print-outs of binary computer data, its odd fit with the now ubiquitous 8-bit byte (conspicuous with 2-byte words), makes octal now a relic, outmoded by †hexadecimal notation.

octane number [refers to the chemical used] *automotives* A measure of the 'anti-knock' or resistance to compressive ignition of a fuel, being the percentage by volume of the chemical *iso*-octane (trimethylpentane) that, when blended to *n* heptane, gives a like performance. Octane rating applies to fuel for petrol engines, which, unlike diesel engines, depend on a spark for ignition and need to avoid compressive ignition of fuel. For adequate performance in modern high-compression engines, an octane rating well above 80 is typically required. For superior engines, even 100% *iso*-octane is inadequate; the scale has therefore been extended, by adding tetra-ethyl lead to the *iso*-octane. *Compare* cetane number.

octave *music* The †interval from any frequency to its precise double or half.

octavo (8vo) *paper and printing* Symbol 8vo. A †paper size, being $\frac{1}{8}$ of a full sheet, the actual size depending on that of the full sheet.

octet *informatics See* byte.

octillion *See* Table 55 under thousand.

o.d. [Lat: omni die, 'every day'] *medicine* once per day; multiples thereof are †b.i.d., †t.i.d., and †q.i.d. *See* q.h. for hourly-based units; p.r.n.

Oe *See* oersted.

oersted [H. C. Ørsted; Denmark 1777–1851] *magnetic field strength*. Symbol Oe. *Metric-c.g.s.-e.m.u.* and *-Gaussian 1930* Identically the strength 1 centimetre from a (notional) unit magnetic pole, $= \frac{1}{4}\pi$ abampere·turn·cm^{-1}; technically defined in a three-dimensional system, it corresponds in the four-dimensional electromagnetic sector of the †SI system to $\frac{1000}{4\pi}$A·turn·m^{-1}. It was called the gauss before the International Electrotechnical Committee agreement in 1930 (when that name was applied to flux density),[71] during which time oersted was sometimes the unit of magnetic reluctance. The oersted is a very small unit, only about four times Earth's magnetic field strength at the surface. The corresponding †practical unit is the even smaller praoersted, $= 10^{-3}$ oersted.

oerstedt *electrics* An early, barely used name for the ampere.

ohm [G. S. Ohm; Germany 1787–1854] *electric resistance*. Symbol officially Ω; sometimes, to avoid an alien graphic, R. The ohms of resistance between two points of a conductor carrying a †steady current equals the ratio of potential difference in volts across these points to the amperes of current flowing in the conductor (the conductor not being the seat of any electromotive force). $\Omega = V \cdot A^{-1} = W \cdot A^{-2}$.
SI, Metric-m.k.s 1948 ($= m^2 \cdot kg \cdot s^{-3} \cdot A^{-2}$ in base terms). The following are among the coherent derived units:

- $\Omega \cdot m^2 \cdot m^{-1}$ ($= \Omega \cdot m$) for electric resistivity;
- ohms per square for surface resistivity – the resistance, between perfect conducting strips, of any square on a surface being independent of size (the mutliplicative effect of any change in distance between the wires exactly offsetting the reciprocal multiplication of the width of the conducting surface).

The reciprocal of the ohm in the SI is the siemens, in older m.k.s. the mho.

Metric-c.g.s. See abohm; statohm. *See also* practical unit.

History

The ohm, initially the **ohmad**, was the original unit defined specifically for electrical usage, in 1838, with 1 ft of #11 copper wire being the standard. The burgeoning telegraph, the pioneer of telecommunications, was critically affected by the resistance of its wires, which set the interval at which repeater stations had to be provided. Resistance measurements based on familiar wires were natural, but laboratory specifications and more theoretical approaches were a necessary parallel. The abohm (the ohm of the absolute †e.m.u. system) was one form, but a very small unit. This gave the **practical ohm**, agreed in 1881 at the first International Electrical Conference[33] as being 10^9 abΩ.

That Conference agreed that the ohm should be represented by a column of mercury of cross-section 1 mm^2 and at 0°C, the length to be determined, making it a base unit instead of a derived unit; an accuracy of one part in 2 000 was expected. The initial determination for the established ohm gave a length of 106 cm for what has often been called the **mercury ohm** or **Hg ohm**. The specification was subsequently shown to have made the ohm smaller than intended by about 0.3%, prompting the IEC of 1891 to change the specification to 106.3 cm of mercury, elaborated in 1893 to 106.300 cm of mercury with a mass of 14.452 1 g, at the temperature of melting ice, i.e. precise now to six significant figures, plus defining the more important conditions.[5] In 1908 the IEC, still beset with discrepancies between the e.m.u.-based units and the experimental specifications, adopted the unadorned qualifier 'international' for the latter, giving the **international ohm**, lacking reference to being absolute or practical (though it was the latter). Because of experimental vagaries, the value for conversions is normally referred to as the mean international ohm, $\Omega_M = 1.000\,49 \sim \Omega$.[162] The **US international ohm** $= 1.000\,495 \sim \Omega$.

By the 1930s it was easier to reproduce the absolute-based ohm than the mercury ohm, allowing definition of the common ohm from abohm, as done in 1881. World War II delayed implementation until 1948, when it became simply a derived unit. The modern SI ohm, sometimes called the **absolute ohm** and identically that of the Metric-m.k.s.A. system, has reverted to the original value (of the practical ohm). Discovery of the Josephson effect, then of the quantum Hall effect, applying at very low temperatures with superconductors, together with subsequent development of the moving-coil balance and related work with the volt, improved accuracies about a thousand-fold for the ampere, volt, ohm, etc.[7] Revised conventional values for the constants involved in laboratory realizations using these effects, adopted internationally beginning in 1990, resulted in reducing the resistance value of many laboratory standards by about 0.000 15%.[130]

> **1946** CIPM '*Ohm* (unit of electric resistance) The ohm is the electrical resistance between two points of a conductor when a constant potential diference of 1 volt, applied to these points, produces in the conductor a current of 1 ampere, the conductor not being the seat of any electromotive force.'[8]

acoustics See acoustic ohm; mechanical ohm.

ohma *electromagnetics* An early, briefly used name for the volt.

ohmad *electromagnetics* Initial name for ohm, used briefly in the 1800s.

omega [Anglicized name of Ω, the last letter in the Greek alphabet] *electromagnetics See* ohm.

open window unit (o.w.u.) *acoustics* An early name for †sabin.

optical density *photics* A logarithmic expression of resistance to light transmission or absorbency, being cumulative rather than per unit (as applies to the more familiar use of the term 'density'); quantitatively defined, if L_i is the quantity of incident light and L_t the quantity transmitted through, as

$$\log_{10} L_i/L_t = -\log_{10} L_t/L_i = -\log_{10} e^{-a}$$

where a is the product of absorbtivity and distance. Thus, for a given substance and thickness, an optical density of 1 means that the emerging transmitted light is $\frac{1}{10}$ of the incident, a value 2 that it is $\frac{1}{100}$ of the incident, each unit increase representing a reduction in the light transmitted of 90%. Two optical filters each of optical density 1 operating in series have an aggregate optical density of 2, the sum of their individual values.[131] The unit can be applied similarly to reflected light.

order of magnitude An indication of the relative numeric sizes of two items of very different size, being the exponent of ten that separates the numbers geometrically, e.g. from 7 to 700, being a multiplication by 10^2, is a difference of two orders of magnitude, from 7 to 7 000 a difference of three. Since the essence of the expression is simple approximation using integer values, from 7 to 1 000 would also be seen as a difference of three orders. Effectively, the order is the number of places between the first significant digit in each of two numbers written in normal decimal notation, with a reduction of one where the larger number starts with a much smaller digit than the smaller number. Technically, the order equals the †rounded value of the †logarithm to base 10 of the ratio of larger to smaller, but common usage is far too loose to warrant detailed concern.

Technically, as discussed for logarithms, numbers other than 10 could be used, for instance 16 in some computer contexts, which would require a statement that the difference is 'hexadecimal orders of magnitude' rather than the generally implied decimal order.

ordinal number The †number expressing order, e.g. first, second, as distinct from the †cardinal numbers one, two, and so on.

orthogonal [Gk: 'right' + 'angle'] *mathematics* For two lines, being at a right angle within their common plane; for a line relative to a plane, being at a right angle to any line in the plane.

osmolality *chemistry* For a given solution of a non-dissociating substance, the †molality of an ideal solution of it that would exert the same osmotic pressure; this is usually expressed as **osmole**s of the original.

osmolarity *chemistry* For a given solution of a non-dissociating substance, the †molarity of an ideal solution of it that would exert the same osmotic pressure; this is usually expressed as **osmole**s of the original.

osmole [osmotic mole] *chemistry* See osmolality; osmolarity.

Ostwald system [W. F. Ostwald; Latvia, Germany 1853–1932] *colorimetry* A scheme for specifying †colours relative to black, white, and a given set of coloured pigments.[38, 132] *Compare* Munsell system.

ounce *mass*. Symbols oz, oz av, oz avdp. Distinctively the ounce avoirdupois: $\frac{1}{16}$lb av = 437$\frac{1}{2}$gr = 35.274 0~ g. See pound for more precise values and Table 34 for scale.

See ouncedal for the derived unit of force.

See apothecaries' scale and troy scale for the distinct ounce used for medications and for precious metals.

Table 34

Bl-avdp, US-C-av			Internat values:		SI
grain	...	...	...		64.8 μg
27.3~	dram	...	...		1.77 g
437.5	16		ounce		28.3 g
7 000	256		16	pound	454 g

See hundredweight for upward extensions.

weight, force (**ounce-force**) *See* gravitational system; weight.

volume See fluid ounce.

ouncedal [parallel with poundal] *force*. Symbol ozl. *UK* The force that can accelerate at the rate of 1 ft s^{-2} the mass identified with 1 oz, i.e. the mass accelerated at the rate of $g = 32.1740\sim$ ft s^{-2} by the force that is the true ounce (more specifically the oz-f, the 'ounce- force'), hence 1 ozl $= 8.640\,938\sim$ mN $(0.031\,081\,0\sim$ oz-f) $= \frac{1}{16}$ poundal. *See* gravitational system for wider discussion.

o.w.u. *acoustics See* open window unit.

oz. *mass See* ounce.

P, p *Metric* As an upper-case prefix, **P-** , *see* peta-, e.g. Pg = petagram.
Metric As a lower-case prefix, **p-**, *see* pico- , e.g. pg = picogram.
   mechanics Metric-c.g.s. As **P**, *see* poise.

P-, p- *Metric* As symbol, *see* P, p.

Pa *mechanics Metric* See pascal; *see also* SI alphabet for prefixes.

p.a. Per †annum.

paH *chemistry* See pH.

paper size Paper is usually manufactured on a continuous basis, as
huge rolls. Except for newsprint, which is normally used directly from
such a roll, the product is generally unrolled and progressively cut into
lesser entities, characteristically into very large sheets. These in turn can
be used directly for artwork, used in a printery to produce a mosaic of
printed pages that are folded, cut, then, if for a book, bound and usually
trimmed, else they can be similarly processed by a paper supplier into
the various sizes of general stationery. Whether a leaf of a book or a sheet
of stationery, the size typically reflects a systematic repeated halving of a
large sheet. The familiar terms **quarto** and **octavo** represent this
process, carried out respectively twice, giving four sheets, and three
times, giving eight sheets. While these terms are readily associated with
particular sizes, e.g. quarto with $8\frac{1}{2}$ in × 11 in, they are really relative
terms applicable to any original sheet size, very many different sizes
having been produced. The 'demy' is the full sheet that gives rise to the
familiar 'quarto' (**letter size** in North America), of $17\frac{1}{2}$ in × $22\frac{1}{2}$ in.
Foldings of the demy progress as shown in Table 35, the ratio repeating
since any pair of consecutive folds produces like-shaped quarters of any
beginning sheet. Less obviously, the alternating values are reciprocals
relative to $\sqrt{2}$. Were the initial proportions equal to that figure, i.e.
1.414∼, the ratio of long side to short side of successive products would
remain the same. Such a ratio, being an irrational number, is not
precisely convenient in practice. The ISO values[133] come

Table 35

	in × in		mm × mm		Ratio
full sheet	17.5	22.5	444.5	571.5	1.286
folio	11.25	17.5	285.75	444.5	1.556
quarto (4to)	8.75	11.25	222.25	285.75	1.286
octavo (8vo)	5.625	8.75	142.875	222.25	1.556
16mo	4.375	5.625	111.125	142.875	1.286

as close as practicable, the reference 'A' full sheet of 1 square metre being 841 mm × 1 189 mm. The rounded-down theoretical related sizes of this 'A Series' are shown in Table 36, the ratios all effectively the same.

As can be seen, the A4 of this series is very close in size to the old British quarto and the North American letter size, being half an inch narrower and half an inch longer, a balanced adjustment to meet the $\sqrt{2}$ standard. Specified trimmed sizes in this A Series are the whole mm figures. Clearly this is incompatible, for the smaller sheets at least, with the realities of repeated halving, never mind real trimming. To allow for practicalities, larger sizes are specified for the base material, in the form of a general 'RA Series' for normal trimming, with RA0 being 860 mm × 1 220 mm, and a special 'SRA Series' that provides a greater trimming margin by having SRA0 at 900 mm × 1 280 mm.

The dimensions 841 mm × 1 189 mm of the parent 'A' sheet, were chosen to make it 1 m². The related 'B Series', related more to posters, and 'C Series', to envelopes, have similar rectangularity but parent areas of $\sqrt{2}$m² = 1.413 8~m² and $\sqrt[4]{2}$m² = 1.189 2~m², giving different grid sizes, as shown in Table 37, the last of these being just comfortably

Table 36

	mm × mm		in × in	
2A	1 189	1 682	46.8	66.2
A0 or A	841	1 189	33.1	46.8
A1	594	841	23.4	33.1
A2	420	594	16.5	23.4
A3	297	420	11.7	16.5
A4	210	297	8.2	11.7
A5	148	210	5.8	8.2
A6	105	148	4.1	5.8
A7	74	105	2.9	4.1

Table 37

	mm × mm		in × in	
B0	1 000	1 414	39.37	55.67
B4	250	353	9.84	13.90
C0	917	1 297	36.10	51.06
C4	229	324	9.02	12.76

larger than A4, its successors the same proportionally larger than A5 et seq., equally of A4 folded centrally once, twice, et seq. However, to meet the widely established practice of folding letters into thirds, there is also the special size:

	mm × mm		in × in	
DL	108	219	4.25	8.62

A fractional prefix can be used to indicate a sheet derived by fractioning the length, e.g. $\frac{1}{4}$A3 means 297 mm by 105 mm.

Paris point $\frac{20}{3}$ mm, previously used for footware; *see also* French scale.

parsec [parallax second] *astronomy*. Symbol pc. The elementary length unit of stellar and galactic astronomy, being the distance at which 1 AU subtends an angle of 1 arcsec, 206 264.8∼ AU (30.856 78∼ × 10^{12}km, 3.261 631∼ ly, 19.173 51∼ × 10^{12} mi). Often used in the form of the megaparsec, denoted Mpc. 1 Mpc = $\frac{1}{5}$ siriusweit. (Since the tangent of a very small angle, as occurs with parallax observations, is virtually equal to the angle expressed in radians, parsecs can be obtained directly from angular readings.)

particle size *geology* While the range of sizes of soil, rock, and similar particles varies through a continuum, it is convenient to classify the sizes, and hence the particles, using common names in a defined way. Various schemes have been used, all broadly †geometric to focus on relative size. The finest particles are clay (a term misleadingly associated in vernacular usage with the clumps formed by the highly adhesive fine particles). A clay is usually defined as particles less than 2 μm in diameter. Sand forms the central band of sizes, from various minima around 50 μm invariably to a maximum of 2 mm, qualified progressively

Table 38

US sieve	USDA	U-W	ϕ	mm	International	British
	boulder	boulder	−12	4 096	cobble	boulder
			−11	2 048		
			−10	1 024		
			−9	512		
		cobble	−8	256		
	cobble			254		
				203.2		cobble
			−7	128		
	gravel			76.2		
		pebble	−6	64		
				60	coarse pebble	gravel
			−5	32		
				20	medium pebble	
			−4	16		
			−3	8		
				6	fine pebble	
		granule	−2	4		
10	very coarse sand	very coarse sand	−1	2	coarse sand	coarse sand
18	coarse sand	coarse sand	0	1		
				0.6		medium sand
35	medium sand	medium sand	1	0.5		
60	fine sand	fine sand	2	0.25		
				0.2	medium sand	fine sand
		very fine sand	3	0.125		
140	very fine sand			0.105 8~		
	silt	coarse silt	4	0.062 5		
				0.06	fine sand	silt
		medium silt	5	0.026 25		
				0.02	silt	
		fine silt	6	0.015 6		
		very fine silt	7	0.007 8		
	clay	coarse clay	8	0.003 9		
				0.002	clay	clay
		clay	9	0.001 95		

from very fine up to very coarse. Silt is between clay and sand, while coarser than sand comes a variety of terms: gravel, pebbles, cobbles, then ultimately boulders. The **phi scale**, in contrast, avoids names, merely uses numbers to represent size, increasing values for decreasing size; the size is the negative of the †logarithm to the base 2 of the size in millimetres, written with the Greek letter following. Thus 2ϕ represents a size up to $2^{-2} \times 1\,\text{mm} = 250\,\mu\text{m}$, and -2ϕ represents a size up to $2^2 \times 1\,\text{mm} = 4\,\text{mm}$.

Table 38, organized in decreasing size, shows limiting size for each label used in the various schemes; for instance, the label 'cobble' is limited to 200 mm in the British scheme but 256 mm (-8ϕ) in the Uddan-Wentworth and US Department of Agriculture schemes, and effectively unlimited in the International scheme, where it is the largest labelled particle. Within any one scheme, the label applies to all sizes down to the figure for the next label below, e.g. the cobble of U-W must exceed 64 mm else it would be a pebble, while in the International and British schemes it would remain a pebble if >60 mm. The left-hand column gives the US †sieve numbers for gradations of sand within the USDA scheme. (Note that the U-W scheme is based on the SI but employs binary subdivisions, to produce a true †geometric scale; the International scheme, endeavouring to be decimal, uses the square root of 10 as its ostensible multiplier step, but compromises the geometric pattern by using alternate multipliers of $\frac{3}{1}$ and $\frac{10}{3}$ to achieve an overall decimal progression.)

pascal [B. Pascal; France 1623–62] *pressure, stress*. Symbol Pa. *SI 1971* Identically $\text{N}\cdot\text{m}^{-2}$, e.g. the result of a force of 1 newton spread over an area of 1 square metre ($= \text{m}^{-1}\cdot\text{kg}\cdot\text{s}^{-2}$ in base terms).

Although the Pa is defined to be †coherent with the other SI units, the round 100 kPa is as representative of atmospheric pressure at Earth's surface as the round 30 in (of mercury) has long been. For this reason, 100 kPa is also called the bar; the millibar is then 100 Pa, i.e. the hectopascal, the symbol for which, hPa, thus appears on many weather maps in place of the once-familiar millibar. The †standard atmosphere of the SI is defined at 101.325 0 kPa (29.53~ in of mercury).

The pascal is a small unit for everyday use, as can be seen above; the kilopascal is much more convenient, though even that is small; 10 kPa is little more than 1 p.s.i., and 250 kPa is the typical pressure for a car tyre.

The pascal was introduced in 1950s, but was not accepted fully until its adoption for the SI at the 14th †CGPM of 1971.[8]

pascal · second *rheology*. Symbol Pa·s. *SI 1971* The unit of dynamic viscosity, $= N \cdot m^{-2} \cdot s$ ($= m^{-1} \cdot kg \cdot s^{-1}$ in base terms).

pastille dose *radiation physics* A unit of radiation dose, being that required to turn a pastille of platinocyanide from Tint A (apple green) to Tint B (reddish brown); equals about 500 roentgens. Initially called a **B dose**.

pc *See* per cent.

pc. *astronomy See* parsec.

pct *See* per cent, but note the common practice in US sports to so label a column that is actually not written as a percentage. Usually written as a fraction to three decimal figures (hence even less like a percentage), then read as a three-digit integer. A true 50% is represented by a value of 0.500 or 'five hundred'; as this represents a balanced performance, i.e. equal number of wins and losses, 'better than five hundred' means winning more than you lose. A 'batting percentage' of 250 in baseball means 25.0% or one hit per four visits to the batting plate.

Pe, *Pe** *rheology See* Péclet number.

pea-sized As applied to hail and other items, though not a formal unit of †particle size, it implies a globular object of about 1 cm or 0.5 in diameter.

pebble *geology* A †particle size, typically 3 to 64 mm or 0.1 to 2.5 in.

pebi- [peta- binary] *informatics*. Symbol Pi-. The 1 125 899 906 842 624 $= 2^{50}$ multiplier, as in pebibytes (PiB) and pebibits (Pib). *See* kibi-.

peck *volume*. Symbol pk. *British/American* $\frac{1}{4}$ bushel.
BI Primarily for dry goods, 2 BI gallon (9.092 2~ L).
    The peck was removed from official UK measures in 1970.[28]
US-C dry Solely for dry goods, $\frac{1}{4}$ US bushel (8.809 8~ L).

Péclet number [J. C. E. Péclet; France 1793–1857] *rheology*
As *Pe*, relating to heat transport, a measure of transference of thermal energy in a fluid, the dimensionless ratio of the product of speed and a representative length to thermal diffusivity.[61] Identically the product of the †Prandtl number and the †Reynolds number.

As *Pe**, the **Péclet number for mass transfer**, relating to transport of matter in a binary mixture, the dimensionless ratio of the product of the mass transfer coefficient and a representative length to the diffusion coefficient.[61] Identically the product of *Pe* and the †Lewis number, and of the †Reynolds number and the †Schmidt number.

pennysize *USA* A scheme for expressing the size of nails, from finishing nails to spikes, expressed as a number followed by the symbol d for penny in the English tradition. The lowest value is 2d, representing a nail of length 1 in; to value 8d the length increments by $\frac{1}{4}$ in, making the latter value correspond to $2\frac{1}{2}$ in. The only further values are 10d for a length of 3 in, 12d for $3\frac{1}{4}$ in, 16d for $3\frac{1}{2}$ in and 20d for 4 in. The scheme appears to reflect the cost of one hundred nails of the size, at some past point of time, with 'penny' relating to the vernacular name for cent.

pennyweight *mass.* Symbol dwt, pwt. *See* troy scale.

penta- [Gk: 'five'] e.g. prefix for five, **pentagon** = five angles (hence five sides).

per [Lat: 'for each'] Occurs widely in expressing speed and other ratios of one item to every or each one of another and, in the form of †per cent, to a multiplicity. Though long included in the expression m.p.h. for miles per hour (but just as widely omitted when codifying pounds per square inch as p.s.i.), the term is deprecated except in normal full English. The language-independent solidus is preferred, as in km/h for kilometres per hour, or the use of a negative exponent, e.g. $km \cdot h^{-1}$.

per cent ['per hundred'] Symbols pc, pct, pph, %, the last being usual. A measure of proportion per 100 of the whole. A doubling of any entity is an increase *to* 200%, but an increase *of* only 100%.

Note that when expressing such things as proportional content in a mixture, such as alcohol and water, the dimension of measure must be included, e.g. by volume, or by mass, since the different components can have different †specific masses. The qualifiers 'mass fraction' and 'volume fraction' can serve such a purpose. Often, particularly when expressed as pph and for **ppm** (parts per million), it is necessary to give two units of measure, the proportional ratio being indirect, e.g $mg \cdot L^{-1}$ for the amount of substance dissolved in some liquid; the terms pph and ppm then should be omitted.[134] (It is also necessary to indicate whether the proportion is to the solvent else to the solution.)

See also percentage point; pct; proof.

percentage point A measure of the change of a part in terms of the whole. As an example, let us take an interest rate of 4 pence on the pound, i.e. 4%. If this is increased to 6 pence, it would have increased by 50% relative to itself, but, being from 4% to 6% relative to the capital figure against which the interest rate is expressed, increased by two percentage points. See †basis point for a related expression.

percentile *statistics* A form of †quantile.

perch *See* rod.

perche [Fr: 'perch', i.e. rod] *length Canada 1879* For lands within Quebec originally granted under the old seigniorial system, this old French unit was retained, along with the **arpent** and the **pied** (**foot French measure, Paris foot**), the latter specified as 12.79 in (324.866 mm) — a rounded value of the Paris unit when metric was introduced. The scale is 1 perche = 18 pied = $\frac{1}{10}$ arpent = 19.185 ft (5.847 588 m).
   area Canada 1879 324 sq. pied = 34.194 3~ m^2 (40.896~ yd^2) = $\frac{1}{100}$ arpent.

period *geology* The third-largest unit (following era) of the †geochronologic scale, and the secondary one into which it, excluding the pre-Cambrian, is divided; examples include Devonian, Jurassic, and Quaternary. Typical size is about 50 000 000 years. The next smaller unit is epoch.

perm *engineering* A measure of the permeability of natural reservoirs, e.g. for oil fields; at 0°C 1 perm = 57.214~ × 10^{-12} kg·Pa^{-1}·m^{-2}·s^{-1}. Hence **perm·in** = 1.453 2~ × 10^{-12} kg·Pa^{-1}·m^{-2}·s^{-1}.

permeability *See permittivity.*

permicron *physics* Waves per micron, i.e. per 10^{-6} m.[65]

per mill, per mille *See* permille.

permil, permille ['per thousand'] Symbols per mill(e), pro mill(e), $\frac{0}{00}$. A measure of proportion per 1 000 of the whole. See note under percentage.

permittivity *physics.* Symbol ε_0. *SI* The two elements electric permittivity (dimensionally farads per metre) and magnetic **permeability** (henrys per metre) of a vacuum are immeasurable, but their product equals the reciprocal of the square of c, the †speed of light in vacuum. Within this constraint, the values can be chosen arbitrarily.

With the †c.g.s. system two basic competing choices were made; the †e.m.u. system had permeability set at 1, the †e.s.u. system had permittivity set at 1. The †practical units derived from the former and which became those of the pioneer †m.k.s. system, effectively had permeability set at 10^{-7}.

The equations of electromagnetic theory routinely involve these two elements and the irrational number †pi, with 4π as a multiplier of permittivity, and a divider of permeability. To avoid unnecessary proliferation of irrational numbers, for the SI an offsetting 4π was built into the adopted definitions of the elements (making it a 'rationalized' system), giving

permittivity $\varepsilon_0 = (4\pi \cdot c^2)^{-1} \times 10^7$ F·m^{-1} = 8.854 188~ $\times 10^{-12}$ C^2·m^{-1}·J^{-1}

permeability $\mu_0 = 4\pi \times 10^{-7}$ H·m^{-1} = 1.256 637~ $\times 10^{-6}$ N·A^{-2}

now called, respectively, the †electric constant and the †magnetic constant.

pet- *SI* Contracted form of †peta-.

peta- [Lat: penta 'five'] Symbol P-. *SI* The $10^{15} = 1\,000^5$ multiplier, e.g. 1 petagram = $1\,\text{Pg} = 10^{15}$ g; contractable to pet- before a vowel.
   informatics Sometimes 2^{50}, but *see* pebi- then kibi-.

pH *chemistry* An index of acidity/alkalinity of a solution, being an expression of concentration of hydrogen ions, quantitatively defined as

   $= -\log_{10}$ hydrogen activity of the solution

which, except for extremes,

   $= -\log_{10}$ hydrogen concentration in mol·dm^{-3}

giving a value in the range 0 to 14, with neutral at 7 and increasing acidity descending from there, alkalinity ascending. Most acids, even at 0.1 N solution (*see* normal solution), have a pH less that 4, though boric acid is an exception at 5.2. Most 0.1 N alkali solutions are above 10, though borax is 9.2 and sodium bicarbonate only 8.4. Drinking water and most human bodily fluid is around 7.5, but the stomach operates below 3. Many fruits are of that same level of acidity, and most foodstuffs are acidic; even the bland potato usually has a pH below 6. The notable exception is the egg, whose pH can be 8 when fresh.

The pioneer 1909 scale of Sørensen, labelled **psH**, and its successor **paH** were used prior to greater understanding of the electrochemistry and the definition of pH; various other scales have been proffered.[135]

ph *photics* Metric-c.g.s. *See* phot, also prefixed, as in cph = centiphot.

phi scale *geology See* particle size.

phon *acoustics*. Symbol P. A measure of loudness level, as agreed in 1937 at the first International Acoustical Conference,[136] the decibels of sound pressure above 20 μPa of a pure tone of 1 000 Hz subjectively judged by a group of listeners as equally loud as the sound being measured.[137] (The reference pressure was 316 μPa earlier.) If p and s represent the loudness respectively in phons and sones, then $p = 40 + 10 \log_2 s$.

phot [Gk: 'light'] *illuminance* Symbol ph. *Metric-c.g.s. 1921* Identically candela·steradian per square centimetre[101] ($cd \cdot sr \cdot cm^{-2} = 10^4$ lx). The milliphot (mph) was more common than the phot.

photo-absorption unit *chemistry* In calorimetry, the concentration, in grams per millilitre, required for a reduction of 1% in light transmitted through 1 centimetre of the solution.

photographic emulsion speed indicators *photography See* film speed.

photon *See* quantum.

pi *mathematics* Symbol π. The proportionality constant of the circle – of its circumference to its diameter and of its area to one quarter the area of the square within which it is precisely inscribed. A precise valuation for pi, albeit with very slow convergence, is given by

$\pi = 4\ (1 - \frac{1}{3} + \frac{1}{5} - \frac{1}{7} + \ ...)$

$= 3.141\ 592\ 653\ 589\ 793\ 238\ 462\ 643\ 383\ 279\ 502\ 884\ 197\ 169\ 399\sim$

with reciprocal

$\pi^{-1} = 0.318\ 309\ 886\ 183\ 790\ 671\ 537\ 767\ 526\ 745\ 028\ 724\ 068\ 919\ 291\sim$

The number π is irrational, i.e. it cannot be expressed precisely as the quotient of two integers (and, further, it is transcendental – *see* number). It can, of course, be approximated to as closely as desired, starting with the rough estimate of 3 through the familiar $\frac{22}{7}$ then onward. Some of the historic approximates, with discrepancy from true value, are:

3	$= 3.0$	$-4.51\sim\%$
$\sqrt{10}$	$= 3.162\ 277\ 7\sim$	$+0.66\sim\%$
$\frac{256}{81}$	$= 3.160\ 493\ 8\sim$	$+0.602\sim\%$
$\frac{25}{8}$	$= 3.125$	$-0.528\sim\%$
$\frac{157}{50}$	$= 3.14$	$-0.044\ 6\sim\%$
$\frac{22}{7}$	$= 3.142\ 857\ 1\sim$	$+0.040\ 2\sim\%$
$\frac{355}{113}$	$= 3.141\ 592\ 9\sim$	$+0.000\ 008\ 49\sim\%$

Pi- *informatics See* pebi-.

pica *paper and printing British* One specific size or, when qualified, one of two else more sizes of a type body size, of height 11 points for **small pica**, 12 for **pica**, plus various fractions and multiples. Also used, in the sense of the pica †em, to mean virtually $\frac{1}{6}$ in (4.233~ mm). Applied to a typewriter of fixed pitch, it refers to a type size equivalent to pica, with 6 lines per inch and, horizontally, 10 characters per inch; *compare* elite.

pico- [Span: 'small'] Symbol p-. *Metric* The 10^{-12} multiplier, e.g. 1 picogram = 1 pg = 10^{-12} g.

pied *length Canada See* perche.

pièze *pressure.* Symbol pz. *Metric-m.t.s. 1919* That which, applied over an area of 1 square metre, produces a force of 1 sthene; = 1 kPa. The hectopièze (100 pz) is the bar.

pint *volume.* Symbol pt.
BI (**Imperial pint**) $\frac{1}{8}$ BI gallon = 568.26~ mL (1.201 0~ US pt).
US-C liq $\frac{1}{8}$ US gallon = 28.875 in³ (473.18~ mL).
US-C dry (**dry pint**) $\frac{1}{64}$ US bushel = 33.600 31~ in³ (550.61~ mL).

pipe (**butt**) A large bulk-measure cask, with established volumes and quantities for various commodities in historic marketplaces, = $\frac{1}{2}$ tun.

Pisces *astronomy See* zodiac; right ascension.

pitch *music* The vibrational frequency of a note, and the fundamental frequency in reality because each instrument and voice produces a complex of frequencies (both a narrow range about the fundamental and their harmonics). The pitch can be expressed in hertz (30 to 3 000 for audible notes), but is not so expressed in a musical context.

Because of simultaneous generation of multiples of the basic frequencies, any note sounds very similar to notes of twice and half its frequency. Termed 'harmonious' because of its pleasant sound to the ear, this provides the reason for the repeating pattern of musical scales. The span of the repetition is the **octave**, extending from any note to the note twice that frequency. A similar but declining pleasantness applies to the nearby ratios of small integers, e.g. 2:3, 3:4, 4:5, 5:6, 5:8, etc. (*See* interval.) The octave can be subdivided accordingly, first to produce the note that has its frequency $\frac{3}{2}$ that of the initial note, and so on, to produce a set of notes, a scale, of harmonious relationships. This is the 'scale of just intonation', in which seven notes per octave make a good set of distinct and harmonious notes, but adjacent spacings are erratic, and awkward for many instruments. The 'well-tempered scale' or 'scale of equal temperament' adopts a consistent two-sized spacing: the tone and its exact half, the semitone. Then 2 semitones and 5 tones must give the octave, i.e. effectively 12 semitones must produce a doubling. So the semitone is $\sqrt[12]{2}$ ($= 1.059\sim$), the tone its square ($1.122\sim$). The result, alongside the 'just' ratios, is shown below, along with the actual frequencies if f corresponds to middle C set for modern **concert** or **international pitch** (440 Hz for the A above it), and putting • to represent the (geometrical) midpoint of the double spacings, identically the black keys of the piano.

pK *chemistry* A measure of the strength of an acid. If K_a is the acid dissociation constant, then

$pK = -\log_{10} Ka.$

This †geometric scale is used for comparison of acids.

pk, pk. *See* peck.

planck [M. K. E. L. Planck; Germany 1858–1947] *power* A rate of working of one joule per second, equal to the watt.

Planck constant (also **elementary quantum of action**) *radiation physics*. Symbol h. The fundamental constant ratio of the energy of a quantum radiating from a black body to its frequency, $= 6.626\,068\,76(52)$

Table 39

Just intonation		Equal temperament	Hz	Note
(1f/1	$= 1f$)	1.0f	261.626$\sim$	C
		1.059 5$\sim f$	277.183$\sim$	•
(9f/8	$= 1.125\,0f$)	1.122 5$\sim f$	293.665$\sim$	D
		1.189 2$\sim f$	311.127$\sim$	•
(5f/4	$= 1.250\,0f$)	1.259 9$\sim f$	329.628$\sim$	E
(4f/3	$= 1.333\,3\sim f$)	1.334 8$\sim f$	349.228$\sim$	F
		1.414 2$\sim f$	369.994$\sim$	•
(3f/2	$= 1.500\,0f$)	1.498 3$\sim f$	391.995$\sim$	G
		1.587 4$\sim f$	415.305$\sim$	•
(5f/3	$= 1.666\,7\sim f$)	1.661 8$\sim f$	440.	A
		1.781 8$\sim f$	446.164$\sim$	•
(15f/8	$= 1.875\,0f$)	1.887 7$\sim f$	493.883$\sim$	B
(2f/1	$= 2f$)	2.0f	523.251$\sim$	C

$\times 10^{-34}$ J·s with †relative standard uncertainty 7.8×10^{-8}.[4] (J·s $= m^2 \cdot kg \cdot s^{-1}$ in base terms.)

The **rationalized Planck constant** or **Planck constant over 2 pi** (i.e. 2π) $= 1.054\,571\,596(82) \times 10^{-34}$ J·s; denoted by $\hbar$, it is the †atomic unit and †natural unit of action.

It has been proposed[138] that the name planck be accorded to the m.k.s. unit of angular momentum, i.e. $1\,kg \cdot m^2 \cdot s^{-1}$, which is dimensionally identical to $1\,J \cdot s$, although the latter relates to action, not momentum.

Planck length, etc., *sub-atomic physics* Units of length, mass, and time chosen to give †normalized values for the †Newtonian constant of gravitation (G), the †speed of light (c) and the †Planck constant over 2π ($\hbar$). Similar to those for †atomic units but seemingly very disparate in size (though perhaps understandably so[139]), they are, based on the latest valuations of the underlying constants:

Planck length	$l_P = \sqrt{G \cdot \hbar / c^3}$	$= 1.616\,0(12) \times 10^{-35}$ m
Planck mass	$m_P = \sqrt{\hbar \cdot c / G}$	$= 2.176\,7(16) \times 10^{-8}$ kg
Planck time	$t_P = \sqrt{G \cdot \hbar / c^5}$	$= 5.390\,6(40) \times 10^{-44}$ s

each value having †relative standard uncertainty 7.5×10^{-4}.[4] Planck saw the trio of constants they normalize as the absolute fundamentals connecting length, mass, and time, and the resulting units as the truly fundamental ones for the respective dimensions, surely comprehensible even by extra-terrestrials.[140] The units are hopelessly sized for †base units, but can readily be scaled – decimally for human use.

pm Picometre, i.e. 10^{-12} m.

PM, p.m. [post meridian, i.e. after meridian] *time* Indicative of a time after noon, i.e. after the Sun has nominally crossed the meridian, so the time is after the meridian. Thus 12:30 p.m. identifies the moment 30 minutes after noon. Similarly 12:30 a.m. identifies the moment 30 minutes after midnight. Technically 12:00 can be neither a.m. nor p.m.; it should be qualified as midnight else as noon, when the number can be just 12. (The 24-hour clock avoids all qualification, whether by a.m. else p.m., or by noon else midnight. Its ambivalence is whether to have midnight as 24:00 in the day it ends, else 00:00 in the day it initiates; the latter is preferable.) *See* Universal Time re †leap seconds delaying midnight.

point Deriving from the Latin for 'pierce' or 'prick', the measurement term 'point' characteristically refers to the smallest and indivisible entity, an atom, one might say, in the classical meaning of that word. It originally meant the mark left by a sharply pointed stylus or other tool. The theoretically dimensionless point on a drawing, as distinct from a line or larger entity, is the literal manifestation, while perhaps the most literal usage in measurement is the expression of saw fineness; a seven-point saw has seven sharp points per linear unit (usually per inch), with the repeating point included in the count. Writing 'in point form' or discussing the good or other points of someone or something, or scoring in sport, exemplifies the atomistic sense. Atomistically, it has dimension; if regarded as dimensionless, the point obviously is of no utility in measuring rather than marking; it serves for an instant, not an extent, of time, for instance. As a punctuation mark separating integral part from fractional, it has a mighty role in the expression of measure (called decimal point, but with a role that is identical whether the number is written in decimal, †binary, †octal, else to any other radix).

See also basis point; percentage point.

length USA, UK trad $\frac{1}{12}$ line = $\frac{1}{144}$ in = 176.39$\sim$ μm.

mass (**jewellers' point**) $\frac{1}{25}$ diamond grain = $\frac{1}{100}$ carat.

Internat 1907, including USA, UK (though not implemented strictly in UK until 1913) As **metric point**, $\frac{1}{100}$ metric carat = 2 mg (0.030 864 7$\sim$ gr).

plane angle As a compass point, $\frac{1}{32}$ revolution (= $\pi/16$ rad, 11° 15′). Hence any specific point of such fraction around the compass from the North. Specifically North, East, South, or West if a **cardinal point**, the succession of 32 cardinal and lesser points, starting at N and circling clockwise, grouping them by quadrant with cardinal points repeated, are:

N	N-by-E	NNE	NE-by-N	NE	NE-by-E	ENE	E-by-N	E
E	E-by-S	ESE	SE-by-E	SE	SE-by-S	SSE	S-by-E	S
S	S-by-W	SSW	SW-by-S	SW	SW-by-W	WSW	W-by-S	W
W	W-by-N	WNW	NW-by-W	NW	NW-by-N	NNW	N-by-W	N

On this basis, one point equals an eighth of a right-angle or quadrant, but it is also sometimes used to mean a twelfth of such.

astronomy For expressing the proportion of linear obscuration in an eclipse, a twelfth of the transverse radius, else diameter, of an obscured body.

paper and printing British/American As the **printers' point**, $\frac{1}{12}$ the width of the letter m in some reference texts, and approximately $\frac{1}{72}$ inch. That m letter is called just its generic name **em**, its half the **en**.

UK, USA, etc. Often stated as $\frac{1}{72}$ in = 0.013 889$\sim$ in, but not precisely so.

The em measures 0.166 044$\sim$ in, often regarded as exactly 0.166 in, so

the point equals 0.013 837~ in (351.460~ μm) else 0.013 833~ in
(351.667~ mm). There are thus 72.27~ else 72.29~ points per inch.

poise [J. L. M. Poiseuille; France 1799–1869] *dynamic viscosity*. Symbol
P. *Metric-c.g.s.* Identically dyne·second per square centimetre
$(dyn \cdot s \cdot cm^{-2}) = 10^{-1} \ N \cdot s \cdot m^{-2} = 10^{-1} \ Pa \cdot s$ ($= cm^{-1} \cdot g \cdot s^{-1}$ in c.g.s. base
terms). Hence

• $P \cdot (g \cdot cm^{-3})^{-1}$ = stokes for kinematic viscosity.

The poise is a large unit for most purposes, water at 20°C having a viscosity
of 10.020~ mP and most free-flowing liquids typically being similar; gases
are around 1 μP. There is no equivalent special term in the SI.

poiseuille *dynamic viscosity*. Symbol Pl. *Metric-m.k.s.* Identically
newton·second per square metre or kilogram per metre·second,
$= N \cdot s \cdot m^{-2} = Pa \cdot s$; there is no use of it else any equivalent term in the SI
($= m^{-1} \cdot kg \cdot s^{-1}$ in m.k.s. base terms). It has rarely been used outside of
France.

Even more than its predecessor, the poise of c.g.s., the poiseuille is a
large unit for most purposes, water at 20°C having a viscosity of
1.002 0~ mPl and most free-flowing liquids typically being similar, with
gases being below 1 μPl.

pole *length* British See rod.

poncelet [J.-V. Poncelet; France 1788–1867] *power* France 1919
Identically the work done in 1 second by a force, acting over 1 metre, that
accelerates 100 kg by 1 metre per second, per second, = 100 W
(0.134 102~ hp).

pond *force* Metric See gram-force.

poumar [pounds per million yards] *textiles* See yarn units.

pound *mass*. Symbols lb, lb av, lb avdp. Distinctively the pound
avoirdupois:

Internat, BI, US-C, Australia, Canada, New Zealand, etc., 1959 453.592 37 g, so
1 kg = 2.204 622 62~ lb.[141]

Canada 1951 453.592 43 g, so 1 kg = 2.204 622 33~ lb.[82]

BI 1898 By Order-in-Council it was declared that, on the basis of
comparative measurement of the †prototype pound and the
international prototype kilogram, 1 kg = 2.204 622 3 lb, so 1 lb =
453.592 436~ g (making the prototype pound obsolescent, but not
displacing it).

US-C 1893 Defined by the †Mendenhall Order, $= \frac{1}{2.20462234}$ kg $= \frac{1}{453.5924277} \sim$ g, i.e. 1 kg = 2.204 622 34 lb.

US-C 1866 Specified by the statute that spawned the Mendenhall Order (but not enacted to displace the extant prototype definition) = 1/2.204 622 78 kg = 453.592 338~ g, i.e. 1 kg = 2.204 622 78 lb.

BI 1855 The prototype pound avoirdupois, carefully created[73] to succeed the master prototype troy pound, was damaged, along with the prototype yard, in the conflagration at the Houses of Parliament in 1834.

BI 1825, US-C 1838 defined as $\frac{7000}{5760}$, the prototype troy pound.

See ounce and hundredweight for scales.

See poundal for the derived unit of force.

See apothecaries' scale and troy scale for the distinct pound used for medications and for precious metals.

History

Always = 7 000 gr (although the grain was discontinued officially in the UK in 1985), the pound avoirdupois has for centuries been very close to its current international value, probably the same to at least six significant figures.[73] However, the unit of modern times differed minutely between the UK (lb avdp), the USA (lb av),[142] and Canada until the agreed international standard, with 1 gr = 64.798 91 mg, was implemented in 1959 (though in the UK not exclusively until 1964). Careful measurement in 1960 of the prototype BI pounds still in practical use showed the British master copy as being 453.592 338~ g, and the Canadian as being 453.592 43~ g.[142]

weight, force (**pound-force**) *See* gravitational system; weight.

poundal *force*. Symbol pdl. *BI-f.p.s.* The derived †coherent unit of force in the non-gravitational form of the system (i.e. where lb is the pound-mass), identically $ft \cdot lb \cdot s^{-2}$, i.e. the force of which one unit gives to a mass of 1 pound-mass an acceleration of 1 foot per second, per second.

1 pdl = 0.138 255~ N (0.031 081 0~ lb-f).

Hence, for instance,

- pdl·second per square foot for dynamic viscosity = 0.671 968~ $N \cdot s \cdot m^{-2}$.

See †gravitational system for a wider discussion.

pound·foot *engineering*. Symbol lb.ft. *BI-f.p.s.* The unit of torque in the gravitational form of the system, being the torque produced by one pound-force acting tangentially at a radius of 1 foot.

The foot·pound, although having the same units, is preferably used distinctively for work. With work, the applicable foot is measured in the same line as the force; with torque, the two are †orthogonal.

pound-force, pound-mass *See* gravitational system.

pp [parts per] Except for **ppm** for parts per million, such expressions are generally undesirable, if not ambiguous. They include **ppb** (per billion), **ppc** (per cent, i.e. hundred), **pph** (per hundred), **pphm** (per hundred million), **ppm** (per million), **ppq** (per quadrillion), and **ppt** (either per thousand else per trillion). As discussed under †percentage, some elaboration is usually necessary; **ppmv** indicating ppm measured in volume terms is an example in use.

Pr *rheology See* Prandtl number.

practical unit *electromagnetics Metric* Any of the various units of more practical size derived, by multiplying with powers of ten, from the absolute units of the †e.m.u. system. These were accorded the unqualified familiar names volt, etc., instead of the abvolt (and statvolt of the e.s.u. system).

ampere	$= 10^{-1}$ abampere
coulomb	$= 10^{-1}$ abcoulomb
farad	$= 10^{-9}$ abfarad
henry	$= 10^{9}$ abhenry
mho	$= 10^{-9}$ abmho
ohm	$= 10^{9}$ abohm
pramaxwell	$= 10^{8}$ maxwell, $= 10^{9}$ abvolt·second
volt	$= 10^{8}$ abvolt

which survive, at like size in SI, the mho as the siemens, the pramaxwell effectively as the weber, plus the absolete.

pragilbert	$= 10^{-1}$ gilbert, $= \frac{1}{40\pi}$ abampere·turn
praoersted	$= 10^{-3}$ oersted, $= \frac{1}{4000\pi}$ abampere·turn·cm^{-1}

History

The members of the †absolute system of units developed under the title electromagnetic units, forming the metric †e.m.u. system, were mostly far from the amounts occurring in everyday electrical work; most were minute. This prompted the first International Electrical Conference[33] to define, simultaneously, 'practical units' that were multiples, by appropriate powers of ten, of the absolute c.g.s. ones.

To make the realization of the practical units more truly practical, it was agreed to specify key ones in practical experimental terms, giving the Hg ohm expressed in terms of the resistance of a specified column of mercury, and a volt set by a Clark cell. These two definitions, plus †coherent derivatives for current (ampere), for quantity of electricity (coulomb), and for capacitance (farad), were established in 1881. By 1900 these specifications were found to be at minor variance with the previous ones, and not mutually self-consistent. The consequence was the agreement in 1908 on the distinctively labelled †international units, which varied slightly from the intended practical units, based on the Ag amp expressed as a depositional rate of silver.

The e.m.u. system within which the abvolt, etc., were defined was itself based on the setting of the value of magnetic permeability to 1; in 1901 it was realized that, were permeability set at 10^{-7} rather than 1, the whole array of practical terms shown above fitted unchanged into the metric †m.k.s. system without any multipliers. Thus, what began as part of a c.g.s. system became very readily part of the m.k.s. system, which was increasingly favoured, and ultimately the like-structured †SI system.

pragilbert [practical gilbert] *magnetomotive force* Metric-prac Identically $\frac{1}{4\pi}$ ampere·turn = 0.079 577 47~ A·turn (10^{-1} gilbert, $\frac{1}{40\pi}$ abampere·turn).

pramaxwell [practical maxwell] *magnetic flux* Metric-prac 10^8 maxwell = Wb.

Prandtl number [L. Prandtl; Germany 1875–1953] *rheology*. Symbol Pr. Characterizing a particular matter, the dimensionless ratio of the kinematic viscosity to the thermal diffusivity, identically the ratio of the †Schmidt number to the †Lewis number.

praoersted [practical oersted] *magnetic field strength* Metric-prac Identically $\frac{1}{4\pi}$ ampere·turn per metre = 0.079 577 47~ A·turn·m^{-1}(10^{-3} oersted, $\frac{1}{4000\pi}$ abampere·turn·cm^{-1}).

preece [W. Preece; UK] *electric resistivity* Metric A derived unit appropriate to insulators, identically 1 megohm per quarter-turn.

preferred numbers, preferred values *engineering* A (near-) †geometric series of numbers, providing a pre-defined count of essentially consistent multiplicative steps to a set objective, typically an overall multiplication by 10. The formalization of this simple mathematical process is attributed to a Capt. Renard in France, who applied it to the mooring ropes for balloons in the 19th century,

reportedly reducing the variety stocked from over 400 to under 20.
(Modern usage is most common for electronic components.)

A series of preferred numbers could be truly geometric, e.g. each
double the previous, but the decimal world favours series with decimal
compounding, as with the international scale for †particle size with two
alternating multipliers $\frac{3}{1}$ and $\frac{10}{3}$ to achieve an overall decimal
progression. The International Standards Organization established
several series in 1953,[143] prefixed by R (for †Renard number), and
qualified by a number showing the count of steps for each decimal
multiplication. As the 10^n is irrational for any integer $n > 1$, all series
involve numbers rounded to some extent away from the true geometric
pattern. The R80 series is shown in folded Table 40, for the range 1 to 101,
other values being obtained by multiplication.

For coarser scales there are parallel series with a gradation of
rounding, indicated by successive hash marks. The series for
$n = 5, 10, 20,$ and 40 are given in the next table, again for the range 1 to
10. Table 41 shows at the left hand the step numbers for the densest
range given (R40) and centrally, in italics, the values for the true

Table 40

1.00	1.25	1.60	2.00	2.50	3.15	4.00	5.00	6.30	8.00
1.03	1.28	1.65	2.06	2.58	3.25	4.12	5.15	6.50	8.25
1.06	1.32	1.70	2.12	2.65	3.35	4.25	5.30	6.70	8.50
1.09	1.36	1.75	2.18	2.72	3.45	4.37	5.45	6.90	8.75
1.12	1.40	1.80	2.24	2.80	3.55	4.50	5.60	7.10	9.00
1.15	1.45	1.85	2.30	2.90	3.65	4.62	5.80	7.30	9.25
1.18	1.50	1.90	2.36	3.00	3.75	4.75	6.00	7.50	9.50
1.22	1.55	1.95	2.43	3.07	3.87	4.87	6.15	7.75	9.75

Table 41

	R5	R10	R20	R40	true	R40′	R20′	R10′	R5″	R10″	R20″
0	1.0	1.0	1.0	1.0	*1.0000*	1.0	1.0	1.0	1.0	1.0	1.0
1				1.06	*1.0593~*	1.05					
2			1.12	1.12	*1.1220~*	1.1	1.1				1.1
3				1.18	*1.1885~*	1.2					
4		1.25	1.25	1.25	*1.2589~*	1.25	1.25	1.25		1.2	1.2
5				1.32	*1.3335~*	1.3					

(contd.)

Table 41 (contd.)

	R5	R10	R20	R40	true	R40′	R20′	R10′	R5″	R10″	R20″
6			1.4	1.4	1.4125~	1.4	1.4				1.4
7				1.5	1.4962~	1.5					
8	1.6	1.6	1.6	1.6	1.5849~	1.6	1.6	1.6	1.5	1.5	1.6
9				1.7	1.6788~	1.7					
10			1.8	1.8	1.7783~	1.8	1.8				1.8
11				1.9	1.8836~	1.9					
12		2.0	2.0	2.0	1.9953~	2.0	2.0	2.0		2.0	2.0
13				2.12	2.1135~	2.1					
14			2.24	2.24	2.2387~	2.2	2.2				2.2
15				2.36	2.3714~	2.4					
16	2.5	2.5	2.5	2.5	2.5119~	2.5	2.5	2.5	2.5	2.5	2.5
17				2.65	2.6607~	2.6					
18			2.8	2.8	2.8184~	2.8	2.8				2.8
19				3.0	2.9854~	3.0					
20		3.15	3.15	3.15	3.1623~	3.2	3.2	3.2		3.0	3.0
21				3.35	3.3497~	3.4					
22			3.55	3.55	3.5481~	3.6	3.6				3.5
23				3.75	3.7584~	3.8					
24	4.0	4.0	4.0	4.0	3.9811~	4.0	4.0	4.0	4.0	4.0	4.0
25				4.25	4.2170~	4.2					
26			4.5	4.5	4.4668~	4.5	4.5				4.5
27				4.75	4.7315~	4.8					
28		5.0	5.0	5.0	5.0119~	5.0	5.0	5.0		5.0	5.0
29				5.3	5.3088~	5.3					
30			5.6	5.6	5.6234~	5.6	5.6				5.5
31				6.0	5.9566~	6.0					
32	6.3	6.3	6.3	6.3	6.3096~	6.3	6.3	6.3	6.3	6.0	6.0
33				6.7	6.6834~	6.7					
34			7.1	7.1	7.0795~	7.1	7.1				7.0
35				7.5	7.4989~	7.5					
36		8.0	8.0	8.0	7.9433~	8.0	8.0	8.0		8.0	8.0
37				8.5	8.4140~	8.5					
38			9.0	9.0	8.9125~	9.0	9.0				9.0
39				9.5	9.4406~	9.5					
40	10.0	10.0	10.0	10.0	10.000	10.0	10.0	10.0	10.0	10.0	10.0

geometric series going from 1 to 10 in 40 steps. (Coincidentally, these steps are almost identical with those of the semitone in musical †pitch.)

When a portion of a series is used, standard practice is to show the end values in the style R40 (14..20).

As can be seen, the coarser series are, except for R5″ and R10″, precise sub-sets of the denser set for the same level of rounding, specifically every second and every fourth, starting at 10. Any such methodical subset can be used, e.g. R20′/3 (10..40) for every third value of R20′ running from 10 to 40, i.e. the values 10, 14, 20, 28, 40.

The USA has an equivalent standard used worldwide for electronic components, but with the number of steps binary multiples of 6 rather than of 5, ranging up to 192 steps. (It has also had a standard using fractions of an inch.[144]) The densest is specified[145] at three levels of tolerance, the others each at one level. Identified by the prefix E, the values are shown in folded Table 42, inclusive of the tolerances.

The E192 series, which can be used with 0.5%, 0.25%, or even 0.1% tolerance specified, is shown in folded Table 43.

The E96 series, with 1% tolerance, is the alternating subseries of E192.

Table 42

E48 ±2%	E24 ±5%	E12 ±10%	E6 ±20%	E48 ±2%	E24 ±5%	E12 ±10%	E6 ±20%	E48 ±2%	E24 ±5%	E12 ±10%	E6 ±20%
1.0	1.0	1.0	1.0	2.15	2.2	2.2	2.2	4.64	4.7	4.7	4.7
1.05				2.26				4.87			
1.10	1.1			2.37	2.4			5.11	5.1		
1.15				2.49				5.36			
1.21	1.2	1.2		2.61	2.7	2.7		5.62	5.6	5.6	
1.27				2.74				5.90			
1.33	1.3			2.87	3.0			6.19	6.2		
1.40				3.01				6.49			
1.47	1.5	1.5	1.5	3.16	3.3	3.3	3.3	6.81	6.8	6.8	6.8
1.54				3.32				7.15			
1.62	1.6			3.48	3.6			7.50	7.5		
1.69				3.65				7.87			
1.78	1.8	1.8		3.83	3.9	3.9		8.25	8.2	8.2	
1.87				4.02				8.66			
1.96	2.0			4.22	4.3			9.09	9.1		
2.05				4.42				9.53			

Table 43

1.0	1.21	1.47	1.78	2.15	2.61	3.16	3.83	4.64	5.62	6.81	8.25
1.01	1.23	1.49	1.80	2.18	2.64	3.20	3.88	4.70	5.69	6.90	8.35
1.02	1.24	1.50	1.82	2.21	2.67	3.24	3.92	4.75	5.76	6.98	8.45
1.04	1.26	1.52	1.84	2.23	2.71	3.28	3.97	4.81	5.83	7.06	8.56
1.05	1.27	1.54	1.87	2.26	2.74	3.32	4.02	4.87	5.90	7.15	8.66
1.06	1.29	1.56	1.89	2.29	2.77	3.36	4.07	4.93	5.97	7.23	8.76
1.07	1.30	1.58	1.91	2.32	2.80	3.40	4.12	4.99	6.04	7.32	8.87
1.08	1.32	1.60	1.93	2.34	2.84	3.44	4.17	5.05	6.12	7.41	8.98
1.10	1.33	1.62	1.96	2.37	2.87	3.48	4.22	5.11	6.19	7.50	9.09
1.11	1.35	1.64	1.98	2.40	2.91	3.52	4.27	5.17	6.26	7.59	9.20
1.13	1.37	1.65	2.00	2.43	2.94	3.57	4.32	5.23	6.34	7.68	9.31
1.14	1.38	1.67	2.03	2.46	2.98	3.61	4.37	5.30	6.42	7.77	9.42
1.15	1.40	1.69	2.05	2.49	3.01	3.65	4.42	5.36	6.49	7.87	9.53
1.17	1.42	1.72	2.08	2.52	3.05	3.70	4.48	5.42	6.57	7.96	9.65
1.18	1.43	1.74	2.08	2.55	3.09	3.74	4.53	5.49	6.65	8.06	9.76
1.20	1.45	1.76	2.13	2.58	3.12	3.79	4.59	5.56	6.73	8.16	9.88

prime ['first'] Usually as the symbol ′ in any graphic form or, as the
†minute, any unit at the first layer of fractioning.

length Using the symbol, a common denotation for the foot, the
first subdivision of yard; as **prime**, $\frac{1}{10}$ Rathborn †chain, = 19.8 in
(502.92 mm).

geometry Using the symbol, a common denotation for the minute of
plane angle, the first subdivision of degree.

prime number A natural †number greater than 1 that has no divisor
between 1 and itself is said to be prime, hence called a prime number or
simply a **prime**. Every natural number greater than 1 has at least the two
distinct divisors 1 and itself; a prime has no others.

The number 2 is a prime, there being no candidate divisors between 1
and itself; from it, all even numbers thereafter are non-prime, i.e. 50% of
all subsequent numbers. The numbers 3, 5, and 7 are all prime, meaning
that, of the first six such subsequent numbers, precisely half are prime,
half non-prime. However, of any subsequent six consecutive numbers, at
least one of the odd values must be divisible by 3; including the three
even numbers this means that at least 66% must be non-prime. So the
trend goes; as we look further afield, with an accumulating collection of
primes to be divisors, the density of primes declines progressively. But,

no matter how far up the numbers we travel, we never exhaust the primes, nor is there any known point above which all further primes are spaced by more than the minimal value of 2 (representing consecutive odd numbers).

The number of primes is infinite. This is easily proved by postulating otherwise, namely that there is a finite set of primes, then constructing the integer that is 1 greater than the product of all of them. Clearly it is not divisible by any of the postulated ones, since every such candidate leaves a remainder of 1. Hence our constructed integer must be either a new prime or be the product of two integers both not divisible by our postulated primes, pointing ultimately to at least one prime excluded from our postulated set. Clearly, then, the number of primes must thus be unlimited, as must the size of them.

For any natural number N, the number of primes not greater than it is of the order of the logarithm of N. It can be proved also that for any prime p, the next prime is less than $2p$. There is no consistency, however; for instance, the nearby numbers 86 629 and 86 677 are both primes, and the virtually adjacent numbers 8 004 119 and 8 004 121 are both primes, called 'paired primes'. Primes appear to be distributed generally without pattern, but the **Mersenne primes** provide something of a patterned subset. These develop the fact that $3 = 2^2 - 1, 7 = 2^3 - 1, 31 = 2^5 - 1$, and $127 = 2^7 - 1$ to suggest that $2^n - 1$ is a prime if n is a prime. But the prime $n = 11$ fails, as do many others. However, the formula holds true for an extended if not unlimited range, for four three-figure primes, for eight four-figure primes, and at least to $n = 216\,091$ (giving a Mersenne prime with over 65 000 decimal digits); it provides one relatively economical means for the esoteric exercise of seeking ever larger prime numbers.

Huge prime numbers have acquired significant utility in the modern world, as keys to encryption.

prism diopter, prism dioptre *photics* For a prism, the measure of deviating power, $= 100 \tan \phi$, where ϕ is the angle of deviation.

p.r.n. [Lat: pro ra nata] *medicine* 'As occasion requires'. *See also* s.o.s.

pro-maxwell *electromagnetics* *See* pramaxwell.

pro mill, pro mille *See* permille.

proof, proof spirit *alcoholometry* Schemes for indicating alcohol content of a mixture by the comparison of relative volumes of alcohol to water (part of a complex scenario discussed under †relative volumic mass). The term came from the use of a simple test to decide whether a liquor fell into the 'strong' category, which carried a higher excise duty than the 'weak' one. Floating (or not) on oil and igniting the mixture, else a rag soaked with it, were two early techniques, but the term 'proof' came from the historic test of mixing some of the liquor with a small amount of gunpowder and igniting; rapid burning was 'proof' of it being strongly alcoholic. Many scientific procedures have followed.[117]

UK The value 100 and the term 'proof spirit' correspond to a mixture of ethyl alcohol and water, which, compared to an equal volume of distilled water, has $\frac{12}{13} = 0.923\,076\,9\sim$ the weight when both are weighed in air at 51°F (10.56∼°C). That corresponds to a proportion of alcohol in the mixture of about 49.3% by weight, 57.1% by volume. For any other mixture, the proof figure is its relative proportion of alcohol by volume, multiplied by 100, i.e. its percentage alcohol divided by 57.1 and multiplied by 100 (= percentage multiplied by 1.753 5, pointing to 175.35 proof as the maximal figure, relating to pure alcohol). The terms 'under proof' and 'over proof' refer to the proof spirit of 100, and the figure applied relative to that, e.g. 35 under proof means $(100 - 35) = 65$ proof. The figures are expressed interchangeably as **degrees proof** (°) or as **percentage proof** (%), the latter sometimes producing a disconcerting impression when it exceeds 100%.

USA Twice the percentage alcohol by volume. The maximum is thus 200. The value may be expressed as above or below proof. As in the UK, percentage proof figures, being based on volume proportions, equal the change in overall volume that would produce a mixture of proof 100, a useful factor for proportional taxation.

proper *mathematics* Literally such, e.g. a proper subset is less than the whole set (but not empty); a proper fraction is less than 1 (rather than e.g., $\frac{3}{2}$); a proper interval is not a single value.

proton *sub-atomic physics* Values[4] of associated fundamental constants, with †relative standard uncertainty, are

> **proton charge** = †elementary charge,
> **proton gyromagnetic ratio** (γ_p) $2.675\,222\,12(11) \times 10^8$ $s^{-1}\cdot T^{-1}$ 4.1×10^{-8},
> **proton magnetic moment** (μ_p) $1.410\,606\,633(58) \times 10^{-26}$ J·T^{-1} 4.1×10^{-8},
> **proton mass** (m_p) $\quad$ $1.672\,621\,58(13) \times 10^{-27}$ kg $\quad$ 7.9×10^{-8},
> $\quad = 1\,836.152\,6675(39)\, m_e$ $\quad$ 2.1×10^{-9}.

prototype Relative to units of measure, a prototype is a physical model that enshrines the size: in medieval days perhaps a crude iron bar or an etched brick set in a wall of the town hall or market building, more recently an article made of an elaborate alloy, preserved and used under very particular conditions, e.g. for the metre and kilogram, and similarly for the yard and the †troy pound. Unless multiple accurate copies exist, relying on an arbitrary prototype instead of a †natural unit is hazardous.[41] (Since 1960 the metre has been defined by a reproducible laboratory process based on natural phenomena, though the kilogram remains defined by prototype. The yard and pound are now defined by metric values.)

Even carefully protected prototypes lose material.[146] *See also* pound.

prout [W. Prout; UK 1786–1850] *sub-atomic physics* A rarely used unit of binding energy, equal to one twelfth of that of deuteron, = 0.185 MeV = 0.000 195 dalton.[56]

psH *chemistry* See pH.

p.s.i. *pressure* Pounds per square inch, which remains a common unit of pressure in North America, even where metric is used otherwise. 1 p.s.i. = 6.894 76~ kPa. (Hence 100 kPa = 14.503 8~ p.s.i., which is within the normal range for the pressure of the atmosphere.)

puff [a pronunciation of 'pF'] *electromagnetics* See farad.

puncheon (**tertian**) A bulk-measure cask, with established volumes and quantities for various commodities in historic marketplaces, = $\frac{1}{3}$ tun.

pwt *mass* pennyweight. *See* troy scale.

pyron [Gk: 'fire'] *astrophysics* Metric-c.g.s. 1942 Of solar radiation, identically the energy-flux density of 1 15°C-gram-calorie per square centimetre else, as used originally, the rate of the said amount per minute. The amount = 41.840 kJ·m^{-2}, the rate = 697.33~ W·m^{-2}, about half the average solar radiation rate for Earth. Effectively the renamed langley.

QF *radiation physics* See sievert.

Q factor [quality factor] *electrics* Indicating overall conservational quality of an electric circuit,

$$= \frac{2\pi \cdot f \cdot L}{R} = \frac{(L \cdot C)^{1/2}}{R \cdot C}$$

where f is frequency, L is inductance, C is capacitance, and R is resistance.[147] Sometimes called **magnification**, and abbreviated as **m**. (The reciprocal is called the **dissipation factor**, shown as **d**.) The unit effectively indicates the ratio of energy stored to the energy lost.

    mechanics The term can be extended to any vibrational situation.

q.h. [Lat: quaque hora] *medicine* 'Every hour'; multiples thereof are †q.2h., †q.3h. et seq. *See* o.d. for day-based units, also p.r.n.

q.i.d. [Lat: quater in die] *medicine* Four times per day. *See also* o.d.

q.s. [Lat: quantum sufficiat] *medicine* As much as may be required.

qt. *See* quart.

qtr. *See* quarter.

q.2h., q.3h., et seq. *medicine* 'Every 2 hours', 'every 3 hours', et seq. *See* q.h. for basis.

quad [quadrillion] *engineering* A generic measure of energy for fuel reserves, = 1 quadrillion B.t.u., i.e. 10^{15} B.t.u. = 1.0554$\sim \times 10^{18}$ J, but apparently used sometimes to mean 10^{18} B.t.u., i.e. the Q unit. For natural gas it equates almost exactly to a trillion m.c.f. For other units *see* b.o.e.

quadrant [a quarter of a revolution] *length* A quarter of Earth's circumference, particularly from Equator along a meridian, nominally 10 000 000 m, by the original definition of the metre (6 213.7$\sim$ mi), but actually slightly greater, which is more of a problem to coordinate systems than were it less.

plane angle The right angle, $= 90°$ (1.570 796~ rad).
electrics The original name of the henry.

quadrillion *See* Table 55 under thousand.

quantile *statistics* A subset of adjacent-valued members of a set of observed values for a statistic. If the members of the total set are ordered by increasing size, the set can be divided progressively, beginning at the smallest and ending at the largest, into two or more subsets of equal or nearly equal counts of members. For a statistical variable that can take any value within some overall range, hence representable by a continuous graph, the division can be into precisely equal subsets, by dividing the graph into equal sub-intervals. For a discrete or finite set (e.g. the heights of people in some group), equality of the subdivisions can usually be closely approximated if the count of members is large compared with the number of subdivisions.

If the division is into four bands or groups, the quantile is a **quartile**, if into ten it is a **decile**, if into 100 it is a **percentile**. In each case the usual ordinals can be used to indicate the specific one of the four, ten else 100, graduating conventionally from the smallest values of the statistic to the largest. Thus a member in the third quartile is larger than at least half the members and smaller than at least a quarter of the members; a member in the third decile is larger than at least 20% of the members and smaller than at least 70% of the members; a member in the third percentile is larger than at least 2% of the members and smaller than at least 97% of the members. The demarcation between the second and third quartile, the fifth and sixth decile and the 50th and 51st percentiles, is the **median** (*see* average).

quanta (pl.), **quantum** *physics* A packet of radiation possessing energy equal to $h \cdot f$ joules, where h is the †Planck constant and f the frequency of the radiation, $= (\frac{h \cdot f}{160.2177\sim}) \times 10^{21}$ eV. For light, sometimes also **photon**.

quantum of action *physics See* Planck constant.

quantum of circulation *physics* Half the ratio of the †Planck constant to the mass of the †electron, $= 3.636\,947\,516(27) \times 10^{-4}$ m$^2 \cdot$s^{-1} with †relative standard uncertainty 7.3×10^{-9}.[4]

quart ['quarter'] *volume* Symbol qt.
BI (**Imperial quart**) $\frac{1}{4}$ BI gallon $= 1.136\,5\sim$ L.
US-C liq $\frac{1}{4}$ US gallon $= 57.75$ in^3 (0.946 33~ L).

US-C dry (**dry quart**) $\frac{1}{32}$ US bushel = 67.200 625 in^3 (1.101 2$\sim$ L).

quarter *length* Symbol qtr. *UK* For textiles, $\frac{1}{4}$ yard = 9 in (0.228 6 m); also, in a nautical context, a quarter of a fathom = 18 in (0.457 2 m). *North America* For dressed lumber $\frac{1}{4}$ in (6.35 mm) of thickness.

   area Canada Notably for land as the quarter-†section (160 ac, 64.749 7$\sim$ ha).

   volume 8 bushels. Originally a quarter of a wey.

   mass A quarter of a hundredweight:
BI 28 lb (12.701$\sim$ kg, but *see* pound for precise values). The finest measure to which truck weighings were expressed in UK pre-metric practice, the quarter was removed from official UK measures in 1985.[28]
US-C 25 lb (11.340$\sim$ kg).

   time Notably for the first 3 months (**first quarter, Q1**) and succeeding 3–month units of the year (**Q2, Q3, Q4**), whether the calendar year or such other year as a fiscal year; hence also lesser periods of such nominal years as a 10-month school year. Also used for a quarter of an hour in duration and the quarter-hour time-point.

   plane angle Notably in a nautical context, a quarter of a compass point, $= \frac{1}{4} \times \frac{1}{32}$ revolution $= \frac{1}{128}$ revolution (= $\pi/64$ rad, 2° 48′ 45″).

quartile *statistics* A form of †quantile; divides the range into quarters.

quarto (4to) *paper and printing* Symbol 4to. A †paper size, being $\frac{1}{4}$ of a full sheet, the actual size depending on that of the full sheet. Used very commonly in the UK to mean specifically the quarto or folio post or demy trimmed, at 8.5 in × 11 in (215.9 mm × 279.4 mm), termed †letter size in North America.

quaternary *geology* Although etymologically meaning of four parts, the quaternary in the †geochronologic scale is the fourth part in sequence of what was originally the Cambrian period. It began 1.5 million years ago and is still continuing; hence it can be taken to mean the current time (its second epoch, the Holocene, being the latest 11 000 years).

quintal *mass BI* The hundredweight, i.e. 112 lb (50.802$\sim$ kg); by either name it was removed from official UK measures in 1985.[28]
Canada 1870 The short †hundredweight of 100 BI lb.

quintillion *See* Table 55 under thousand.

Q unit [quintillion unit] *engineering* A generic measure of energy for fuel reserves, = 1 quintillion B.t.u., i.e. 10^{18} B.t.u. = 1.055$\sim$ × 10^{21} J. *See* quad.

R *physics* As **R**, *see* molar gas constant.

   electromagnetics *Metric* As **R** (for resistance), sometimes used in lieu of the alien Ω (capital Greek letter omega) as the symbol for ohm.

   engineering As a unit of thermal insulation, *see* R value.

R₀ *astronomy* See galactocentric distance.

Ra *rheology* See Rayleigh number.

rad *plane angle* *Metric See* radian.

   radiation physics [radiation absorbed dose] Energy imparted to a dosed material, per unit mass, identically defined in 1953 within Metric-c.g.s. as:

$$1\,\text{rad} = 100\,\text{erg}\cdot\text{g}^{-1} = 10^2 \times 10^{-7}\,\text{J}\cdot(10^{-3} \times \text{kg})^{-1} = 10^{-2}\,\text{J}\cdot\text{kg}^{-1}$$

then redefined in the last form for the SI in 1970. Not †coherent in either system, it was succeeded in 1975, within the SI, by the coherent gray, $1\,\text{rad} = 0.01\,\text{Gy}$. The 1978 decision of the †CIPM considering it acceptable to continue to use the rad with the SI still stands.

History

The rad began, in 1918, as the dose required to kill a mouse, clearly an imprecise but usefully indicative amount. It received sharper definition in 1953, the International Commission on Radiology defining it in metric-c.g.s. terms as above. In 1956, it was accepted as the general unit of dose, displacing the roentgen in its use for x-rays; the rad is about 10% smaller than the roentgen, the precise relationship depending on the dosed material. Also expressed as $1\,\text{rad} = 62.4 \times 10^6\,\text{MeV}\,\text{g}^{-1}$.

radian *plane angle.* Symbol rad. *Metric* The angle subtended by a section of circumference equal in length to the radius (hence slightly less than the 60° angle subtended by the secant of the same length that forms a side of the readily constructed hexagon). Quantitatively the ratio of the circumferential length to the radial length. Since the circumference of the complete circle is 2π times the radius, a complete revolution equals $2\pi\,\text{rad} = 6.283\,185\sim\,\text{rad}$; $1\,\text{rad} = \frac{360°}{2\pi} = 57.295\,78\sim\,°$. Hence

- rad·s^{-1} for angular frequency, angular speed, revolution speed;
- rad·s^{-2} for angular acceleration.

Creating an equivalent †base unit in the SI, giving dimensionality to the plane angle, has been proposed.[148] (*See also* spat.) But the radian of the SI, once a supplementary unit, since 1980 is a dimensionless derived unit.[149]

With the irrational quantity 2π per complete turn or revolution, the radian is awkward in various ways. It has influenced the setting of conventional values for units (*see* permittivity). While the degree is preferable for most working purposes, the radian is the natural unit in any mathematical equations, e.g. for map projecting despite the degree being the convenient expression on the map.

The radian is sometimes divided into 100 centrads, and into 1 000 mils.

1960 11th CGPM

1980 CIPM: '*considering*

— that the units radian and steradian are usually introduced into expressions for units when there is need for clarification, especially in photometry where the steradian plays an important role in distinguishing between units corresponding to different quantities,

— that in the equations used one generally expresses plane angle as the ratio of two lengths and the solid angle as the ratio between an area and the square of a length, and consequently that these quantities are treated as dimensionless quantities,

— that the study of formalisms in use in the scientific field shows that none exists which is at the same time coherent and convenient and in which the quantities plane angle and solid angle might be considered as base quantities,

considering also

— that the interpretation given by the CIPM in 1969 for the class of supplementary units introduced in Resolution 12 of the 11th CGPM in 1960 allows the freedom of treating the radian and the steradian as SI base units,

— that such a possibility compromises the internal coherence of the SI based on only seven base units, *decides* to interpret the class of supplementary units in the International System as a class of dimensionless derived units for which the CGPM allows the freedom of using or not using them in expressions for SI derived units.'[8]

radiation constant *fundamental constant* Values, with †relative standard uncertainties:[4]

first	c_1	$3.741\,771\,07(29) \times 10^{-16}\,\text{W}\cdot\text{m}^2$	7.8×10^{-8}
second	c_2	$1.438\,775\,2(25) \times 10^{-2}\,\text{m}\cdot\text{K}$	1.7×10^{-6}

radiation length, radiation unit *astrophysics See* cascade unit.

Ramsden's *See* chain; link.

rank [W. J. M. Rankine; UK 1820–72] *energy See* clausius.
temperature See Rankine.

Rankine *temperature*. Symbols deg R, degree R, °R. A scale and a unit of temperature, being the 'absolute' scale relating to †Fahrenheit, i.e. having its zero at the †thermodynamic null and its degree such that there are 180 between the freezing point and the boiling point of water. Thus the degree Rankine equals the degree Fahrenheit and 0°R = −459.67∼ °F. Readings on the scale are expressed usually as °**R**, temperature intervals preferably as **deg R** or **degree R**, sometimes **R**°. (Note use of R also for †Réaumur scale.)

    See temperature for other scales and conversions between scales.

Rathborn's *See* chain; link.

rational number Any †number that can be expressed as the ratio of two †natural numbers.

rayl [J. W. Strut, 3rd Lord Rayleigh; UK 1842–1919] *acoustics Metric* A unit of specific impedance, the ratio of the effective sound pressure to the effective particle velocity at the surface, identically the product of volumic mass of a gas and the speed of sound therein. The rayl has the dimension of force·second per volume, and is defined †coherently, but thereby of different size, in both c.g.s. and m.k.s.
Metric-c.g.s. $\text{dyn}\cdot\text{s}\cdot\text{cm}^{-3}$ (= $\text{cm}^{-2}\cdot\text{g}\cdot\text{s}^{-1}$ in m.g.s. base terms) = $10\,\text{N}\cdot\text{s}\cdot\text{m}^{-3}$.
Metric-m.k.s. $\text{N}\cdot\text{s}\cdot\text{m}^{-3}$ (= $\text{m}^{-2}\cdot\text{kg}\cdot\text{s}^{-1}$ in m.k.s. base terms).
    For air, the specific impedance is around 400 $\text{rayl}_{\text{m.k.s.}}$.

rayleigh [4th Lord Rayleigh; UK 1875–1947] *astrophysics*. Symbol R. A very small unit of luminous intensity commensurate with the moonless night sky, equal to 10^6 photons or †quanta per square centimetre per second.

Rayleigh number *rheology*. Symbol *Ra*. Relating to heat transport, the dimensionless ratio of the product of the acceleration of free fall, the cubic expansion coefficient, a representative temperature difference and the cube of a representative length to the product of the kinematic viscosity and thermal diffusivity.[61] Identically the product of the †Grashof number for momentum transport and the †Prandtl number.

RBE *radiation physics See* sievert.

RE [Retinol Equivalent] *nutrition* The preferred unit (to †IU) for the expression of the effective content of Vitamin A in foods, retinol being the active form of this vitamin. For animal-derived foods, where the Vitamin A content is retinol, 10 IUs represent about 3 REs. Where the content is provided as carotene from plants, 10 IUs represent only about 1 RE. The recommended daily intake is near 1 000 REs for adult males and pregnant women, less for others.

Re *rheology See* Reynolds number.

real number *mathematics See* number.
  informatics Used to describe a field that can hold fractional real numbers in traditional style within a modest fixed range, in contrast on the one hand to the much more restrictive †integer fields that hold only integer real numbers and on the other to †floating-point fields that hold fractional numbers over an astronomical range by a distinctive two-part notation. Single-precision real-number fields are usually assigned 32 bits, equating with nine decimal digits, the decimal point often being locatable at the pleasure of the user; double precision real-number fields are usually assigned twice this.

Réaumur [R. A. F. de Réaumur; France 1683–1757] *temperature* Symbols deg R, degree R, °R but not to be confused with †Rankine. *France* A scale and a unit of temperature, its defining points being 0 at the freezing point of water and 80 at its boiling point. Readings on the scale are expressed usually as °R, temperature intervals preferably as deg R or degree R, sometimes R°. There is no equivalent 'absolute' scale. The Réaumur scale ranks after only Celsius and Fahrenheit in terms of everyday usage. It provided the temperature standard for the †Baumé scale of hydrometry.
  See temperature scale for other scales and conversions between scales.

History
The use of 80 degrees between the reference points derives from graduations for expansion of the alcohol mixture in a thermometer; with a setting of 1 000 corresponding to the freezing point, each gradation represented a unit of expansion as the alcohol warms. The

precise value 80 for boiling, reflecting an increase of 8.0% from 1 000 to
1 080 in the length of the alcohol column, may, as well as being rounded,
have been extrapolated from lesser readings.

Red, Red I, Red II *engineering* See Redwood.

reduced frequency *rheology* See Strouhal number.

Redwood *engineering.* Symbol Red, specifically Red I and Red II. *UK*
A scheme for measuring viscosity, being the seconds required for a
defined volume of fluid to pass through a specified orifice, there being
scales I and II; for lighter oils 1 sec Red I = 4 to 7 centipoise; for heavier
oils 1 sec Red II is about ten times the former.

Reech number [F Reech; France 1805–84] *rheology* See Froude
number.

register ton For maritime use, usually 100 ft^3 (2.831 7$\sim$ m^3). *See*
shipping ton.

relative atomic mass The ratio of the mass per atom of an element to
one twelfth of the mass of one atom of ^{12}C. For naturally occurring forms
of the element, usually involving a mixture of isotopes, the ratio applies
to the average mass per atom. Previously and widely known as atomic
†weight.

relative density, relative mass density *See* relative volumic mass.

relative molecular mass The ratio of the mass per molecule of an
element or compound to one twelfth of the mass of one atom of ^{12}C. For
naturally occurring forms of the material, usually involving a mixture of
isotopes, the ratio applies to the average mass per molecule. Previously
and widely known as molecular †weight.

relative standard uncertainty Symbol u_r. The established standard
for the evaluations of the physical constants, reflecting the complications
at the quantum level arising from the Heisenberg uncertainty principle,
i.e. that measuring one quantity renders others uncertain.

For the values themselves, the standard short form of expression, used
in this work, shows in brackets the related standard deviation (σ),
applicable at the last digit position of the stated value; the implication is a
68% confidence of the true value being within $\pm\sigma$ of that stated.

relative volumic mass (also **relative density, relative mass
density**) The dimensionless ratio of the mass of a volume of an object

substance to the mass of the same volume of a reference substance, i.e. the ratio of their respective mass densities. Usually the reference substance is water at its densest (close to 4°C) with the object substance being measured at 20°C, but dry air at †s.t.p. is typical when measuring gases. (Though commonly called specific gravity, it is independent of gravitational acceleration.) Besides the temperature of measurement and, particularly for gases, the pressure being significant to the result, the proportion of particular isotopes of constituent elements affects the more precise readings.

In addition to the simple numeric ratio, various scales have been adopted for the expression of relative volumic mass, often to express it in a manner more easily converted to some consequent use but just as often in a manner whose only convenience seems to be the construction of the instrument.[150] Many are utterly arbitrary from a user's viewpoint. Most of them are expressed as degrees, including using the familiar ° symbol. Thus any measurement in such a generic unit is meaningless unless qualified as to the scale employed, though there is some favour for its general use meaning 100 times the relative volumic mass (with a better symbol than ° or degree to signify it).

The measurement of aqueous solutions and mixtures has a long history relative to the taxation of alcoholic drinks, the distinct difference in the relative densities of water and alcohol providing a ready means for assessment of alcoholic content in a beverage. More subtly, the measurement of the relative volumic mass of the mash from which a beer is made provided a fair indication of its alcohol potential, while the difference between readings at the start and end of brewing translated well into alcohol content. (The retention of solids precludes a simpler measurement.) This last, dealing with solutions denser than water, was the pioneering area of such densitometry.

An aqueous solution of sugar may have a relative volumic mass close to 1.5 (1 L can contain over 1.3 kg of dissolved sugar). Typical brewing mashes start around 1.05 to 1.09, and are stopped 0.02 to 0.05 lower. Interestingly, the drop in value multiplied by a hundred is close to the resulting percentage alcohol content by volume. That percentage is wanted to one decimal place, so the relative volumic mass is taken to three places. The term **degree of gravity** is used for the specific gravity (i.e. relative volumic mass) difference multiplied by one thousand, either the difference from 1 for a mash or other solution (i.e. 55° means a relative volumic mass of 1.055) or a drop (e.g. 55° when from 1.085 to 1.030). Divided by 3.86, the degree of gravity gives a measure of percentage solids by mass. The †Brix scale is just such measure, specifically the percentage by mass of sucrose in water solution, at 17.5°C.

For sugar solutions more generally, and similar denser-than-water solutions, the **Twaddell** scale measures specific gravity by degrees that equal a half percentage point relative to water at its densest (4°C), e.g. 1.12 becomes 24°Tw. The twin †Baumé scales provide for unlimited application, with arbitrary values escalating by about 1 per step of 0.07 away from the reference value of 1 in specific gravity, from 0° Baumé in the heavier scale, and from 10° for the lesser volumic mass of water; at 10° Réaumur (12.5°C, 54.5°F). The scale of the **degree API** (†API gravity) is structured identically with the lighter Baumé but for a temperature of 60°F; readings are about 1% greater.

Wines and distilled spirits have specific gravities of less than 1. The best-known measure for hard liquor is †proof. This term relates any sample to a standard that is regarded as 100, the resulting measure being expressed as either **percentage proof** or **degree proof** in absolute terms, or, as under or over proof, relative to 100. In the UK the historic standard is a mixture with 57.1% alcohol by volume; in the USA it is 50%. Proof figures equal the change in overall volume that would produce a mixture of proof 100, a useful factor for proportional taxation. Various hydrometers have been used as measuring tools for proof, each typically with a bevy of alternative weights and its own scale.[117] The **Sikes**, developed by the Royal Society, was probably the first to encompass the problems of temperature by incorporating tables for converting its readings into proof, but its scale was peculiar in having its zero at 66.7% over proof (UK); its 100 was at the maximal specific gravity, i.e. pure water in this context. Subsequent extension above 66.7% brought the Sikes 'A' and 'B' scales.

For wines, several competing schemes have persisted into the contemporary world. The most preferable of these is the **Gay-Lussac**. Widely used in the French-speaking world, this scale is effectively the percentage of alcohol by volume, expressed in degrees that are percentage points, so is just half the US proof and 0.571 times the UK proof; that is, with the traditional Gay-Lussac scale used at 15°C, just 1 degree Fahrenheit below the 60°F that applies to proof. The modern consensus favours using the Gay-Lussac internationally at 20°C, which increases the figures by about 0.2 in mid-range. The **Tralles** scheme used in Russia is essentially the same as the traditional Gay-Lussac, but set at 60°F. The **Windisch** scale of Germany differs distinctly in being percentage point by mass. The **Cartier** scale of Spain is, in contrast, utterly arbitrary in every way, going from 10° at 0% alcohol to 44° at 100%, in irregular steps, with about 19.5° for 50% alcohol by volume (US 100 proof) and 26.5° for 70% (US 40 over proof).

rem *radiation physics* [roentgen equivalent man/mammal] An obsolescent unit of radiation dose equivalent used in radiobiology, derived by appropriate weighting factors from the absorbed energy. It was succeeded in 1979 by the sievert; 1 rem = 0.01 Sv. Derivation of dose equivalent in rems from the absorbed dose measured in rads was identical with that for the sievert from the gray. The 1978 decision of the †CIPM considering it acceptable to continue to use the rem with the SI still stands.

The rem was defined initially, in the 1950s, as the amount of ionizing radiation that causes damage in the human equal to that caused by 1 roentgen of X-rays from a source of 200–250 kilovolts.

Re$_m$ *rheology* *See* Reynolds number.

Renard number [C. Renard; France 1849–1905] *See* preferred numbers.

rep *radiation physics* [roentgen equivalent physical] An obsolete unit of absorbed dose of radiation, being a generalization of the roentgen to accommodate particulate as well as electromagnetic ionizing radiation, and to apply to other materials besides air, biological tissue being the prime but not exclusive interest. It is equated for other ionizing rays in air with the roentgen in terms of energy per unit mass, $= 8.38\sim \times 10^3$ Gy. For tissue, it is adjusted with volumic mass, to range from 6.5 to 100 mJ·kg^{-1} (the latter being bone), with a typical value near 10 mJ·kg^{-1}. It has also been called the **tissue roentgen** and **equivalent roentgen** (er).[44]

revolution *plane angle* (also in some senses **turn**). The complete revolution or one turn around the circle, $= 2\pi$ rad. The turn is fundamentally important in the magnetic effects of an electric current in a coil (e.g. as ampere·turn for magnetomotive force), and suggestions have been made that it be represented by a base unit in the SI (suggested name spat); the radian is a derived unit. See Table 44 for fractioning of the revolution or turn, then degree and grad.

reyn [O. Reynolds; UK 1842–1912] *rheology* *BI-f.p.s.* The unit of dynamic viscosity, for lubricants, $=$ poundal·second per sq. ft., $= 6\,894.76\sim$ Pa·s.

Reynolds number *rheology*. Symbol *Re*. Relating to momentum transport, the dimensionless ratio of the product of representative speed

Table 44

						SI	
grad	...	...	...	...		15.7~ mrad	
	degree	...	...	...		17.5~ mrad	
60		hexangle	...	...		1.047~ rad	
100	90	3:2	right angle	...		0.14~ rad	$= \pi$ rad
400	360	6	4	2	revolution	6.28~ rad	$= 2\pi$ rad

and length elements to kinematic viscosity (the ratio of the inertia forces to the viscous forces in a flowing fluid).[61] For flow through pipes, turbulence rather than laminar flow is indicated by a value greater than 2 000. *See also* Dean number; Taylor number.

The **magnetic Reynolds number** (Re_m, R_m), for fluids flowing in a magnetic field, is the ratio of mass transport diffusivity to magnetic diffusivity, identically the ratio of the product of fluid speed, permeability, and electrical conductivity to a representative length.

rhe [Gk: 'flow'] *rheology* The unit of fluidity, the reciprocal of dynamic viscosity, defined as the reciprocal of the poise, $= 10\ (Pa \cdot s)^{-1}$, though originally defined as the reciprocal of the centipoise, $= 10^3\ (Pa\,s)^{-1}$, and sometimes similarly applied to the stokes or centistokes as kinematic fluidity.

rhm *radiation physics* [roentgen per hour at 1 metre] A unit of effective strength of a gamma-ray source, quantitatively the number of roentgens dose per hour at 1 metre, roughly equivalent to the curie.

Richter scale [C. F. Richter; USA 1900–85] *geophysics* A †geometric scale for magnitude (total energy) of an earthquake, ranging from 1 upwards, with each increment of 1 equalling a 60-fold increase in energy; originally specified in 1935.[151] In contrast to the †Mercalli scale, which expresses received intensity locally with 12 discrete levels, the Richter scale is single-valued for one earthquake and is open-ended with a nominal continuum of values (though usually expressed to just one decimal place, and with no earthquake within historic times having reached 9).

Valuations on the Richter scale are derived from standardized seismographs, and relate to the subterranean point of the initial disturbance – the hypocentre; the epicentre is the surface point immediately above that.[109]

Ridgeway *See* hardness numbers.

Ridgway *See* colour.

right ascension (RA) *astronomy* A form of longitude that expresses the angular position of a star or other celestial body, referenced to the celestial extension of Earth's equatorial plane and axis, with zero value at the †First Point of Aries, and representing the additional time until ascension. (*Compare* celestial †latitude; longitude.) The sidereal day is the time of one rotation of Earth relative to the stars, which is $\frac{1}{366.24\sim}$ less than the normal apparent solar day; its distinctive hour and minute are each the same proportion of the more familiar units. As Earth rotates, the stars repeat their longitudinal position every 24 sidereal hours, ascending from the eastern horizon just as the Sun does.

Observationally, the overhead position is a more convenient functional marker than the horizon. Right ascension is the time, expressed in sidereal units, from the passage of the First Point of Aries to that of the star being measured, and, unlike most forms of longitudinal measure, is unidirectional, the values going from 0 to virtually 24 (sidereal) hours. (For navigational usage, degrees rather than sidereal hours are used, with values 0° through to virtually 360°.)

The corresponding latitudinal measure is **declination**, which uses the angular degrees, etc., qualified by North else South (hemisphere).

As discussed under †zodiac, the First Point of Aries is not fixed, so the specification data for a star position must refer to a year.

The twelve constellation names used notionally in the zodiac, to divide the year into twelve equal elements, are used even more notionally to divide the day likewise as Earth rotates; Aries is the segment of the sky from 0 to 2 h (0 to 30°), Taurus from 2 to 4 h (30 to 60°), et seq. through Gemini, Cancer, Leo, Virgo, Libra, Scorpio, Sagittarius, Capricorn, and Aquarius to Pisces at 22 to 24 h (330 to 360°).

Greenwich Hour Angle (GHA) is a similar concept to right ascension, being the west longitude (0° to 360°) of the point directly below the celestial body at any instant. Since the GHA of any body equals the GHA of the First Point of Aries plus (360° minus RA of the body), the latter referred to as the **Sidereal Hour Angle** (SHA) of the body, which is virtually constant for stars, navigators can easily gain one GHA from the other. (Astronomers use the same terminology, but with the angles expressed in sidereal hours rather than degrees.)

While the pattern of the stars recurs virtually exactly every 24 sidereal hours, it advances every 24 normal hours by an amount corresponding to nearly 1°, specifically by $\frac{360°}{365.24\sim}$, which figure apparently prompted our 360 degrees in a circle; *see* degree.

ringing equivalent number *telecommunications* UK A conductance of 250 μΩ.

R_K *electromagnetics* See von Klitzing constant.

Rm *rheology* See Reynolds number.

RMS, r.m.s. [root-mean-square] *electromagnetics* The statements of simple relationship between volt, ampere, ohm, and other electromagnetic units (e.g. resistance in ohms = $V \cdot A^{-1}$, electric charge = $A \cdot s$) apply basically only to direct current. For alternating current, with its inherent oscillation of voltage between some matched positive and negative figures of maximum amplitude, the same statements can apply if an average is used – an average not of the voltage itself, but relative to its effective power level – to give a figure for voltage equal to that for a steady direct current with the same heating effect. Allowing that power is proportional to the square of voltage, that average is the square root of the mean of the voltage squared. Termed **root-mean-square** or simply r.m.s., for a smooth sinusoidal oscillation it equals the maximum amplitude divided by $\sqrt{2}$; identically the maximum amplitude = $\sqrt{2} \cdot$r.m.s. The stated voltage on systems and equipment is usually the r.m.s., so the 110 V a.c. of North America oscillates between $\pm 110 \cdot \sqrt{2}\,V \approx \pm 156\,V$ and the 230 V a.c. of Europe oscillates between $\pm 230 \cdot \sqrt{2}\,V \approx \pm 325\,V$.

The concept of r.m.s. value is applied similarly to sound and other oscillatory phenomena.

roc [reciprocal ohm·centimetre] *Metric-c.g.s.* $(\Omega \cdot cm)^{-1}$.[152]

Rockwell hardness number *See* hardness numbers.

rod (**perch**, **pole**) *length* BI, US-C $\frac{1}{4}$ chain = $5\frac{1}{2}$ yd (5.029 2~ m); *see* inch for greater details, including reference to the **coast rod** or **survey rod** of $5.5 \times 36/39.37\,m = 5.029\,210\,06\sim m$.

Canada See also perche.

Roemer [O. Rømer; Denmark 1644–1710] *temperature* The precursor of the †Fahrenheit scale, with zero at the freezing point of brine and 60° at the boiling point of water. Taking simplified fractioning, the freezing point of water was 7.5° ($\frac{1}{8} \times 60°$), human body temperature was 22.5° ($\frac{3}{8} \times 60°$).

roentgen [W. K. Röntgen; Germany 1845–1923] *radiation physics*
Symbol R, röntgen. An obsolescent unit of ionizing electromagnetic
radiation, being the quantity of x-rays or gamma rays that, through
ionization, produces 2.58×10^{-4} coulombs of electricity per kilogram of
dry air at †s.t.p.

The roentgen was originated relative only to x-rays, being agreed in
1928 as the amount that would produce 1 †electrostatic unit of electric
charge from 1 cubic centimetre of standard dry air. In 1937, for practical
reasons, mass replaced volume for the reference amount of air, making
the amount 1.293~ mg of air. The 1978 decision of the †CIPM considering
it acceptable to continue to use the roentgen with the SI still stands.

Such ionization displaces electrons from individual atoms to produce
negatively charged free electrons and matching positively charged ions;
1 roentgen produces $1.61\sim \times 10^{15}$ of each per kilogram. The energy
required for such displacing is $8.69\sim$ mJ·kg^{-1} or $11.2\sim$ mJ·m^{-3}.

Using this energy figure of $8.69\sim$ mJ·kg^{-1}, the roentgen was extended
to apply to materials other than air, and to the particulate radiation of
alpha and beta rays. In this extended form it was called the **tissue
roentgen** or, much more usually, the **roentgen equivalent physical**,
abbreviated to and universally called the **rep**. However, the roentgen
itself was often so used; until 1956 it was used in a radiological context
not only as a measure of exposure (which it correctly is) but also as a
measure of absorbed dose; the rad took over the latter role.

Until 1952 the British authorities used a version about 8% less.

roentgen equivalent mammal, roentgen equivalent man *See*
rem.

roentgen equivalent physical *See* rep.

rom [reciprocal ohm·metre] *Metric-c.g.s.* $(\Omega \cdot m)^{-1}$.[152]

Roman numerals The graphic characters used as numerals by the
ancient Romans, and thereby by many other peoples since, are:

    I and i for 1 (also j; see note below)

    V and v for 5

    X and x for 10

    L for 50

    C for 100

    D for 500

    M for 1 000.

The fashion of using lesser ones subtractively, as in IV for 4, IX for 9 and

XC for 90, was introduced only in medieval Europe; it was not practised by the Romans. (On clocks 4 is invariably the classical IIII.) A line over the top, indicating a thousand-fold increase, was used for larger values, essential over the M to increase its meaning to be a million.

Where the symbol i occurred multiply, it was common medieval practice to substitute j for the last one, as a clear terminator, e.g. iij for 3.

See numerals for comparative discussion.

röntgen *radiation physics See* roentgen.

rood *area* British $\frac{1}{4}$ acre = 40 sq. rods = $1\,210\,\text{yd}^2$ $(1\,011.714\sim \text{m}^2)$. The rood was removed from official UK measures in 1985.[28]

root-mean-square *electromagnetics See* r.m.s.

root-mean-square error *See* standard deviation.

Rossby number [C. A. Rossby; Sweden 1899–1959] *rheology* The ratio of inertial forces to Coriolis force.

rounded, rounding off *mathematics* Rounding off numbers involves the termination of a string of digits at some incomplete point, but, unlike truncating, taking into account the digits beyond that point. It is well illustrated with the simple fraction $\frac{2}{3} = 0.666\,6\sim$ with the digit 6 occurring endlessly. To terminate as shown is merely to truncate; to write the number as 0.666 7, which is the closest value at four decimal places to the real value of the fraction, is to have it rounded.

Some form of termination, i.e. rounding else truncation, must be induced for all decimal fractions that do not terminate naturally, which is true for all irrational †numbers as well as many rational numbers. All non-terminating rational numbers fall into a repeating pattern (shorter in digit count than the value of the denominator), so they can be expressed by terminating at the completion of the first pattern, then indicating its repetition by putting a dot above the first and last digits of the pattern (which is a single digit in the above case). But naturally terminating numbers must often be shortened to fit circumstances, e.g. $\frac{1}{16} = 0.0625$ (exactly) may have to be entered into a column that allows for only three decimal places. In such circumstances there are two numbers equally close to the true value; to avoid an undue overall bias, the convention is to end the rounded value with an even digit, 0.062 in this case.

rowland [H. A. Rowland; USA 1848–1901] *astrophysics* A unit of the order of an angstrom used in the 1980s for expressing a discrepancy in published data on solar spectra, arising from a pervasive misunderstanding of a reference length.

RPM, r.p.m. *engineering* The number of completed revolutions per minute of a revolving shaft.

RSI, RSI value *engineering* The SI-expressed unit of thermal insulation corresponding to the earlier R value, i.e. expressed in terms of $K \cdot m^2 \cdot W^{-1}$ instead of $°F \cdot ft^2 \cdot h \cdot Btu^{-1}$, the RSI value = $0.176\sim \times$ R value. Hence R12 becomes RSI 2.1, R20 becomes RSI 3.5.

rundlet A cask, with established volumes and quantities in historic marketplaces, for wine, $== \frac{1}{2}$ pipe or butt = $\frac{1}{14}$ tun.

R unit, r unit *radiation physics* A unit of the 1920s that expressed X-ray intensity, existing in different forms. One **Solomon R unit** gave the ionization equivalent of 1 gram of radium placed 2 cm from an ionization chamber with an intervening platinum screen 0.5 mm thick. One **German R unit** was about 2.5 that size.[110]

rutherford [E. Rutherford, 1st Baron Rutherford; NZ, Canada, UK 1871–1937] *radiation physics* A unit of radioactivity, 1 rutherford corresponding with 10^6 disintegrations per second = 10^6 Bq.

R value *engineering* A unit of thermal insulation, expressed in terms of $°F \cdot ft^2 \cdot h \cdot Btu^{-1}$. Its reciprocal, for a given time and area, is †U factor. Under the SI, this is expressed as RSI value = $0.176\sim \times$ R value.

rydberg [J. R. Rydberg; Sweden 1854–1919] *mechanics* An initial name for kayser.

Rydberg constant *physics.* Symbol R_∞. A constructed quantity fundamental to wave number for all atomic spectra (hence their frequencies), $2\pi^2 m_e e^4/c \cdot h = \mu_0^2 m \cdot e(e \cdot c/2\,h)^3 = $ 10 973 731.568 549(83) m^{-1} with †relative standard uncertainty 7.6×10^{-12}.[4] The reciprocal of the frequency term $4\pi R_\infty \cdot c$ is the †atomic unit of time (= $24.188\,8\sim$ as).

For **double Rydberg**, *see* Hartree.

S

S, s *electromagnetics* As **S**, *see* siemens.
   volume Metric As **s**, the original symbol for stere.
   time Metric As **s**, *see* second.
   astronomy As **S**, mass of the Sun; *see* astronomical unit.

sabin [W. C. Sabine; USA 1868–1919] *acoustics* A measure of sound absorption. If T is the number of seconds for a characteristic sound to drop 60 dB to inaudibility in a room of volume v cubic feet, then the sound-absorptive power of the room can be expressed as $a = 0.049\,5\ v/T$, the numeric factor varying with the speed of sound (*see* Mach number). Ignoring the minor contribution of the air within the room and letting S be the total surface area of walls, floor, and ceiling, the total absorption represents an average absorption of a/S per square foot of surface. But it can just as easily be represented by a square feet of total absorption and $(S - a)$ square feet of non-absorption.

   Earlier called both **open window unit** (an open window being a total absorber of sound) and **total absorption unit**, the sabin (a name adopted in 1937) is the unit applicable to a, i.e. the example room is said to have a sound absorption of a sabins. The unit has areal dimensionality; the numeric factor, which embeds the reciprocal of length, becomes 0.162 when using metres instead of feet.

SAE [Society of Automotive Engineers, USA] *engineering* A scheme for grading lubricating oils according to their viscosity under varying temperature conditions. Approximate equivalents in centipoises are in Table 45.

Saffir–Simpson Hurricane Scale [J. H. Saffir; USA: R. H. Simpson; USA] *meteorology* USA 1975 a Code for forecasting the characteristics of and damage from an imminent hurricane.[153] A synopsis is shown in Table 46 (the surge height to be adjusted to a particular coast, the damage that of the wind alone).

Sagittarius *astronomy* See zodiac; right ascension.

Table 45

at:	0°F (−17.8°C)	100°F (−37.8°C)	210°F (−98.9°C)
SAE 10W	1 950	41	6.0
SAE 20W	6 550	71	8.5
SAE 30		114	11.3
SAE 40		173	14.8
SAE 10W-30		76	12.7
SAE 20W-40		93	13.7

Table 46

Category	Central pressure mbar	Max wind m.p.h.	surge ft	damage
1 weak	≥ 980	75 to 95	4 to 5	Mostly to trees and loose objects; no real damage to building structures
2 moderate	965 to 979	96 to 110	6 to 8	Flimsy structures damaged, trees down, some damage to roofing, windows, and doors
3 strong	945 to 964	111 to 130	9 to 12	Mobile homes, etc., and signs destroyed, some structural damage to small buildings
4 very strong	920 to 944	131 to 155	13 to 18	Extensive roof, window, and door failures, some structural damage to better buildings
5 devastating	<920	>155	>18	Very extensive roof and glass failure; some buildings blown down, over or away

sand *geology* A †particle size, of maximum diameter 2 mm (0.787~ in), ranging to about a sixth of that depending on the scheme employed.

saros *time* The inherent period for recurrence of eclipses, essentially the least common multiple of eclipse year and †synodic month = 6 585.32 days, = 18.998~ eclipse years = 223.000 synodic months = 18 y 11.32 d. A period for cruder recurrence of eclipse sequences has 1 386.38 days (3 y 291.38 d), close to 4 eclipse years and to 47 synodic months.

The fractional part of the day count causes each solar eclipse of any one sequence episode in a given series to be centred about a third Earth's rotation westward from the corresponding one of the previous episode (115° for the saros, 137° for the lesser period). *See also* Metonic cycle.

A sequence of eclipses separated by 1 saros is called an eclipse series. A solar eclipse series contains about 70 eclipses spread over about 1 300 years; a lunar eclipse series extends over about 900 years. Many eclipse series of each type are running at any one time.

Sander's theater cushion [Sander's Theater, Boston, USA] *acoustics* The first unit of absorption,[154] literally a theatre cushion of the 1890s.

savart [F Savart; France 1791–1841] *music* A †geometric scale for measuring †intervals. 1 octave = 1 000 ($\log_{10} 2$) savarts = 301.03~ savarts, though it is usually taken as precisely 300; hence an increase in frequency of 1 savart represents a 1.002 305~ multiplication. For frequencies f_1 and f_2, the latter the higher, the difference in savarts is officially

$$1\,000 \times \log_{10}(f_2/f_1)$$

or, modifying to give a round 300 savarts to the octave,

$$996.578\sim \times \log_{10}(f_2/f_1).$$

Saybolt Furol Second [fuel and road oil] *rheology*. Symbols SSF, SFS. *See* Saybolt Universal Second.

Saybolt Universal Second *rheology*. Symbols SSU, SUS. *USA* A scheme for measuring viscosity, being the seconds required for 60 mL of fluid to pass through a specified orifice. The **Saybolt Furol Second** is a variant used for heavier oils, being about ten times the SUS. The usual conversion from SUS to kinematic viscosity in centistokes is, for reading S,

$$0.226 \times S - 195 \times S^{-1} \text{ for } S \le 100.$$
$$0.220 \times S - 135 \times S^{-1} \text{ for } S \ge 100.$$

Thus 10 cS is about 58.8 SUS, and 100 cS is about 48.5 SFS at 122°F.

sb *photics* Metric-c.g.s. *See* stilb. Also prefixed, as in csb = centistilb.

Sc *rheology See* Schmidt number.

Scheiner [J. Scheiner; Austria 1858–1913] *photography See* film speed.

Schmidt number [E. W. H. Schmidt; Germany 1892–1975]. *rheology*. Symbol *Sc*. Characterizing a particular matter, the dimensionless ratio of the kinematic viscosity to the diffusion coefficient, identically the product of the †Prandtl number and the †Lewis number.

scleroscope hardness *See* hardness numbers.

Scorpio *astronomy See* zodiac; right ascension.

scruple [Lat: scripulus 'small stone'] *mass See* apothecaries' scale.

s.d. *statistics See* standard deviation.

sea mile *length British* One minute of latitude locally, but sometimes a synonym for nautical mile; *see* geographic mile.

sec. *time See* second.

secohm Ohm·sec = ohm·second, i.e. $\Omega \cdot$s.

second The ordinal (also **2nd**), being the element of an ordered set preceded by just one other, and indirectly from that, the element of a second layer of subdividing. In a hierarchical scheme of subdivided units, the unit of the second rung or layer. Often a subdivision of minute, and then rarely other than a sixtieth of a minute, but also a subdivision of prime, when it is often a tenth. Being the second subdivision, it is commonly represented by a double prime mark, e.g. 3″ for 3 seconds of time, of angle, etc.

length British Using the symbol ″ but not its name, a denotation for inch, the second subdivision (following foot) of yard. Also as link, $= \frac{1}{10}$ prime $= \frac{1}{100}$ Rathborn chain $= \frac{99}{50}$ in $= 1.98$ in (50.292 mm).

time. Symbol s in the SI. Traditionally $\frac{1}{60}$ of a minute, thereby $\frac{1}{86400}$ of a †day (a unit varying in size depending on the qualifier) and sized by such fractioning. Since 1967, however, the second of normal usage (derived from the mean solar day), the base unit for time in the SI and other †metric systems, has been defined as the duration of 9 192 631 770 periods of the radiation corresponding to the transition between the two hyperfine levels of the ground state of the caesium-133 atom.[155] It has since been affirmed that this applies to a caesium atom at rest at 0 K, so unperturbed by black-body radiation, hence that the frequencies of primary frequency standards should be corrected for ambient radiation.[156] The scale on second of time in any definition is:

second			
60	minute		
3 600	60	hour	
86 400	1 440	24	day

to which can be added week, fortnight, †month, †year, †century, and †millennium in their varying senses. All these figures can be varied by one or more †leap seconds whenever astronomical observations indicate the need for such 'correction'.

1952 8th General Assembly of the International Astronomical Union: '...the second of ephemeris time (ET) is the fraction $\frac{12960276813}{408986496} \times 10^9$ of the tropical year for 1900 January 0 at 12 h ET i.e. effectively the traditional $\frac{1}{86400}$ of a day, but that day being the mean solar day based on the length of a specific year.'

1956 CIPM: '...*considering* the above, *decides* The second is the fraction $\frac{1}{31556925.9747}$ of the tropical year for 1900 January 0 at 12 hours ephemeris time.'

1960 11th CGPM: '*ratifies* the above.'

1964 CIPM: '*declares* that the "molecular or atomic frequency" standard to be employed is the transition between the hyperfine levels F = 4, M = 0 and F = 3, M = 0 of the ground state $^2S_{\frac{1}{2}}$ of the caesium 133 atom, unperturbed by external fields, and that the frequency of this transition is assigned the value 9 192 631 770 hertz i.e. indirectly defining the second as the time for the said number of such transitions.'

1967–68 13th CGPM: 'The second is the duration of 9 192 631 770 periods of the radiation corresponding to the transition between the two hyperfine levels of the ground state of the caesium 133 atom.'

1997 CIPM: '*affirms* that this definition refers to a caesium atom at rest at a temperature of K.'

geometry (also **arcsec**, **second of arc**) Also as **sec** and often as the symbol ", the second layer of fractioning (after minute) of the degree, the traditional measure of plane angle; $= \frac{1}{60}$ min $= \frac{1°}{3600} = \frac{\pi}{(3600 \times 180)}$ rad.

astronomy Note: for †right ascension the second of the †sidereal clock, for declination, etc., it is the second of plane angle, a ratio of 15:1.

music (2nd) *See* interval.

other For **second degree**, *see* degree.

see also centesimal second; Redwood; Saybolt Universal Second.

section *area North America* The standard module of division of the vast Western lands upon survey for settlement, usually 1 mile square and bounded by the cardinal directions N, E, S, and W (so not precisely 1 mile along north and south perimeters), and embedding 1-chain-wide allowances for several roads, varying the assignable area from the nominal 640 acres. Settlement was originally on the basis of a †quarter-section per family, but soon proved markedly inadequate to support a family in less-favoured regions. Ontario and some other districts had bigger sections, often not cardinally oriented.

semitone *music See* interval.

sensation unit, sensation level *acoustics* The sensation unit is effectively the bel, applied to acoustic pressure as to electric pressure (i.e. amplitude).[157] Sensation level is expressed in the unit equivalent to the decibel[158] (q.v. for related discussion).

septillion [Lat: 'seven'] *See* Table 55 under thousand.

serving *nutrition* A representative amount of a particular foodstuff, used mainly for guidance for healthy living. The US Dept. of Agriculture uses the following interpretations:

fats	1 tspn butter or margarine; 1 tspn mayonnaise else salad dressing.
farinaceous foods	1 slice of bread; 1 oz breakfast cereal; $\frac{1}{2}$ cup cooked rice or pasta.
vegetables	1 cup raw solid; $\frac{3}{4}$ cup juice; $\frac{1}{2}$ cup cooked.
fruit	1 medium piece; $\frac{3}{4}$ cup juice; $\frac{1}{2}$ cup chopped, cooked or canned.
dairy	1 cup milk else yoghurt; $1\frac{1}{2}$ oz natural or 2 oz processed cheese.
meat	2 to 3 oz cooked lean red meat, poultry or fish; 1 egg.
vegetarian 'meat'	$\frac{1}{2}$ cup cooked dried beans; 1 tablespoons peanut butter.

Dietary recommendations for daily intake of the average adult are: 6 to 11 servings of bread, etc., 3 to 5 vegetables, 2 to 4 fruit, 2 to 3 dairy plus meat.

seven *Metric* For use applied as suffix to units, *see* tenth gram.

seventh *music See* interval.

seventh gram, seventh metre *Metric See* tenth gram.

sexagesimal [Lat: 'sixtieth'] With divisor/multiplier steps of 60, in contrast with the steps of 10 for †decimal, of 12 for †duodecimal, of 20 for †vigesimal, etc. The sexagesimal approach was dominant in Ancient Mesopotamia, and survives today on the clock and in the †degrees for measurement of plane angles. **Babylonian numerals** had a composite two-part structure, the first counting 0 to 4 dozens, the second part 0 to 11 units, the composite 4,11 meaning 59, i.e. 1 short of 60.

sextillion *See* Table 55 under thousand.

S.F.S. *rheology See* Saybolt Furol Second.

Sh *rheology See* Sherwood number.

SHA [Sidereal Hour Angle] *astronomy, navigation See* right ascension.

shake *time* Rare, informal term in sub-atomic physics for 10^{-8}s.[15]

shed *sub-atomic physics* Obsolete unit of area, $= 10^{-24}$ barn.

Sherwood number [T. K. Sherwood; USA 1903–76] *rheology*. Symbol Sh. *See* Nusselt number *Nu** relating to transport of matter in a binary mixture.

shipping cubic inch *volume* Actually the cuboid having a height of 1 in but a base of 1 ft × 1 ft $= 144$ in^3 (2 359.7$\sim$ cm^3). Identical to the board-foot.

shipping ton (measurement ton) The prime tariff unit for general maritime cargo, typically 40 ft^3 (1.133$\sim$ m^3), usually with concomitant weight restrictions, e.g. 1 long ton, reflecting the fact that, while a ship's hold capacity is limited by volume, the ship's buoyancy is limited by weight. See Table 47.

Shore hardness number *See* hardness numbers.

shower length, shower unit *astrophysics*. Symbol s. Applicable to cosmic rays, the mean path length required in a given medium to reduce the energy of a charged particle by 50%; e.g. 35 mm in lead, 300 mm in water, 230 m in air. s $= (\ln 2) \times$ cascade units.

SI [Fr: Système International d'Unités] The international system for units of measurement, the now generally accepted international form of the

Table 47

UK					SI	US-C
tea chest	...	...	...	...	125.$\sim$ dm^3	4.44$\sim$ ft^3
	barrel bulk	...	...	...	141.$\sim$ dm^3	5 ft^3
	7	displacement ton		...	0.991$\sim$ m^3	35 ft^3
9	8		shipping ton	...	1.13$\sim$ m^3	40 ft^3
		5:2		register ton	2.83$\sim$ m^3	100 ft^3
		32		twenty-foot unit	36.2$\sim$ m^3	1 280 ft^3

†metric system, and indeed that normally referred to by the latter phrase within a contemporary context. The name was adopted by the †CIPM in 1956. For an extended discussion see the following entries:

SI alphabet for an alphabetic presentation of the many symbols adopted within the SI;

SI system for a general discussion of the system;

SI units for a presentation of realms of measurement with matching units, ordered first alphabetically by realm, then algebraically relative to the base units.

SI alphabet The characters used as symbols for units in the †SI system are as shown in Table 48, the meanings followed by a hyphen, e.g. femto-, being usable only as prefixes to the others.

Table 48

		a-	= atto-
A	= ampere,	a	= are
Å	= angstrom		
Bq	= becquerel		
		c-	= centi-
C	= coulomb		
Ci	= curie (temporarily)		
	see also °C below		
		cd	= candela
		d-	= deci-
D	= dalton (unofficially)	d	= day
		da-	= deca-, deka-
E-	= exa-		
		eV	= electronvolt
		f-	= femto-
F	= farad		
G-	= giga-		
		g	= gram
Gy	= gray		
		h-	= hecto-
H	= henry	h	= hour
Hz	= hertz		
J	= joule		
		k-	= kilo-
K	= kelvin		
L	= litre	l	= litre
		lm	= lumen
		lx	= lux

Table 48 (contd.)

M-	= mega-	m-	= milli-
		m	= metre
		min	= minute of time
		mol	= mole
		n-	= nano-
N	= newton		
P-	= peta-	p-	= pico-
Pa	= pascal		
		rad	= radian
S	= siemens	s	= second
		sr	= steradian
Sv	= sievert		
T-	= tera-		
T	= tesla	t	= tonne
		u	= unified atomic mass unit
V	= volt		
W	= watt		
Wb	= weber		
Y-	= yotta-	y-	= yocto-
Z-	= zetta-	z-	= zepto-
		μ-	= micro-
Ω	= ohm		
°	= degree		
°C	= degree Celsius		
′	= minute (of arc)		
″	= second (of arc)		

Only one prefix should be used for any one unit to produce a compound unit; such expressions as millimicrogram are outmoded and unacceptable. The one-time millimicrogram is now correctly called the nanogram, to be written ng; note that any prefix directly adjoins the elementary measure-unit symbol, making an unpunctuated composite symbol for the compound unit of measure. The symbol for each unit of measure, whether a simple unqualified symbol else a composite, should be separated from others, by individual spaces or the elevated point or, where appropriate, by the solidus indicating 'per', and the resulting string of them should be separated by a space from any preceding number (else a hyphen in English adjectival form, but then applicable only for one unit, e.g. 'a 2-mm clearance'). For

Resistance = Resistivity × L/A
[Ω] = [Ω-cm] × L/A

example, for 8 kilopascals, which is identically 8 kilonewtons per square
metre, any of the following can be used:

$8\,kPa$ $8\,000\,Pa$ $8 \times 10^3\,Pa$ $8\,kN\,m^{-2}$ $8\,kN \cdot m^{-2}$ $8\,kN/m^2$.

Note that, though a pluralizing 's' is added to the full words in text in
accordance with normal rules of the language, it is never added to any
symbol. The letter 's' has its own distinct role of representing the
second of time; mostly, this appears in a quotient role, 'per second'. As
part of the endeavour to make the scheme invariable across languages,
the term 'per' should not be used with the symbols; any such quotient
should be written without controlling words, either by using the '/' or by
using negative indices, e.g. s^{-1}. Likewise the words 'square', 'squared',
'cubic', etc., and their abbreviations should not be used except within
normal English text; most notably, †cc for what is properly mL is
deprecated.

Technically any prefix can be added as needed to any unit. For mass
this means to gram, not to kilogram, although the latter is the actual base
unit in the SI (relative to †coherent derived units); however, it is not usual
to apply a larger prefix than kilo- to the gram. It is likewise incorrect to
apply them to the tonne, which is the megagram, but it is common
practice to do so (this application being to a unit that does not
incorporate a prefix). The alternative of explicit multiplicative powers of
10 is always available. Positive and negative indices of any magnitude can
be used as appropriate with any unit. Powers of ten should be placed
immediately in front of the space before the first (unqualified else
composite) measure-unit symbol in the string. The preferred practice
now is to restrict such uses of powers to exponents that are multiples of
3, i.e. to integer powers of 1 000. Thus a value that might be written as
1.732×10^{11} should be written as 173.2×10^9 else $0.173\,2 \times 10^{12}$. For mass,
such multipliers should apply to the kilogram; for most other
circumstances they should preferably apply directly to an unqualified
unit else to a string of them.

Unless there is special cause, a compound expression should be
resolved into standard unit terms and prefixed by powers of 10^3
as needed, but, if the numerator is not compound, this can
reasonably be prefixed to remove the powers, as in the kilopascal
example above.

sidereal [Lat: 'star'] *astronomy* A version of a unit with size determined
by reference to the stars. For **Sidereal Hour Angle** *see* right ascension,
otherwise *see. . .* , e.g. for **sidereal year**, *see* year.

[handwritten annotation: Conductance = Conductivity × $\frac{A}{L}$ {S} [S/m] (from) S·m⁻¹]

siegbahn unit [K. M. G. Siegbahn; Sweden 1886–1978] *See* X unit.

siemens [E. W. von Siemens; Germany 1816–92] *electrical conductance*
Symbol S. *SI 1971, Metric-m.k.s. 1933* The reciprocal ohm, $= \Omega^{-1} =$
$A \cdot V^{-1} = A^2 \cdot W^{-1} (= m^{-2} \, kg^{-1} \, s^3 \, A^2$ in base terms). Hence

  • S·m for electric conductivity.

The siemens was renamed the mho (q.v. for variations in the value prior
to the SI).

    The siemens was adopted as a derived special name within †Metric-
m.k.s. by the International Electrical Conference in 1933, but not
accepted widely until its adoption for the SI at the 14th †CGPM of 1971.[8]

sieve number (**mesh number**) A number representing the fineness of
a sieve mesh, and hence of the particles allowed through it. Official
standard sizes are common; while expressible as a number per inch or
such, implying a size, the manner of expression and the width of the
wires are of major significance. The old British standard sieve of mesh
number 200 has 200 modules per inch, i.e. a pitch of 0.005 in, but,
because the wire used has a diameter of 0.002 in, the functional aperture
is only 0.003 in, i.e. smaller than $\frac{1}{300}$ in, so is able to pass only particles
below such a size. This proportion, with a mesh number or sieve number
of n giving an aperture of $0.6/n$ in ($15/n$ mm) is reasonably indicative for
the British BSI and the USA standards, and for the **Tyler** mesh numbers.
The IMM scheme of the British Institute of Minerals and Metals has a
somewhat smaller number for a given size, translating roughly to an
aperture of $0.5/n$ in ($12.5/n$ mm).

    Modern terminology relates to hole size, using the R40(3) series of
†preferred numbers based on the millimetre, hence a progressive
increment of $18.85 \sim \%$.[159]

sievert [R. Sievert; Sweden 1896–1966] *radiation physics.* Symbol Sv.
SI 1979 The derived unit for dose equivalent or organ equivalent dose,
$= J \cdot kg^{-1}$, i.e. joules per kilogram ($= m^2 \cdot s^{-2}$ in base terms). Additionally
qualified as for ambient, directional, and personal dose equivalent.

    The effect of radiation depends on its total amount of energy, the type
of radiation, and the energy levels of particular particles. The dose
equivalent in sieverts of radiation is the product of the absorbed dose in
grays and a dimensionless numeric factor, called the quality factor (**QF**)
or relative biological effectiveness (**RBE**), dependent on the type of
radiation. Its unit value represents the effect of photons as x-rays or
gamma rays; electrons and positrons are very close to 1, neutrons,

protons, and alpha particles range up to 10, but heavy recoil nuclei can be 20. (Somewhat surprisingly, the higher RBE values for any particle apply to smaller kinetic energies, reflecting the greater opportunity for damage that occurs with slower-moving particles.) The factor for mixed radiation requires separate conversion for each type, then addition of the individual product values.

Being in basic terms a multiple, using dimensionless quantities, of the gray, the sievert has dimensionality identical with that SI unit of absorbed dose. The distinctive name exists, despite a general reluctance to create extra names, in the interests of minimizing risks to people (as well as the fact that the two differ in what they measure). The gray should be used only for absorbed dose, the sievert only for dose equivalent.

History

The unit was named the **intensity millicurie** when proposed by Sievert in 1932, and was initially defined as the dose delivered in one hour at a distance of 1 cm by a point source of 1 mg of radium enclosed in a platinum case 5 mm thick. That unit was only about a twelfth of the modern unit.

The sievert succeeded the rem as the measure of dose equivalent; 1 Sv = 100 rem. (As 1 gray = 100 rad, the numeric QFs apply identically when using those older units.)

1979 16th CGPM: '*considering* the effort made to introduce SI units in the field of ionizing radiations, the risk to human beings of an underestimated radiation dose, a risk that could result from a confusion between absorbed dose and dose equivalent, that the proliferation of special names represents a danger for the International System of Units and must be avoided in every possible way, but that this rule can be broken when it is a matter of safeguarding human health, *adopts* the special name *sievert*, symbol Sv, for the SI unit of dose equivalent in the field of radiation protection. The sievert is equal to the joule per kilogram.'[8]

sigma [Anglicized name of σ, the Gk letter 's'] *See* σ.

sign *plane angle* $\frac{1}{12}$ revolution of $360° = 30°$ ($0.523\ 598\ 8{\sim}$ rad), being the portion of the circle pertaining to each sign of the †zodiac.

Sikes *See* relative volumic mass.

silt *geology* A †particle size, typically 4 to 60 μm or 0.000 2 to 0.002 in.

silver amp, silver ampere *See* ampere.

siriometre *astronomy* 10^6 AU (4.848 137~ pc, 149.6 $\times 10^{12}$ km, 92.957 130~ $\times 10^{12}$ mi), being the decimally rounded number of †astronomical units approximating the distance of 0.546~ $\times 10^6$ AU (from Earth) to the star Sirius.

siriusweit *astronomy* 5 parsecs (1.031 324~ $\times 10^6$ AU, 154.283 9~ $\times 10^{12}$ km), being the integral number of parsecs approximating the †siriometre (but virtually double the 2.65~ pc actual distance from Earth to Sirius).

SI system The current international form of the †metric system, and indeed that normally referred to by the latter phrase within a contemporary context. The name was adopted by the †CIPM in 1956, for a system founded on six base units adopted by the 10th †CGPM in 1954, these being the metre, kilogram, second, ampere, kelvin, and candela. (Hence it is an m.k.s. system rather than a †c.g.s. system as long applied routinely, though not exclusively, to the metric system. Thus, for example, the derived unit for force is the newton, $= \mathrm{m \cdot kg \cdot s^{-2}}$, rather than the dyne, $= \mathrm{cm \cdot g \cdot s^{-2}}$. Because of being based on the ampere, rather than the volt or some other electrical unit, the SI system is sometimes referred to as an †m.k.s.A. system.) The name and the international standard label SI were ratified by the 11th CGPM in 1960, along with a repertoire of basic, supplementary, and derived units.[8] The current repertoire, with their purposes and, for all derived units, their expression in base-unit terms, is thus as shown in Table 49 (the derived units being methodically indented to indicate their relative dependence).

Table 49

Base units:
    length: m = metre
    mass: kg = kilogram
    time: s = second
    electric current strength: A = ampere
    thermodynamic temperature: K = kelvin
    luminous intensity: cd = candela
    amount of substance: mol = mole
Derived units (the first two being ratios):
    rad = radian = $\mathrm{m \cdot m^{-1}}$ for plane angle:
    sr = steradian = $\mathrm{m^2 \cdot m^{-2}}$ for solid angle:

(*contd.*)

Table 49 (*contd.*)

degree C = degree Celsius = K for temperature
C = coulomb = s·A for electric charge, quantity of electricity
N = newton = m·kg·s^{-2} for force
Pa = pascal = N·m^{-2} for pressure, stress
J = joule = N·m for energy, work, quantity of heat
W = watt = J·s^{-1} for power, radiant flux
V = volt = W·A^{-1} for voltage, e.m.f., potential difference
F = farad = C·V^{-1} for electric capacitance
Ω = ohm = V·A^{-1} for electric resistance
S = siemens = Ω^{-1} for electric conductance
Wb = weber = V·s for magnetic flux
T = tesla = Wb·m^{-2} for magnetic flux density
H = henry = Wb·A^{-1} for electric inductance
Gy = gray = J·kg^{-1} for absorbed radiation
Sv = sievert = J·kg^{-1} for dose equivalent
lm = lumen = cd·sr for luminous flux
lx = lux = lm·m^{-2} for illuminance
Hz = hertz = (cycles)·s^{-1} for frequency
Bq = becquerel = (disintegrations)·s^{-1} for activity of a radionuclide
kat = katal = mol·s for catalytic activity.
Other established units acceptable temporarily where already used:
length: nautical mile = 1 852 m
speed: knot = 1 nautical mile per hour = $\frac{1852}{3600}$ = 0.514 444 4~ m·s^{-1}
length: Å = angstrom = 0.1 nm (10^{-10} m)
area: a = are = 100 m^2, hence ha = hectare = 10 000 m^2

Reference should be made to the individual unit names for relevant definitions, plus details of adjustments in values.

The familiar range of prefixes was expanded and the upper/lower case distinction of D- for deca- and d- for deci- (both largely shunned outside Europe) replaced by da-versus d-. Hecto- (of similar limited use) and kilo- are now routinely and officially lower case in their symbol form; myria has been discarded. The larger multipliers than kilo- continue the tradition of being capitalized as symbols (and of being derived from Greek, but their opposites all wander from the Latin tradition). The full range of SI prefixes, with etymological derivation and symbol (old in brackets) is as shown in Table 50. The inner terms, all in use by 1951, were redefined in 1960 by the 11th †CGPM, the outer as shown in the table.

Table 50

Symbol	Prefix	Value	Etymological source	Approved
y	yocto-	10^{-24}	oct-'eight', for $1\,000^{-8}$	1990 CIPM
z	zepto-	10^{-21}	sept-'seven', for $1\,000^{-7}$	1990 CIPM
a	atto-	10^{-18}	'eighteen' in Danish	1964 12th CGPM
f	femto-	10^{-15}	'fifteen' in Danish	1964 12th CGPM
p	pico-	10^{-12}	'small' in Spanish	
n	nano-	10^{-9}	'dwarf' in Greek	
μ	micro-	10^{-6}	'small' in Latin	
m	milli-	10^{-3}	'thousand' in Latin	
c	centi-	10^{-2}	'hundred' in Latin	
d	deci-	10^{-1}	'ten' in Latin	
da (D)	deca-	10^1	'ten' in Greek	
h (H)	hecto-	10^2	'hundred' in Greek	
k (K)	kilo-	10^3	'thousand' in Greek	
M	mega-	10^6	'great' in Greek	
G	giga-	10^9	'giant' in Greek	
T	tera-	10^{12}	'monster' in Greek	
P	peta-	10^{15}	pent-'five', for $1\,000^5$	1975 15th CGPM
E	exa-	10^{18}	hex-'six', for $1\,000^6$	1975 15th CGPM
Z	zetta-	10^{21}	sept-'seven', for $1\,000^7$	1990 CIPM
Y	yotta-	10^{24}	oct-'eight', for $1\,000^8$	1990 CIPM

See metric system for further historical background.

SI unit While many combinations of the base units of the †SI system have specific names, those for many distinct purposes do not, e.g. the ampere·second is called the coulomb, but the metre per second squared, despite its common occurrence for acceleration, has no special name. The SI units for realms of measurement, as defined by the †CGPMs, are by subject as follows, showing for each the relevant powers of base units (and any exceptional factor counted).

- absorbed dose, †kerma,
  specific energy (imparted): $J \cdot kg^{-1} = Gy = gray = m^2 \cdot s^{-2}$;
- absorbed dose rate: $Gy \cdot s^{-1} = (m^2 \cdot s^{-2}) \cdot s^1 = m^2 \cdot s^{-3}$;
- acceleration: $m \cdot s^{-2}$;
- activity of a radionuclide: (disintegrations)$\cdot s^{-1} = Bq = becquerel = s^{-1}$;
- amount of substance: mol = mole, a base unit;
- angular acceleration: $= rad \cdot s^{-2} = (m^1 \cdot m^{-1}) \cdot s^{-2} = s^{-2}$;

- angular speed: $= \mathrm{rad \cdot s^{-1}} = (\mathrm{m^1 \cdot m^{-1}}) \cdot s^{-1} = s^{-1}$;
- area: $= \mathrm{m^2}$, also a = are = $100 \, \mathrm{m^2}$;
- capacitance: $\mathrm{C \cdot V^{-1}} = \mathrm{F} = \mathrm{farad} = (\mathrm{s \cdot A}) \cdot (\mathrm{m^2 \cdot kg \cdot s^{-3} \cdot A^{-1}})^{-1}$
  $= \mathrm{m^{-2} \cdot kg^{-1} \cdot s^4 \cdot A^2}$;
- catalytic activity: $\mathrm{mol \cdot s^{-1}} = \mathrm{kat} = \mathrm{katal} = \mathrm{s^{-1} \cdot mol}$;
- catalytic (activity) concentration: $\mathrm{kat \cdot m^{-3}} = (\mathrm{s^{-1} \cdot mol}) \cdot \mathrm{m^{-3}}$
  $= \mathrm{m^{-3} \cdot s^{-1} \cdot mol}$;
- dose equivalent,[i] organ equivalent dose: $\mathrm{J \cdot kg^{-1}} = \mathrm{Sv} = \mathrm{sievert} = \mathrm{m^2 \cdot s^{-2}}$
  (*see* Sievert for qualified variants);
- dynamic viscosity: $\mathrm{N \cdot m \cdot s^{-1}} = (\mathrm{m \cdot kg \cdot s^{-2}}) \cdot \mathrm{m \cdot s^{-1}} = \mathrm{m^{-1} \cdot kg \cdot s^{-1}}$;
- electric current density: $\mathrm{A \cdot m^{-2}} = \mathrm{m^{-2} \cdot A}$;
- electric charge: *see* quantity of electricity;
- electric charge density: $\mathrm{C \cdot m^{-3}} = (\mathrm{A \cdot s}) \cdot \mathrm{m^{-3}} = \mathrm{m^{-3} \cdot s \cdot A}$;
- electric conductance: $\Omega^{-1} = \mathrm{V^{-1} \cdot A} = \mathrm{W^{-1} \cdot A^2} = \mathrm{S} = \mathrm{siemens}$
  $= (\mathrm{m^2 \cdot kg \cdot s^{-3} \cdot A^{-1}})^{-1} \cdot \mathrm{A} = \mathrm{m^{-2} \cdot kg^{-1} \cdot s^3 \cdot A^2}$;
- electric current strength: A = ampere, a base unit;
- electric field strength: $\mathrm{V \cdot m^{-1}} = (\mathrm{m^2 \cdot kg \cdot s^{-3} \cdot A^{-1}}) \cdot \mathrm{m} = \mathrm{m \cdot kg \cdot s^{-3} \cdot A^{-1}}$;
- electric flux density: $\mathrm{C \cdot m^{-2}} = (\mathrm{A \cdot s}) \cdot \mathrm{m^{-2}} = \mathrm{m^{-2} \cdot s \cdot A}$;
- electric potential difference, electromotive force, voltage:
  $\mathrm{W \cdot A^{-1}} = \mathrm{V} = \mathrm{volt} = (\mathrm{m^2 \cdot kg \cdot s^{-3}}) \cdot \mathrm{A^{-1}} = \mathrm{m^2 \cdot kg \cdot s^{-3} \cdot A^{-1}}$;
- electric resistance: $\mathrm{V \cdot A^{-1}} = \mathrm{W \cdot A^{-2}} = \Omega = \mathrm{ohm} = (\mathrm{m^2 \cdot kg \cdot s^{-3} \cdot A^{-1}}) \cdot \mathrm{A^{-1}}$
  $= \mathrm{m^2 \cdot kg \cdot s^{-3} \cdot A^{-2}}$;
- electric resistivity: $\Omega \cdot \mathrm{m^2 \cdot m^{-1}} = (\mathrm{m^2 \cdot kg \cdot s^{-3} \cdot A^{-2}}) \cdot \mathrm{m^2 \cdot m^{-1}}$
  $= \mathrm{m^3 \cdot kg \cdot s^{-3} \cdot A^{-2}}$;
- electricity (quantity of): *see* quantity of electricity;
- electromotive force: *see* electric potential difference;
- energy, work, quantity of heat: $\mathrm{N \cdot m} = \mathrm{J} = \mathrm{joule} = (\mathrm{m \cdot kg \cdot s^{-2}}) \cdot \mathrm{m}$
  $= \mathrm{m^2 \cdot kg \cdot s^{-2}}$;
- energy density: $\mathrm{J \cdot m^{-3}} = (\mathrm{m^2 \cdot kg \cdot s^{-2}}) \cdot \mathrm{m^{-3}} = \mathrm{m^{-1} \cdot kg \cdot s^{-2}}$;
- entropy, heat capacity: $\mathrm{J \cdot K^{-1}} = (\mathrm{m^2 \cdot kg \cdot s^{-2}}) \cdot \mathrm{K^{-1}} = \mathrm{m^2 \cdot kg \cdot s^{-2} \cdot K^{-1}}$;
- exposure to X- or gamma rays: $\mathrm{C \cdot kg^{-1}} = (\mathrm{S \cdot A}) \cdot \mathrm{kg^{-1}} = \mathrm{kg^{-1} \cdot s \cdot A}$;
- force: $\mathrm{kg \cdot m \cdot s^{-2}} = \mathrm{N} = \mathrm{newton} = \mathrm{m \cdot kg \cdot s^{-2}}$;
- frequency: (cycles) $\cdot \mathrm{s^{-1}} = \mathrm{Hz} = \mathrm{hertz} = \mathrm{s^{-1}}$;
- heat capacity: *see* entropy;
- heat (quantity of): *see* energy;
- heat-flux density, irradiance: $\mathrm{W \cdot m^{-2}} = (\mathrm{m^2 \cdot kg \cdot s^{-3}}) \cdot \mathrm{m^{-2}} = \mathrm{kg \cdot s^{-3}}$;
- illuminance: $\mathrm{lm \cdot m^{-2}} = \mathrm{lx} = \mathrm{lux} = \mathrm{m^{-2} \cdot cd}$;
- inductance: $\mathrm{Wb \cdot A^{-1}} = \mathrm{H} = \mathrm{henry} = (\mathrm{m^2 \cdot kg \cdot s^{-2} \cdot A^{-1}}) \cdot \mathrm{A^{-1}}$
  $= \mathrm{m^2 \cdot kg \cdot s^{-2} \cdot A^{-2}}$;
- irradiance: *see* heat-flux density;

- kerma: *see* absorbed dose
- kinematic viscosity: $= m^2 \cdot s^{-1}$;
- length: m = metre, a base unit;
- light (quantity of): *see* quantity of light;
- luminous flux: lm = lumen = $cd \cdot sr = cd \cdot (m^2 \cdot m^{-2}) = cd$;
- luminous intensity: cd = candela, a base unit;
- magnetic field strength: $A \cdot m^{-1} = m^{-1} \cdot A$;
- magnetic flux: $V \cdot s = Wb = weber = (m^2 \cdot kg \cdot s^{-3} \cdot A^{-1}) \cdot s = m^2 \cdot kg \cdot s^{-2} \cdot A^{-1}$;
- magnetic flux density: $Wb \cdot m^{-2} = T = tesla = (m^2 \cdot kg \cdot s^{-2} \cdot A^{-1}) \cdot m^{-2}$
  $= kg \cdot s^{-2} \cdot A^{-1}$;
- mass: kg = kilogram, a base unit;
- mass density: $kg \cdot m^{-3} = m^{-3} \cdot kg$;
- molar energy: $J \cdot mol^{-1} = (m^2 \cdot kg \cdot s^{-2}) \cdot mol^{-1} = m^2 \cdot kg \cdot s^{-2} \cdot mol^{-1}$;
- molar entropy,
  molar heat capacity: $J \cdot (mol \cdot K)^{-1} = (m^2 \cdot kg \cdot s^{-2}) \cdot (mol \cdot K)^{-1}$
  $= m^2 \cdot kg \cdot s^{-2} \cdot K^{-1} \cdot mol^{-1}$;
- moment of force: $N \cdot m = (m \cdot kg \cdot s^{-2}) \cdot m = m^2 \cdot kg \cdot s^{-2}$;
- organ equivalent dose: *see* dose equivalent;
- permeability: $H \cdot m^{-1} = (m^2 \cdot kg \cdot s^{-2} \cdot A^{-2}) \cdot m^{-1} = m \cdot kg \cdot s^{-2} \cdot A^{-2}$;
- permittivity: $F \cdot m^{-1} = (m^{-2} \cdot kg^{-1} \cdot s^4 \cdot A^2) \cdot m^{-1} = m^{-3} \cdot kg^{-1} \cdot s^4 \cdot A^2$;
- plane angle: rad = radian, a supplementary unit prior to 1980,
  now $= m^1 \cdot m^{-1}$, so dimensionless;
- potential difference = electromotive force;
- power, radiant flux: $J \cdot s^{-1} = W = watt = (m^2 \cdot kg \cdot s^{-2}) \cdot s^{-1} = m^2 \cdot kg \cdot s^{-3}$;
- pressure, stress: $N \cdot m^{-2} = Pa = pascal = (m \cdot kg \cdot s^{-2}) \cdot m^{-2} = m^{-1} \cdot kg \cdot s^{-2}$
- quantity of electricity, electric charge: $A \cdot s = C = coulomb = s \cdot A$;
- quantity of heat: *see* energy;
- quantity of light: $lm \cdot s = cd \cdot s = s \cdot cd$;
- radiance: $W \cdot m^{-2} \cdot sr^{-1} = (m^2 \cdot kg \cdot s^{-3}) \cdot m^{-2} \cdot (m^2 \cdot m^{-2})^{-1} = kg \cdot s^{-3}$;
- radiant flux: *see* power;
- radiant intensity: $W \cdot sr^{-1} = (m^2 \cdot kg \cdot s^{-3}) \cdot (m^2 \cdot m^{-2})^{-1} = m^2 \cdot kg \cdot s^{-3}$;
- solid angle: sr = steradian, a supplementary unit prior to 1980,
  now $= m^2 \cdot m^{-2}$, so dimensionless;
- specific energy: $J \cdot kg^{-1} = (m^2 \cdot kg \cdot s^{-2}) \cdot kg^{-1} = m^2 \cdot s^{-2}$;
- specific energy (imparted); *see* absorbed dose
- specific entropy,
  specific heat capacity: $J \cdot (kg \cdot K)^{-1} = (m^2 \cdot kg \cdot s^{-2}) \cdot (kg \cdot K)^{-1} = m^2 \cdot s^{-2} \cdot K^{-1}$;
- speed: $= m \cdot s^{-1}$;
- stress: *see* pressure;
- surface tension: $N \cdot m^{-1} = (m \cdot kg \cdot s^{-2}) \cdot m^{-1} = kg \cdot s^{-2}$;

- temperature: K = kelvin, a base unit, also °C = degree Celsius = K;
- thermal conductivity: $W \cdot (m \cdot K)^{-1} = (m \cdot kg \cdot s^{-3}) \cdot K^{-1}$;
- thermodynamic temperature: K = kelvin, a base unit;
- time: s = second, a base unit;
- velocity; *see* speed;
- voltage; *see* electric potential difference;
- volume: cu-metre = m^3;
- wave number: waves per metre = (wave) $\cdot m^{-1}$;
- work; *see* energy.

In descending order of the successive powers of the base units these are shown in Table 51.

Table 51

	m	kg	s	A	K	cd	mol	
volume	3							cu metre
energy, work, quantity of heat	2	1	−2					J joule
moment of force	2	1	−2					newton·metre
molar energy	2	1	−2				−1	joule per mole
entropy, heat capacity	2	1	−2		−1			joule per kelvin
molar entropy, molar heat capacity	2	1	−2		−1		−1	joule per mole·kelvin
magnetic flux	2	1	−2	−1				Wb weber
inductance	2	1	−2	−2				H henry
apparent power	2	1	−3					volt·ampere
power, radiant flux	2	1	−3					W watt
radiant intensity	2	1	−3					watt per steradian
electromotive force, voltage, potential difference	2	1	−3	−1				V volt
electric resistance	2	1	−3	−2				Ω ohm
area	2							sq metre
kinematic viscosity	2		−1					sq metre per second
specific energy	2		−2					joule per kilogram
dose equivalent	2		−2					Sv sievert
absorbed dose	2		−2					Gy gray
specific entropy, specific heat capacity	2		−2		−1			joule per kilogram·kelvin
absorbed radiation dose rate	2		−3					gray per second

Quantity	m	kg	s	A	K	cd	mol	Symbol	Unit name
force	1	1	−2					N	newton
permeability	1	1	−2	−2					henry per metre
thermal conductivity	1	1	−3		−1				watt per metre·kelvin
electric field strength	1	1	−3	−1					volt per metre
length	1							m	metre
speed	1		−1						metre per second
acceleration	1		−2						metre per second squared
mass		1						kg	kilogram
surface tension		1	−2						newton per metre
magnetic flux density		1	−2	−1				T	tesla
heat-flux density, irradiance		1	−3						watt/sq metre
radiance		1	−3						watt/sq metre·sterad
electric charge, quantity of electricity			1	1				C	coulomb
quantity of light			1			1			lumen·second
time			1					s	second
magnetomotive force: turn^1				1					ampere·turn
electric current strength				1				A	ampere
temperature					1			K	kelvin
luminous flux						1		lm	lumen
luminous intensity						1		cd	candela
amount of substance							1	mol	mole
plane angle								rad	radian
solid angle								sr	steradian
catalytic activity			−1				1	kat	katal
angular speed			−1						radian per second
frequency: cycle^{-1}			−1					Hz	hertz
activity of a radionuclide: disintegration^{-1}			−1					Bq	becquerel
angular acceleration			−2						radian/second-sqrd
exposure to X- or gamma rays		−1	1	1					coulomb/kilogram
dynamic viscosity	−1	1	−1						newton·sec per sq metre
energy density	−1	1	−2						joule per cu metre
pressure, stress	−1	1	−2					Pa	pascal
magnetic field strength	−1			1					ampere per metre
wave number: wave^{-1}	−1								wave per metre
electric flux density	−2		1	1					coulomb per sq metre
illuminance	−2					1		lx	lux

(contd.)

Table 51 (*contd.*)

	m	kg	s	A	K	cd	mol		
electric capacitance	−2	−1	4	2				F	farad
electric conductance	−2	−1	3	2				S	siemens
luminous efficacy	−2	−1	2			1			lumen per watt
volumic mass	−3	1							kilogram per cu metre
electric charge density	−3		1	1					coulomb per cu metre
catalytic concentration	−3		−1				1		katal per cu metre
permittivity	−3	−1	4	2					farad per metre

six *Metric* For use applied as a suffix to units, *see* tenth gram.

sixth *music See* interval.

sixth gram, sixth metre *Metric See* tenth gram.

size *See* drill size; pennysize; tyre size.

slug (gee pound, g pound) *mass BI-f.p.s.* The †coherent unit of mass in the †gravitational system of the Imperial system, identically ft·lb-f⁻¹·s², i.e. the mass accelerated at the rate of one foot per second per second by the force that is the true pound (the 'pound-force'); 1 slug = 14.593 9∼ kg (32.174∼ lb-mass). The latter number is the standardized value, within BI, of Earth's surficial gravitational acceleration, i.e. *g*; hence also gee pound.

SNU *sub-atomic physics See* solar-neutrino unit.

sol *time* The Martian day, ≈ 1.025 Earth †day.

solar mass *astrophysics* The usual reference mass for the comparative indication of star masses, being the mass of the Sun, 1.989∼ ×10²⁷ t (1.989∼ × 10³⁰ kg, 2.192∼ × 10²⁷ short ton) ≈ 333 000 times the mass of Earth.

solar-neutrino unit *sub-atomic physics*. Symbol SNU. A unit of infinitesimally small frequency, created to reflect the rarity of neutrinos reacting with ordinary matter = one event per second per 10³⁶ target atoms.

Solomon R unit *radiation physics See* r unit.

sone *acoustics* 1 sone = the loudness of a simple tone of 1 kHz at 40 db above the standard listener's threshold of 20 μPa (≈ the hum of a modern refrigerator).[137] A sound twice as loud has value 2, etc.[160] If s and p represent the loudness respectively in sones and phons, then

$s = 2^{(p-40)/10}$.

s.o.s. [Lat: si opus sit] *medicine* When required. *See also* p.r.n.

span *length* Traditionally the span of outstretched fingers and thumb, ≈ 9 in. In architecture, the distance between supports.

spat [Lat: spatium, 'space'] *geometry* A term applied in both the two- and the three-dimensional contexts to mean the complete rotation, i.e. the turn around the circle or about the sphere. Hence, in plane angle it is the revolution, the turn, $= 2\pi$ rad; in solid angle it is the sphere, $= 4\pi$ sr. It is more used in the latter context, but more needed in the former, where the magnetic effects of an electric current in a coil are so fundamentally important (for which it has been suggested it become a base unit of the SI).

astrophysics 1 Tm (10^9 km, $0.621\,37\sim \times 10^9$ mi).

specific *physics* Correctly implies quantity per unit mass, but used otherwise in such obsolete terms as specific gravity (correctly †relative volumic mass) and specific resistance (correctly resistivity).

SPECmark *informatics* A measure of effective speed of a computer workstation, the geometric mean of ten standardized benchmark programs.

speed of light Essentially the speed of radio and other electromagnetic waves too, the speed of light depends on transmission medium. The maximum speed, labelled c and often referred to as the speed of light without qualification, occurs in a vacuum, it equals $299\,792\,458$ m·s^{-1} ($1.079\,252\,85\sim \times 10^9$ km/h, $670.616\,629\sim \times 10^6$ m.p.h.), the first figure being precise since the 1983 re-definition of the metre. (That light has finite speed was accepted after Ole Rømer correctly forecast in 1679 that an eclipse of Io by Jupiter would be seen 10 minutes later than expected.)

Outside of electromagnetic radiations and accelerated atomic and sub-atomic particles, no speeds achieved by the physical creations of mankind are more than a minor fraction of such a speed.

1975 15th CGPM: '*considering* the excellent agreement among the results of wavelength measurements on the radiations of lasers locked on a molecular absorption line in the visible or infrared region, with an uncertainty estimated at $\pm 4 \times 10^{-9}$ which corresponds to the uncertainty of the realization of the metre, *considering* also the concordant measurements of the frequencies of several of these radiations, *recommends* the use of the resulting value for the speed of propagation of electromagnetic waves in vacuum $c = 299\,792\,458$ metres per second.'[8]

SPF [Sun Protection Factor] The fraction of B-type ultraviolet light that sun-screen cream allows to reach the skin, e.g. SPF 15 means reduction to $\frac{1}{15}$.

sphere *solid angle* The equivalent in three-dimensional space to the revolution for the planar degree in two-dimensional space; 1 sphere $= 4\pi$ steradians $= 129\,600\pi^{-1}$ square degrees, $= 160\,000\pi^{-1}$ square grades.

It has been suggested that it, maybe as the spat, be a base unit of the SI.

square *area Britain, North America, Australia* For constructional floor area and roofing $= 100\,\text{ft}^2$ $(9.290\,3\sim\text{m}^2)$ (expressed for a house in North America it includes all floors except the basement); for carpet etc., $= 1\,\text{yd}^2$. Note that, while used as a preceding adjective for deriving areal expressions from linear units (in an English context), 'square' should not be used so for seconds and other non-linear units subject to power-raising; for these the adjective 'squared' should be used after the unit, e.g. an acceleration of 3 metres per second squared.

square degree *solid angle*. Symbols $\square°$ $(°)^2$. The equivalent to the square of the ordinary planar degree, hence $= (\frac{\pi}{180})^2$ steradian.

This simple-looking transformation is misleading, for the two degrees are in spatial dimensions 2 and 3 rather than 1 and 2, and the parameter π is not raised to a power for the steradians in a sphere. Since there are 4π steradians in the sphere, the square degree $= \frac{1}{129600}$ sphere.

square grade *solid angle* Symbol $(g)^2$. The equivalent to the square of the planar grade, $= (\frac{\pi}{200})^2$ steradian $= \frac{\pi}{160000}$ sphere. (*See also* square degree.)

Sr *rheology See* Strouhal number.

sr *geometry SI See* steradian.

ss [Lat] *medicine* Semis, i.e. half.

SSF, SSU [Second Saybolt . . .] *mechanics See* Saybolt Furol Second; Saybolt Universal Second.

St *mechanics Metric-c.g.s. See* stokes. Also prefixed, as in cSt = centistokes.
rheology (usually as **St**) *See* Stanton number.

st *volume Metric See* stere.

standard acceleration of gravity *geophysics See* standard
gravity.

standard atmosphere *physics* SI, *Metric-c.g.s. and -m.k.s.* A
standardized configuration of typical atmospheric conditions, most
notably of pressure. More elaborate standard atmospheres, e.g. for
artillery or meteorological use, define the lapse rate of temperature with
altitude and other particulars, but the main usage of the term relates
just to pressure. As such, it is a laboratory standard corresponding
with typical ground-level atmospheric pressure, now SI-deprecated as a
unit of measure (usually just †atmosphere), but not as a reference
pressure.

On the traditional barometer of mercury, the head of liquid is about
30 in or 760 mm; the latter figure was adopted as the standard,
suitably defined regarding surrounding and intrinsic conditions. Now
the derived (non-†coherent) unit is defined as 101.325 kPa
(14.695 95~ p.s.i.).

The overall picture of the standard atmosphere is shown in Table 52.

Table 52

	SI	BI
temperature	15°C	59°F
pressure	101.325 kPa, 1.01325 bar	2116.2 lb-f·ft^{-2}
gravitational acceleration	9.806 55 m·s^{-2}	32.173 72 ft·s^{-2}
volumic mass	1.2250 kg·m^{-3}	0.076 474 lb·ft^{-3}
viscosity – dynamic	1.7894 ×10^{-5} kg·m^{-1}·s^{-1}	1.2024 ×10^{-5} lb·ft^{-1}·s^{-1}
viscosity – kinematic	1.4607 ×10^{-5} m^2·s^{-1}	15.723 ×10^{-5} ft^2·s^{-1}
mean molecular speed	458.94 m·s^{-1}	1505.7 ft·s^{-1}
speed of sound	340.294 m·s^{-1}	1116.45 ft·s^{-1}

For miscellaneous conversions, *see* atmosphere.

History
Created by the US National Advisory Committee on Aeronautics in 1922
(the **NACA +1922 standard atmosphere**) as 'The atmospheric pressure
registering 760 mm (29.921 3~ in) of head on a mercury barometer at

mean sea level with gravitational acceleration at $9.8066 \, \mathrm{m \cdot s^{-2}}$, assuming the air to be a perfect gas at $15°$ C ($59°$ F) and the mercury having a volumic mass of $13.595 \times 10^3 \, \mathrm{kg \, m^{-3}}$, $= 101.3250{\sim} \, \mathrm{kPa}$ ($14.691{\sim}$ p.s.i.).' In 1924 the International Commission for Air Navigation adopted (as the **ICAN+1924 standard atmosphere**) this definition with a minor change of gravity to $9.8062 \, \mathrm{m \cdot s^{-2}}$. In 1941 the International Civil Aviation Organization did likewise (as the **ICAO+1941 standard atmosphere**), with gravity at $9.80655 \, \mathrm{m \cdot s^{-2}}$. The subsequent definition by the 9th †CGPM of 1948 as a simple derived unit ($1\,013\,250 \, \mathrm{dyn \cdot cm^{-2}}$) and its re-statement in m.k.s. and the SI were only a re-expression of the implied value of the 1941 definition; so any statement in ICAO+1941 terms is equally valid today.

1954　10th CGPM: 'The 10th CGPM, having noted the definition of the standard
　　　atmosphere given by the 9th CGPM when defining the International
　　　Temperature Scale led some physicists to believe that this definition of the
　　　standard atmosphere was valid only for accurate work in thermometry,
　　　declares that it adopts, for general use, the definition:
　　　　1 standard atmosphere = $1\,013\,250$ dynes per square centimetre, i.e.
　　　　$101\,325$ newtons per square metre.'[8]

standard cable *engineering UK, USA 1905* A standard for attenuation used in telecommunications, being a notional cable 1 mile in length, of capacitance $0.054 \, \mathrm{\mu F}$, resistance 88 ohms, inductance 1 mH, and leakance $1 \, \mathrm{\mu mho}$. A signal of 800 Hz (about central in energy terms for telephony) suffered about 20% attenuation over the mile.

The performance of real cables could be measured by equivalence in miles of this standard cable, expressed as **standard miles**, as could the contribution of attenuation-reducing ancillary devices.

This unit was succeeded by the †transmission unit in 1922.
See also cable, cable length, cable's length.

standard deviation *statistics*. Symbol s.d., σ. A measure of the variability in a set of numbers, equal to the square root of the variance (a less convenient measure of variability, as is also the mean deviation). The deviation for each number is the difference between it and the mean value for the set. Clearly, averaging the signed values would produce zero, by the definition of the mean. The **variance** is the mean of the squares of the deviations, the squaring removing any effect of signs, but compounding the scale factor. The standard deviation is the square root of the variance, bringing the measure back to scale (and prompting its other name of **root-mean-square**; *see* r.m.s. for the distinctive usage in electromagnetics).

Together with the mean, the standard deviation gives a first-level indication of the characteristics of any set of numbers. The actual pattern of frequency of the member numbers can be very different for sets with the same mean and standard deviation, but one overall pattern is so common to have been accorded the name 'normal' distribution and its symmetric shape is well known as the bell-curve.

Expressed simply as s.d., the standard deviation is often used to show how far any one member is away from the mean. For a normal distribution, 68.27~% of members lie within 1 s.d. of the mean, 95.45~% within 2 s.d., and 99.73~% within 3 s.d. For a set of laboratory values for a repeated experiment aimed at establishing some value, the standard deviation gives a measure of the consistency of the experiments. Along with the mean to represent the targeted value, the standard deviation is often cited, usually in brackets as an integer to be applied at the last decimal place expressed for the mean, as the 1 standard deviation uncertainty.

If the variance (hence the standard deviation) being computed relates to a full population, then the averaging involves merely dividing by the count of numbers in the set. However, if the set is merely a sample aimed at obtaining a picture of a larger population, then the variance obtained by such division tends to understate the variability of the whole. To compensate and represent the whole, the divisor for a sample has to be one less than its count. The Greek letter σ is used to signify the standard deviation for the whole, σ^2 the variance.

standard gravity (acceleration of free fall, g, standard acceleration of gravity) *acceleration* Metric 1901 Established at the 3rd †CGPM as $9.806\,65$ m·s^{-2} ($32.174\,049\sim$ ft·s^{-2}).[8] Originally conceived as the gravity at mean sea level at latitude 45°N, the term 'standard gravity' has become a label for a conventional value.

From 1907 the 'accepted international base' value for gravity measurements was $9.812\,74$ m·s^{-2}, as recorded at Potsdam, Germany,[161] but this figure was acknowledged at the 9th CGPM in 1948 as 'appreciably in error'.[162] It was subsequently determined that $9.812\,60$ m·s^{-2} ($32.193\,57\sim$ ft s^{-2}) would be more appropriate. However, the 1901 value, being entrenched for units in the †gravitational systems, continues to be acknowledged as the conventional standard gravity, defining the kilogram-force and slug, and in the †standard atmosphere.

The Potsdam figure, with its associated period for the swing of a standard pendulum, provided the base via use of a like pendulum elsewhere, but was superseded by International Gravity Standardization Network (IGSN), now the International Absolute Gravity Base Station Network (IAGBN).[163]

Until 1941 the standard atmosphere for aviation used values relating to the original concept, as favoured by the International Meteorological Committee, being $9.806\,6\,\mathrm{m\cdot s^{-2}}$ initially for the US standard and then, from 1924, internationally $9.806\,2\,\mathrm{m\cdot s^{-2}}$.

The effective gravitational pull at Earth's surface varies consistently by latitude and by altitude, and, because of local geology, erratically too. The downward pull experienced at Earth's surface represents the net effect of the static gravitational pull downwards and the contrary dynamic 'centrifugal' force caused by the rotation of Earth (modified minutely further by the gravitational influence of the Moon and the Sun, and technically other bodies). The former acts vertically downwards towards (but not precisely towards) the centre of Earth and, above the surface, is inversely proportional to the square of the distance from the centre (hence declining with altitude). The centrifugal force acts at right angles to Earth's axis of rotation and outwards from it, increasing proportionally to the distance from the axis. At Earth's surface, the strength of the latter is everywhere well below 1% of the former.

The oblateness of Earth makes the gravitational pull least at the Equator, greatest at the Poles. It is also slightly distorted by Earth's centroid not being exactly at the geographical centre. The offsetting centrifugal force is nil at the Poles (as well as being virtually horizontal in the proximity), and maximal at the Equator, so accentuating the static disparity. The net value of the downward acceleration at sea level varies from $9.780\,4\sim\mathrm{m\cdot s^{-2}}$ ($32.088\sim\mathrm{ft\ s^{-2}}$) at the Equator to an average $9.832\sim\mathrm{m\cdot s^{-2}}$ ($32.26\sim\mathrm{ft\ s^{-2}}$) at the Poles (differing between the two).

The standard value is roughly the mean of these extremes and close to true for the developed inhabited world. Increasing altitude reduces the static factor while increasing the centrifugal factor (for a given speed). For an object travelling at Earth's angular rotational speed, the first 35 000 m (115 000 ft) above sea level reduces the above net figures by little more than 1%, but at 35 788 km (22 238 mi) the net figure is reduced to zero, hence the positioning (in the equatorial plane and orbiting directionally as Earth) at this height of the orbiting 'geostationary' satellites. Table 53 gives sample values for actual gravity, these being finely variable for specific locations.

Table 53

	Latitude:	0°	15°	30°	45°	60°	75°	90°
at	sea level	9.7805	9.7840	9.7934	9.8063	9.8192	9.8287	9.8322
at	1 000 m	9.7774	9.7809	9.7903	9.8032	9.8161	9.8257	9.8291
at	10 000 m	9.7496	9.7531	9.7626	9.7754	9.7884	9.7979	9.8013

For imprecise use, the values 9.81 m·s^{-2} and 32.2 ft s^{-2} are appropriate. The close proximity to 10 m·s^{-2} led to the definition of the leo.

1901 3rd CGPM: in pursuit of clarity for mass vis-à-vis weight '*The Conference declares:*. . .

3. The value adopted in the International Service of Weights and Measures for the standard acceleration due to gravity is 980.665 cm/s², a value already stated in the laws of some countries.'[8]

standard illuminants *engineering, colorimetry* These are reference illuminants of specific spectral composition, established to avoid the problems familiar through the use of fluorescent lights that distort colours relative to natural light. Standard A is the light of an incandescent filament at $2\,848$ K (4666.73°F). Standards B, representing overhead sunlight, and C, representing normal sunlight, are defined from A via filters, each being a double cell made of colourless glass holding in each compartment a solution of specified composition. Standard illuminant B is used along with the three cardinal illuminants red, green, and blue in the †CIE system.[164]

Technically these are standard sources, i.e. lights, but being defined by spectral composition, they are illuminants.

standard mile *engineering See* standard cable.

standard ratio $10^{0.1} = 1.258\,925\,4\sim$, the logarithmic base for decibel, etc.[52]

standard temperature and pressure (s.t.p.) *physics* A temperature of 0°C and a pressure of 101.325 kPa (32°F and $2\,116.2\sim$ lb-f·ft^{-2}), the pressure being that of the †standard atmosphere.

standard volume *fundamental constant.* Symbol V_0. The volume occupied by 1 mole of ideal gas at s.t.p., $= 22.413\,996(39) \times 10^{-3}$m^3 ($22.4\,140\sim$ L) with †relative standard uncertainty 1.7×10^{-6}.[4] *See* molar volume of ideal gas; amagat unit.

The volume occupied by 1 kilogram-mole of ideal gas at s.t.p. is thus $22.4140\sim m^3$.

Standard Wire Gauge (SWG) *UK See* gauge.

Stanton number [T. Stanton; UK 1865–1931] *rheology*

As *St* (also **Margoulis number**), relating to heat transport, the dimensionless ratio of the coefficient of heat transfer to the product of the speed, the volumic mass, and the massic heat content at constant pressure.[61] Identically the ratio of the †Nusselt number to the †Péclet number (both for heat transport). The product of the cube of *St* and the square of the †Prandtl number equals the cube of the **heat transfer factor**.

As *St**, the **Stanton number for mass transfer**, relating to the transport of matter in a binary mixture, the dimensionless ratio of the mass-transfer coefficient to the product of volumic mass and the speed.[61] Identically the ratio of the †Nusselt number to the †Péclet number (both for mass transport). The product of the cube of *St** and the square of the †Schmidt number equals the cube of the **mass transfer factor**.

stat *radiation physics* For radioactivity, 3.63×10^{-27} Ci; obsolete.

statA *See* statampere.

statampere [electrostatic ampere] *electric current*. Symbol statA. *Metric-c.g.s.* The †ampere of the †e.s.u. system and of the †Gaussian system, for a †steady current identically statC/second. 1 statA $= 333.5635\sim pA$.

statC *See* statcoulomb.

statcoulomb [electrostatic coulomb] *electric charge*. *Metric-c.g.s.* The †coulomb of the †e.s.u. system and of the †Gaussian system, defined as the point charge that exerts a force of 1 dyne on an equal point charge 1 cm away *in vacuo*. 1 statC $= 333.564\sim pC$. Also called franklin.

statD *See* statdaraf.

statdaraf [electrostatic daraf] *electric elastance*. Symbol statD. *Metric-c.g.s.* The †daraf of the †e.s.u. system and of the †Gaussian system, a †steady current identically statV/statC and reciprocal statF. 1 statD $= 1\ statF^{-1}$.

statF *See* statfarad.

statfarad [electrostatic farad] *electric capacitance*. Symbol statF. *Metric-c.g.s.* The †farad of the †e.s.u. system and of the †Gaussian system, for a †steady current identically statC/statV and reciprocal statdaraf. 1 statF = 1.112 646~ pF, but often accepted as being $\frac{10}{9}$ pF.

statH *See* stathenry.

stathenry [electrostatic henry] *electromagnetic inductance*. Symbol statH. *Metric-c.g.s.* The †henry of the †e.s.u. system and of the †Gaussian system, for a †steady current identically statV/statA/second. 1 statH = 898.755 2~ GH.

statute [indicating being defined by statute] For statute mile, *see* mile (where the qualifier contrasts with nautical mile, etc.).

statV *See* statvolt.

statvolt [electrostatic volt] *electric potential, electromotive force*. Symbol statV. *Metric-c.g.s.* The electrostatic †volt of the †e.s.u. system and of the †Gaussian system, for a †steady current identically statA·statΩ. 1 statV = 299.793~ V.

statΩ *electromagnetics See* statohm.

Staudinger value, Staudinger molecular weight [J. H. Staudinger; Germany 1881–1965] *chemistry* Notably for expressing size of and thence classifying polymers, being a number representative of molecular size, similar to but usually less than the molecular weight.

std standard.

steady, steady current *electromagnetics* The statements of the simple relationship between volt, ampere, ohm, and other electromagnetic units (e.g. electric charge = A·s) apply basically to an unchanging direct current, along with other stated conditions. For alternating current, *see* r.m.s.

Steel Plate Gage, Steel Wire Gage *USA See* gauge.

Stefan constant, Stefan–Boltzmann constant [J. Stefan; Austria 1835–1893 (*see* Boltzmann constant)] *mechanics* The coefficient in the corresponding law that energy is proportional to the fourth power of the thermodynamic temperature, = 5.670 400(40)~ × 10^{-8} W·m^{-2}·K^{-4} with †relative standard uncertainty 7.0 × 10^{-6}.[4]

stellar magnitude *astronomy* The range of brightness of the different stars prompted the Greeks to introduce a numeric scale for them, with 1 for the brightest and 6 for the faintest they could see, such numbers being called magnitude. The telescope expanded the visible, and hence the scale. The scale was given a mathematical basis in the 19th century, the five steps from 1 to 6 being defined as forming a †geometric scale accumulating to a 100–fold increase in brightness (close to the amount from 1 to 6 in the Greek scale); hence an increment of 1 toward the brighter end of the scale represents a multiplicative increase in brightness of $\sqrt[5]{100} = 2.511\,886$ and fractional values of magnitude can be unambiguously meaningful.

The subsequent introduction of photography allowed a more formal measurement, but confounded the system by its differential response to different colours; since the colour spectra of stars differ, their relative brightnesses changed with the photographic material. This was accommodated by defining an associated colour index. The 20th century brought the discovery of radio stars, X-ray stars, etc., and brought in the ultraviolet and infrared alongside the visible, leaving a complexity of meanings to brightness. Nor is the colour index necessarily a simple parameter. Generally, stellar magnitude means the overall visible light, but can be used appropriately qualified for other radiations. A set of stars near Polaris, the North Pole star, provided the original reference.

The huge variation in distances of individual stars from the observer on Earth (or elsewhere) requires the observed **apparent magnitude** (generically symbolized by **m**) of each star be corrected for distance to give an intrinsic brightness required for scientific comparison of the stars; defined to be equal to the apparent magnitude at the fixed distance of 10 parsecs from the star, this **absolute magnitude** (symbol **M**) is computed as

$M = 5 + m - 5\log_{10} d$

where d is the actual distance away in parsecs. *See also* distance modulus.

steradian (sterad) *solid angle*. Symbol sr. *Metric* For the apex angle of a cone of any shape whose sides are radii of a sphere, 1 sr is the angle subtended by a section of surface of area equal to the square of the radius. Quantitatively defined for the apex angle of any such cone as the ratio of the circumferential surface area enclosed by the cone to the square of the radial length. Since the surface area of a sphere is 4π times the square of the radius, a complete sphere equals 4π sr $= 12.566\,37\sim$ sr; hence 1 sr $= \frac{1}{4\pi}$ sphere $= 0.079\,577\,47\sim$ sphere. The triangular cone defined by a

Pole on Earth and one radian (57.3~ °) of longitude along the Equator has central angle equal to 1 sr.

Creating an equivalent †base unit in the SI, giving dimensionality to the plane angle, has been proposed.[148] (*See also* spat.) However, the steradian of the SI, once a supplementary unit, has been a dimensionless derived unit since 1980.[149]

Unlike plane angle, where degree offers a scheme of essentially †rational number values, there is no rational scheme for solid angle other than as fractions of the †sphere.

For history, *See* radian.

stere *volume Metric* For firewood, 1 m³ (35.315~ ft³); defined in France in 1798 and retained in modern international conventions, but now an SI-deprecated unit except for domestic use.

sthéne *force.* Symbol sn. *Metric-m.t.s.* That which produces an acceleration of $1 \text{ m} \cdot \text{s}^{-2}$ when applied to a mass of 1 t, = 1 kN. Originally called the **funal** when proposed by the British Association in 1876, it was renamed by 1914. It was authorized in France by a statute of 1919 as part of the †m.t.s. system.

stigma [Gk: 'dot'] Symbol σ – the Greek letter called sigma. *See* bicron.

stilb [Gk 'glitter'] *luminous emission.* Symbol sb. *Metric-c.g.s. 1921* For a diffuser, identically candela per square centimetre[101] ($\text{cd} \cdot \text{cm}^{-2}$ $= 10^{-4} \text{ cd} \cdot \text{m}^{-2}$). (There was no corresponding unit in m.k.s., and there are no special units for luminous emission in the SI.) For a perfect diffuser, 1 sb corresponds to the emission of $\pi \text{ cd} \cdot \text{sr} \cdot \text{cm}^{-2} = 3.141\,6\sim$ lambert. The millistilb (msb) was more common than the stilb. (The light of a clear sky at ground level ≈ 1 sb.)

stokes [G. G. Stokes; UK 1819–1903] *kinematic viscosity.* Symbols St, also lentor. *Metric-c.g.s.* Identically poise per gram per cubic centimetre $= 10^{-4} \text{ m}^2 \cdot \text{s}^{-1}$. The kinematic viscosity of water is regarded as 1.003 8 cSt for calibrations.

stone *mass BI* $\frac{1}{8}$ hundredweight = 14 lb (6.350 3~ kg). *See* hundredweight for scales. For centuries the major unit in the UK for expressing a person's 'weight' (with pound the minor), the stone was removed from official UK measures in 1985.[28]

weight, force See gravitational system.

geology A specific †particle size, typically between boulder and gravel in size, of diameter 1 to 8 in (25.4 to 203.2 mm).

storey, story *See* floor.

s.t.p. *physics See* standard temperature and pressure.

striken A qualifier for dry measures, meaning that the contents reach only the plane of the top of the measure, achieved by drawing a stick ('strickle') or roller across the top. The Act of 1824 that created BI specified any strickle to be straight, round, and of a uniform diameter.

strontium unit (sunshine unit) *foods* The number of microcuries of Sr^{90} absorbed per kilogram of calcium (the element strontium primarily replaces).

Strouhal number (reduced frequency) [V. Strouhal; 1850–1922] *rheology.* Symbol *Sr.* Relating to momentum transport, characterizing a taut wire vibrated by a fluid flow, the dimensionless ratio of the frequency of vibration times diameter of the wire to the fluid speed,[61] and hence in the narrow range 0.185 to 0.2 (but sometimes regarded as the reciprocal of this, hence 5 to 5.4). The phenomenon occurs with the aeolian harp; any example is an **aeolian frequency**.

Stubbs Iron Wire Gauge *UK See* gauge.

substance, substance number *paper and printing See* basis weight.

Sumner unit [J. B. Sumner; USA 1877–1955] *chemistry* For enzymes, the amount that liberates 1 mg of ammoniacal nitrogen in five minutes at $20°$ C, = 14.28 enzyme units.

sunshine unit *foods See* strontium unit.

sunspot number (Wolf number) *astrophysics* An indicator of sunspot activity. The sum of the number of spots and ten times the number of disturbed regions, multiplied by a constant, depending on the instrument.

superfoot *See* board foot.

SUS *mechanics See* Saybolt Universal Second.

Sv *radiation physics See* sievert.

svedberg [T. Svedberg; Sweden 1884–1971] *chemistry.* Symbol S. The unit of time 10^{-13} s, used with data from ultracentrifuge separation.

sverdrup [H. U. Sverdrup; Norway 1888–1957] *oceanography* For ocean currents, $10^6 \, m^3 \cdot s^{-1}$ ($35.315\sim \times 10^6 \, ft^3 \cdot s^{-1}$, $15.85\sim \times 10^9 \, US \, gal \cdot min^{-1}$).

SWG Standard Wire Gauge; *see* gauge.

symmetry number Symbol σ. The number of different positions into which a rigid body can be rotated without visible difference.

synodic [Gk: 'together' + 'journey'] *astronomy* An adjectival qualifier normally referring to the period an orbiting body takes to move between successive conjunctions relative to an observer on Earth, i.e. being effectively in line with Earth and our Sun and in same order. The synodic month, for instance, is the period between successive placements of the Moon in line with Earth and the Sun, either consecutive placements outside Earth's orbit (superior conjunction) else both inside (inferior conjunction). These, of course, are precisely full moon and so-called new moon (i.e. in astronomical usage, the point 180° from full moon, actually lacking any visible Moon, and prior to the like-named visible event of religious and everyday terminology), so the synodic month is the commonly observed period of the Moon, typically measured by the new moon rather than strict inferior conjunction. The synodic periods of planets can be measured directly and from each the sidereal period and true orbital period can be calculated.

T

T, t *Metric* As an upper-case prefix, **T-**, *see* tera-, e.g. TV = teravolt.
volume As **T**, *see* tablespoon; as **t**, *see* teaspoon.
mass *Metric* As **t**, *see* tonne.
BI As **t**, ton, but not so used in this work.
electromagnetics As **T**, *see* tesla; *see also* SI alphabet for prefixes.

T- *Metric* As symbol, *see* T, t.

Ta *textiles* *See* yarn units.

tablespoon *volume*. Symbol T, tbs, tbspn. *Metric* 15 mL = $\frac{1}{16}$ cup.

TAI [*Temps Atomique International*, 'International Atomic Time'] *time*
A scheme for measuring †time with a fixed-size second derived from
atomic clocks, rather than the variable natural day or other astronomical
period, and the underlying reference for †Universal Time (UT) and for the
special scheme used for astronomical observations, †Terrestrial Time. In
TAI every day has exactly 86 400 seconds, so it does not stay in exact step
with the real world, i.e. with the rotational realities of Earth; UT is an
adjusted form addressing that problem by allowing †leap seconds, so
does not always have 86 400 seconds in a day.

Modern precision demands a statistical derivation from a multiplicity
of atomic clocks, coordinated through the US Naval Observatory.

1970 CIPM: 'International Atomic Time (TAI) is the time reference coordinate
established by the Bureau International de l'Heure on the basis of the
readings of atomic clocks operating in various establishments in
accordance with the definition of the second, the unit of time of the
International System of Units.'

1971 14th CGPM: '*considering* that the second, unit of time of the International
System of Units, has since 1967 been defined in terms of a natural atomic
frequency, and no longer in terms of the time scales provided by
astronomical motions, that the need for an International Atomic Time (TAI)
scale is a consequence of the atomic definition of the second, that several
international organizations have ensured and are still successfully ensuring

the establishment of the time scales based on astronomical motions, particularly thanks to the permanent services of the Bureau International de l'Heure (BIH) ... that the BIH has started to establish an atomic time scale of recognized quality and proven usefulness, that the atomic frequency standards for realizing the second have been considered and must continue to be considered by the CIPM helped by a Consultative Committee, and that the unit interval of the International Atomic Time scale must be the second realized according to its atomic definition, that all the competent international scientific organizations and the national laboratories active in this field have expressed the wish that the CIPM and CGPM should give a definition of International Atomic Time, and should contribute to the establishment of the International Atomic Time scale, that the usefulness of International Atomic Time entails close coordination with scales based on astronomical motions, *requests* the CIPM

1. to give a definition of International Atomic Time [already done, see above],
2. to take the necessary steps, in agreement with the international organizations concerned, to ensure that the available scientific competence and existing facilities are used in the best possible way to realize the International Atomic Time scale and to satisfy the requirements of users of International Atomic Time.'[8]

talbot [W. H. F. Talbot; UK 1800–77] *luminous energy* Metric-m.k.s. Identically the product of luminous efficiency, in lumens per watt, and the number of joules of input radiant energy (= s·cd·sr in base terms).

tau mass *sub-atomic physics*. Symbol m_τ. The theoretical rest mass of the never-resting particle, $= 3.167\,88(52) \times 10^{-27}$ kg $= 3\,477.60(57)\ m_e$, both with †relative standard uncertainty 1.6×10^{-4}.[4]

Taurus *astronomy* See zodiac; right ascension.

Taylor number *rheology* A dimensionless quantity relating to momentum transport between two rotating cylinders, being the square of the Reynolds number multiplied by the ratio of the difference in diameters of the cylinders to their mean diameter.

tbs, tbspn See tablespoon.

t.c.f. [trillion cubic feet] *volume* North America Particularly for natural gas, 10^9 m.c.f. $= 10^{12}$ ft^3 ($28.317\sim \times 10^9$m^3).

Td *electromagnetics* See townsend.
  textiles See yarn units.

TDT *time See* Terrestrial Dynamic Time.

teaspoon *volume*. Symbol t, tsp, tspn. *Metric* $5\,\text{mL} = \frac{1}{48}$ cup.

tebi- [tera- binary] *informatics*. Symbol Ti-. The $1\,099\,511\,627\,776 = 2^{40}$ multiplier, as in tebibytes (TiB) and tebibits (Tib). *See* kibi-.

technical system of units; *see* gravitational system.

temperature The standard temperature scale of science and most of the world is †Celsius, with its zero at the freezing point of water; the absolute thermodynamic or †kelvin scale has its unit the kelvin (K) of identical size but its zero at the †thermodynamic null. For scientific accuracy over a vast range, the thermodynamic scale exists in a 'practical' form as the †international temperature scale. The †Fahrenheit scale and its 'absolute' version, the †Rankine, are of comparable vintage to the Celsius and kelvin, and have enjoyed major use, but are now little used outside the USA. Their degree is only $\frac{5}{9}$ K or degree C, while the Fahrenheit zero is well below the freezing point of water, which is 32°F. The †Réaumur scale is the next most common of the many scales that have been introduced, but sees no significant use in modern times; it has zero at freezing point but a degree equal to $\frac{5}{4}$ K. Notable comparative points are shown in Table 54.

Relevant conversion formulae between the Celsius, kelvin, Fahrenheit, and Rankine scales, represented respectively by c, k, f, and r, are:

$$c = k - 273.15 \qquad c = (f - 32)5/9 \qquad c = (r - 491.67)5/9$$
$$k = c + 273.15 \qquad k = (f - 32)5/9 + 273.15 \qquad k = (r - 491.67)5/9 + 273.15$$
$$f = 9c/5 + 32 \qquad f = 9(k - 273.15)/5 + 32 \qquad f = r - 459.67$$
$$r = 9c/5 + 491.67 \qquad r = 9k/5 \qquad r = f + 459.67$$

Table 54

Celsius	kelvin	Fahrenheit	Rankine	Réaumur
−273.15	0	−459.67~	0	−218.52~
−40	233.15	−40	419.67~	−32
−20	253.15	−4	455.67~	−16
−17.78~	255.37~	0	459.67~	−14.22~
0	273.15	32	491.67~	0
40	313.15	86	563.67~	32
60	333.15	140	599.67~	48
80	353.15	176	635.67~	64
100	373.15	212	671.67~	80

tempon *sub-atomic physics See* chronon.

ten *Metric* For use applied as a suffix to units, *see* tenth gram.

tenth *length* Vernacularly, a tenth of any implied unit; *see also* tenth gram.

tenth gram, tenth metre *length Metric* Obsolete term for the 10^{-10} multiple of the unit, e.g. tenth metre = 10^{-10} m (the angstrom).

Akin to the modern scheme for †floating-point numbers, G. J. Stoney in the 1860s proposed citing the exponent of a power of ten, verbally, to any unit as a convenient means of expressing decimal multipliers.[165] He used an †ordinal number before the word to indicate a negative power (e.g. †tenth metre), the †cardinal number after for positive powers, e.g. 'gramme-eight' = 10^8 g. The qualifiers were usually hyphenated.

ter- *Metric* contracted form of †tera-, as in teramp = TA.

tera- [Gk: 'monster'] Symbol T-. *Metric* The 10^{12} multiplier, e.g. 1 teragram = 1 Tg = 10^{12} g; contractable to ter- before a vowel, e.g. 1 teramp = 1 TA = 10^{12} A.

informatics Sometimes 2^{40}, but *see* tebi- then kibi-.

Terrestrial Dynamical Time (TDT) *time See* Terrestrial Time.

Terrestrial Time (TT) *time* The current clock, and calendar, of astronomy, using the second of an atomic clock, with no adjustment for the vagaries of the natural day, etc. Until 1991 called **Terrestrial Dynamical Time** (TDT), it succeeded the similar †Ephemeris Time (ET), based on a fixed-size second of $\frac{1}{86400}$ an idealized day, in 1984. It is synchronized with †TAI, offset permanently by the 32.184 s that ET was in advance of TAI at the moment of succession. *See* time, also Universal Time, which, unlike TAI and TT (the preferred name now being the shorter one), allows for adjustment to maintain harmony with the observed day.

tertian ['third'] (**firkin** (for wine), **puncheon**) A bulk-measure cask, with established volumes and quantities in historic marketplaces, = $\frac{1}{3}$ tun.

tesla [N. Tesla; Austro-Hungary, USA 1857–1943] *magnetic flux density, magnetic polarization*. Symbol T. *SI, Metric-m.k.s.* Wb·m^{-2}, i.e. a density of 1 weber of magnetic flux per square metre (= $kg \cdot s^{-2} \cdot A^{-1}$ in base terms).

The tesla was approved by the International Electrical Conference in 1954, and for the SI by the 11th †CGPM of 1960.

tetra- [Gk: 'four'] Prefix for four, e.g. **tetragon** = four angles (hence four sides).

TEU *See* twenty-foot (equivalent) unit.

tex *textiles*. Symbol Tt. *See* yarn units.

TFU *See* twenty-foot (equivalent) unit.

therm *energy BI* The energy required to raise the temperature of 1 000 pounds of water 100 degrees Fahrenheit, = 10^5 Btu (105.505 6~ MJ). The precise amount is dependent on the starting temperature, so use of the term requires an acknowledgement of this; *see* B.t.u. for related discussion.

The British Association originally, in 1888, applied this name to what became the calorie, i.e. the heat required to raise 1 gram of water 1 degree Celsius.

thermal ampere, thermal farad, thermal henry, thermal ohm *thermodynamics* Seeing temperature difference akin to electric potential difference, the current of heat flow can be seen as measurable in thermal amperes, as an entropy flow of 1 watt. One thermal ohm is then the thermal resistance such that unit temperature difference causes an entropy flow of 1 watt, one thermal farad a thermal capacitance such that addition of entropy of 1 joule raises the temperature 1 kelvin. For thermal inductance, one thermal henry can also be defined, relating an entropy flow of 1 watt per kelvin with a hydrodynamic energy of 1 joule.

thermie *heat energy Metric-m.t.s. 1919* The heat energy required to raise the temperature of 1 tonne of water 1 degree Celsius, = 4.185~ MJ (3 967.~ Btu).

thermodynamic null The absolute lowest point achievable on the thermometric scale, being the point at which thermodynamic activity, the essence of temperature, ceases. Commonly called 'absolute zero', it is the zero point on the †kelvin and †Rankine scales.

thermodynamic temperature scale *See* kelvin.

third *music See* interval.
  other For **third degree**, *see* degree.

third gram, third metre *Metric See* tenth gram.

Thomson cross-section [G. P. Thomson; UK 1892–1975] *sub-atomic physics*. Symbol σ_e. $0.665\,245\,854(15) \times 10^{-8}$ m^2 with †relative standard uncertainty 7.6×10^{-12}.[4]

thou [thousandth] (**mil**) *length* UK $\frac{1}{1000}$ in (25.4 μm, 0.001 in).

thousand Ten 100s; $= 1000$ or 1,000 or 1 000, the punctuation usually being omitted until a further digit is involved. The scale of thousands, showing US terms, then the distinctive traditional British and European terms, is set out in Table 55.

It should be noted that the traditional British terms relate the Latin etymology to the powers of a million, while the American relate it to the powers of a thousand that multiply the basic thousand; e.g. the bi- (i.e. 2) of billion relates

for UK to (million)^2, $= 1\,000\,000^2 = 10^{6 \times 2} = 10^{12}$,

for USA to (thousand)$^2 \times$ thousand, $= 1000^{(2+1)} = 10^{3 \times (2+1)} = 10^9$.

Despite the 13th †CGPM in 1948 recommending the British/European forms, the American forms are the more prevalent, even in the UK. However, they are yielding in either sense to expressions formed from the prefixes of the †SI system (e.g. 'megabucks'). The use of the old terms above a million in a scientific context is highly undesirable.

Table 55

		US	UK trad	European trad
$1\,000^1$	10^3	thousand	thousand	millier
$1\,000^2$	10^6	million	million	million
$1\,000^3$	10^9	billion	milliard	milliard
$1\,000^4$	10^{12}	trillion	billion	
$1\,000^5$	10^{15}	quadrillion		
$1\,000^6$	10^{18}	quintillion	trillion	
$1\,000^7$	10^{21}	sextillion		
$1\,000^8$	10^{24}	septillion	quadrillion	
$1\,000^9$	10^{27}	octillion		
$1\,000^{10}$	10^{30}	nonillion	quintillion	
$1\,000^{11}$	10^{33}	decillion		

thousandth mass unit *sub-atomic physics*. Symbol TMU. A unit of energy †coherent with the famous equation $E = m \cdot c^2$, being that relating to a mass of a thousandth of an atomic mass unit, $= 149.2\sim$ fJ.

three *Metric* For use applied as a suffix to units, *see* tenth gram.

Ti- *informatics See* tebi-, e.g. as **TiB** = tebibytes, **Tib** = tebibits (Eib).

t.i.d. [Lat: ter in die] *medicine* 'Three times per day'. *See also* o.d.

tierce ['third'] A bulk-measure cask, with established volumes and quantities in historic marketplaces, $= \frac{1}{3}$ pipe or butt $= \frac{1}{6}$ tun.

time Our world presents three conspicuous units of time: the day, the lunar month, and the year, being respectively the rotation of Earth about its own axis, the orbiting of the Moon about Earth, and that of Earth about the Sun; they can be representatively defined as the time between successive high noons, full moons and midsummers. Mankind has constructed calendars, clocks, and other time matters on the basis of these three apparently unchanging units. However, all three are lacking in constancy within the sensitivities of scientific instruments, and are changing progressively over time. The irregular shape of Earth, for instance, affects its rotation. The fact that gravitational orbiting follows an elliptical path with the 'central' body at an offset focus rather than the geometrical centre complicates the timing of any orbital travel. Additionally, such elliptical travel is inherently not steady in speed. The tilt of Earth's axis relative to its plane of travel (its 'obliquity') and of the Moon's plane of travel relative to that of Earth add further perturbing complications. Gravitational effects between bodies include more than mere orbiting; they tend to slow individual rotations, alter the axial direction of a tilted body (nutation) and, more importantly from the perspective of time measurement, cause the point that is situated at the focus of the elliptical orbit to be a compromise rather than the centre of the commanding body. This is particularly significant for the Earth–Moon pair because of their relative closeness in mass; what follows the orbital path around the Sun is not the centre of Earth but a moving point (their barycentre) about 4 600 km (2 875 mi, 36% of Earth's radius) from Earth's centre towards the centre of the Moon. Exaggerating, the pair travel like a pair of disparate balls tied together with a length of string, then thrown. The other planets add their (varying) influences to these motions. The solar system itself, our Galaxy (the Milky Way) and other elements of the heavens at large are all moving minutely relative to the larger framework, making even the so-called 'fixed stars' not absolutely fixed.

The key natural unit of time for human purposes is the apparent solar day, i.e. the value of the observed time for passage between consecutive high noons or other marks on the sundial (also called sundial day). The standard day of our clocks is the average such, the mean solar day. By definition, this equals 24 (mean solar) hours. Time on this base is referred to as (**mean**) **solar time**. For use within astronomy, however, it is more appropriate to work in terms of the revolution relative to the fixed stars, which gives **sidereal time**, with the sidereal day of 24 sidereal hours. Sidereal differs crucially rather than pedantically from solar time because of the one rotation of Earth that is 'lost' by the traverse of an orbit, a factor readily observable as the progressively changing midnight sky and providing the basis for zodiacal horoscopes, and which puts 366 sidereal days in the normal 365-day year. Midnight in sidereal time is when a designated star is precisely on the local celestial meridian (i.e. the imaginary planar circular arc running between the north and south points on the horizon and through the zenith), which is often during daylight hours! (This makes that one star, and many neighbours, not observable, but the relative time-offsets of a wide range of significant stars are known and catalogued.) As the interval between consecutive overhead circumstances for any one star also varies very slightly, in 1960 astronomers adopted a corresponding artificial regular time called †Ephemeris Time, based on idealized motions for the three key bodies, standardized to the tropical year of 1900. For relative values and further discussion *see* day.

For most of history, the (mean solar) day has been the standard for the measurement of earthly time, divided into 24 hours, each hour containing 60 minutes, and each minute containing 60 seconds. For scientific use the definition was changed drastically in 1967 from the solar base to an atomic one; instead of being a specified fraction of a day the second became the key unit of time, as equalling 9 192 631 770 oscillations of the atom of caesium-133, creating **atomic time**. Atomic clocks are now the primary clocks, their rigid day of 86 400 seconds building †TAI. The various discrepancies between that absolute scheme and the vagaries of nature are attended to within the framework of †Universal Time, and can result in a second being added to else subtracted from the master (secondary) clocks that provide the reference for the world of science and even the street. For astronomical reference,

Terrestrial Time, employing the 'atomic' second, has succeeded Ephemeris Time.

tire size *See* tyre size.

tissue roentgen *radiation physics See* rep; roentgen.

Tj *textiles See* yarn units.

TME [Ger: *Technischer Mass Einheit*] *See* engineering mass unit.[166]

TMU *sub-atomic physics See* thousandth mass unit.

tog *textiles* A unit of thermal resistance applicable to apparel, with value about 1 for normal suitings, ranging beyond 10 for highly insulative outerwear; it is effectively the number of watts required to maintain a temperature differential across the material of 0.1 K.[167] Hence

$$1 \text{ tog} \approx 0.1 \text{ K} \cdot \text{W}^{-1} \cdot \text{m}^2 \ (0.571\sim \text{degF} \cdot \text{h} \cdot \text{ft}^2 \cdot \text{Btu}^{-1}, 0.645\sim \text{clo}).$$

tolerance unit *engineering* The standard applying to the clearance between pistons (and similar cylindrical inserts) and the 'cylinders' into which they fit. For diameter D the fundamental unit is, using millimetres,

$$0.001 (0.45 D^{1/3} + 0.001 D),$$

and using inches,

$$0.001 (0.052 D^{1/3} + 0.01 D).$$

The quality of tolerance is expressed as a multiple of that unit, with notation IT.1 to IT.16 for successive grades of declining fineness.

[handwritten annotations: D = 25 mm : 1.35 + .025 = 1.37 .00137; D = 1 inch : .052 + .01 = .062 .000062 inch = .00155 mm]

ton [tun] *mass SI* As **metric ton** = tonne (1 000 kg, 2 204.6~ lb). *BI* (also **long ton**) 20 hundredweight of 112 lb = 2 240 lb (1 016.~ kg).

The ton was removed from official UK measures in 1985.[28] *US-C* (also **short ton**) 20 hundredweight of 100 lb = 2 000 lb (907.2~ kg). *Canada* (also **short ton**) 20 hundredweight of 100 lb = 2 000 lb. *See* hundredweight for scales.

weight, force See gravitational system.

volume For maritime use typically 40 ft³ (1.412 6~ m³); *see* shipping ton.

For gravel, sand, etc., equated with the cubic yard (27 ft³, 0.954 96~ m³).

For timber, typically 40 ft³, though for hewn timber 50 ft³ (1.765 7~ m³).

engineering For refrigeration, the heat transfer required to freeze completely, in 24 hours, a ton of water at 0°C.

USA (short ton) $288\,000$ Btu/24 hours $= 200$ Btu·min^{-1} ($3.516\,9\sim$ kW).

UK (long ton) $322\,560$ Btu/24 hours $= 224$ Btu·min^{-1} ($3.938\,9\sim$ kW).

UK (metric tonne) $317\,465\sim$ Btu/24 hours ($3.876\,7\sim$ kJ·s^{-1}).

Compare frigorie.

See also assay ton.

tone *music See* interval.

tonne (**metric ton, millier**) *mass.* Symbol t. *Metric* The round metric mass figure very close to the traditional ton, being $1\,000$ kg ($0.984\,21\sim$ long tons, $1.102\,3\sim$ short tons, $2\,204.6\sim$ lb). For the †m.t.s. system the tonne was the base unit for mass. In North American business literature, the tonne is often represented by the letters **mt** for metric ton.

energy The energy of 1 tonne of TNT (trinitrotoluene) $= 4.20\sim$ MJ (1 kW·h).

tor (rarely), **torr** [E. Torricelli; Italy 1608–47] *pressure* For gases, particularly the atmosphere; designed to equate with the millimetre of mercury (*see* head of liquid) but to have a precise value for the standard atmosphere, it is thus defined as $\frac{101325}{760}$ Pa $= 13.585\,53\sim$ Pa. An SI-deprecated unit. (The spelling tor was used briefly for the pascal itself.)

total absorption unit *acoustics See* sabin.

townsend [J. Townsend; UK 1868–1957] *electromagnetics.* Symbol Td. *Metric-c.g.s.* The ratio of electrical field strength to molecular density of a gas.[168]

township *area North America* Usually 6 mi $\times$ 6 mi $= 36$ mi^2 ($93.240\sim$ km^2).

t$_P$ *sub-atomic physics See* Planck time.

trace Inherently an unmeasurable non-zero amount within the measurement method and scheme being used, e.g. visible moisture less than 0.5 mm within a rain gauge read to the nearest millimetre. On a graphical recorder, this is represented by a blip above the base line that likewise falls short of the minimum non-zero reading. However, as applied in 'trace elements', etc., the amount is typically measurable, the inference of the word being that the concentration is below that measurable on a coarser scale, e.g. occurring below 1 g per kilogram, so

usually expressed as milligrams per kilogram or parts per million
(p.p.m.).

Tralles [Switzerland, a member of the first international committee
establishing the †metric system] *See* relative volumic.

transcendental number *See* number.

transmission unit *telecommunications*. Symbol TU. Original name
for what became the decibel.

tri- [Gk 'three'] Prefix for three, e.g. **triangle** = three angles (hence three
sides).

trichromatic unit *colorimetry*. Symbol T unit. *See* CIE system.

trillion *See* Table 55 under thousand.

trimester [tri + Lat: 'month'] *time* Any three-month period, usually a
subdivision of a greater period that is a multiple of such, particularly the
successive parts of the nine-month period of human gestation. Used
erroneously for any period that is a third of some year, e.g. with a year-
round school scheme having three four-month terms.

tristimulus *colorimetry See* CIE system.

troland [L. T. Troland; 1889–1932] (**luxon**) *optometry* A unit of retinal
illumination corresponding to that from a surface producing 1 candela
per square metre, observed through an aperture of 1 square millimetre.
Initially a **photon**.

tropical [deriving from a word meaning 'turning', referring in
astronomy to the apparent turning of the Sun from its northerly passage
to its southerly, and vice versa, and more generally to the Sun] As applied
to year, the period of the repetition of the seasons.

troy scale *mass* The traditional and contemporary scheme for
weighing gold and silver, in recent times extended to platinum and other
precious metals. The widely familiar ounce used for reporting the price
of gold, rarely qualified, is the ounce of the troy scale.

Unlike the more usual avoirdupois scale with its pound of 16 ounces
and 7 000 grains (making its ounce of 437.5 gr), the troy scale, like the
†apothecaries' scales, has a pound of only 12 ounces each of 480 grains,
giving a total of 5 760 grains, the †grain being the one unit common to all
three scales. The distinctive scale for troy units is shown in Table 56.

Table 56

BI-troy, US-C-troy				Internat values:	SI	US-C-av
grain	...	...	...		64.8~ μg	1 gr
24	pennyweight	...	...		1.56~ g	24 gr
480	24		ounce		31.1~ g	1.10~ oz
5 760	288		12	pound	373.~ g	13.2~ oz

As with the avoirdupois units, the troy units have for centuries been very close to their current international value, probably the same to at least six significant figures. Although for many centuries the scale of goldsmiths, etc., the troy scale became officially legal in the UK only with BI in 1825. A prototype troy pound actually provided the standard for all 'weights' in BI until 1855 and in the USA until the †Mendenhall Order of 1893. Current values are based on the international grain, adopted in 1959, of 64.798 91 mg.

The troy pound was removed from official UK measures in 1878. The pennyweight was removed in 1970, and the grain in 1985.[28]

tsp, tspn *See* teaspoon.

TT *time See* Terrestrial Time.

Tt *textiles See* yarn units.

TU *engineering See* transmission unit.

TUC [Fr: *Temps Universel Cordonné*, 'Coordinated Universal Time'] *See* Universal Time.

tun (sometimes its derivative **ton**) A very large bulk-measure cask, with established volumes and quantities for various liquids and other commodities in historic marketplaces, typically weighing about a ton.[73]
    volume BI 210 BI gal (954.68~ L, 252.20~ US gal); see Table 57(a). *US-C* 252 US gal (953.92~ L, 209.83~ BI gal); see Table 57(b).

History
Originally 256 gallons, so utterly compatible with the regular binary structure still persisting for smaller volume units, the tun was specified as 252 (wine) gallons in an English statute of 1425, i.e. as in the USA today. The BI system, introduced in 1825, merely changed the number to compensate for its larger gallon.[73]

T unit *colorimetry* Trichromatic unit; *see* CIE system.

Table 57(a)

UK, US-C				US-C:		SI	US-C liq
firkin* of ale	...	...	...	...		59.7~L	15.8~gal
2	kilderkin	...	...	...		59.7~L	15.8~gal
4	2	barrel	...	...		119.~L	31.5~gal
8	4	2	hogshead	...		239.~L	63.0~gal
16	8	4	2	pipe, butt		477.~L	126.~gal
32	16	8	4	2	tun	954.~L	252.~gal

* a firkin of ale or beer; see Table 57(b) for firkin of wine.

Table 57(b)

UK, US-C				US-C:		SI	US-C liq
rundlet		...		...	...	68.2~L	18.0~gal
	tierce			...	...	159.~L	42.0~gal
	2	firkin,*	puncheon, tertian			318.~L	84.1~gal
7	3	6		pipe		477.~L	126.~gal
14	6	3		2	tun	954.~L	252.~gal

* a firkin of wine; see Table 57(a) for firkin of ale or beer.

turn (**revolution**) *geometry* One complete †revolution, i.e. one turn around the circle, $= 2\pi$ rad. The turn is fundamentally important in the magnetic effects of an electric current in a coil (*see* ampere·turn), and suggestions have been made that it be represented by a base unit in the SI; the radian is a derived unit, seen as dimensionless. *See* radian for discussion.

Twaddell *See* relative volumic mass.

twenty-foot equivalent unit, twenty-foot unit *volume*. Symbols TEU, TFU. The standard element of the containerized shipping scheme, being a box 8 ft by 8 ft in cross-section times 20 ft long, hence of volume 1 280 ft³ (36.245 6~m³, 32 shipping ton). Such a container may be a simple steel box, a steel frame holding a round tank, or any other entity meeting the standard dimensions for size, strength, position of fasteners, etc. Multiples and a 10 ft half unit (actually the pioneer size) exist, reckoned proportionally to the twenty-foot unit for relevant statistics. (A height other than 8 ft is also allowed.)

twenty-twenty Now a general phrase indicating perfection.

medicine (correctly **twenty/twenty**) The representational figure for what is regarded as perfect human vision, the ability to identify an †alphanumeric character $\frac{1}{3}$ in (8 mm) high at a distance of 20 ft (6.1~ m). Applied separately to each eye, the latter figure is increased for sub-normal vision proportionally to the increase in size necessary for identification; thus $\frac{20}{30}$ indicates an eye that can identify at 20 ft what a normal eye can identify at 30 ft, equivalently how much the standard characters would have to be enlarged to be read at 20 ft, i.e. by $\frac{3}{2}$ to $\frac{1}{2}$ in (12 mm) in the given example.

There is a corresponding scale for oridinary print reading, presuming a standard distance of 14 in; then $\frac{14}{18}$ indicates an eye that can read at 14 in distance what a normal eye can read at 18 in.

TW·yr, Tw-yr. *engineering* Terrawatt-year, used in planning electricity supply, and accepted as 31.54 EJ. *See also* b.o.e.

typp *textiles*. Symbol Nt. *See* yarn units.

tyre size *engineering* The modern international standard notation for car-tyre size is an interesting admixture, being an alphanumeric code involving millimetres, inches, and a percentage. These are augmented by letters indicating type of construction and intended load/speed range of use. The dimensional components can be illustrated by a sample:

```
205 / 60   R    15
 ↓     ↓   ↓     ↓
 ↓     ↓   ↓    diameter
 ↓     ↓   construction type
 ↓   aspect ratio
section width
```

The diameter is of the rim, measured at the surface on which the bead of the tyre sits, expressed in inches. The section width is the width of the inflated but unloaded tyre, exclusive of raised lettering, etc., expressed in millimetres. The aspect ratio represents the section height of the tyre, from bead to outside of tread, as a percentage of the width. (So the gross diameter of the equipped wheel depends on the product of width and aspect ratio, and replacing tyres of one width with tyres of another width using rims of the same diameter will change the height of the vehicle unless compensating adjustment is made to the aspect ratio.) Because the aspect ratio characterizes the visible tyre for a given situation, its number is often used as a more generic label, as '60 Series', for instance, indicating a low-profile (i.e. wide for its diametrical depth) tyre.

The embedded R in this illustration indicates radial construction; B represents bias steel construction. This letter may be preceded by a letter representing intended maximal speed for use, but such designation is now usually included, together with a number representing maximal load, as an appended service description, e.g. 85T. The load figure is on a geometric scale, with 100 representing 800 kg (1 764 lb) per tyre, and the unit gradation being just under 3%; examples are:

60	253.1 kg	558 lb
70	337.5 kg	744 lb
80	450 kg	992 lb
90	600 kg	1 323 lb
100	800 kg	1 764 lb
110	1 067 kg	2 352 lb

The speed indicator (referring purely to the ability of the tyre to withstand the thermal and mechanical stresses of speed and not specifying cornering ability or such) has letter codes representing maximal speed:

L	120 km/h	74.5 m.p.h.
M	130 km/h	80.7 m.p.h.
N	140 km/h	86.9 m.p.h.
P	150 km/h	93.2 m.p.h.
Q	160 km/h	99.4 m.p.h.
R	170 km/h	105.6 m.p.h.
S	180 km/h	111.8 m.p.h.
T	190 km/h	118.0 m.p.h.
H	210 km/h	130.4 m.p.h.
V	240 km/h	149.0 m.p.h.

The last of these, or the additional code Z, may appear within the size description, immediately ahead of the construction type code; they then, if not overruled by an explicit service descriptor, indicate tolerance of speeds above that shown, generally the full maximal speed of the car.

The above ISO scheme is being introduced progressively into North America. Initially it is often prefixed with a P (for passenger) on car tyres, and referred to as P-metric. It may have the abbreviation STD (for standard) as a following load indicator. In an earlier form, any speed rating was routinely placed within the size description, immediately ahead of the construction type code.

Up to the 1970s, North America used a different notation, with width being coded alphabetically, e.g.

```
G R 78 – 15  B
↓ ↓ ↓  ↓ ↓
↓ ↓ ↓  ↓ load range
↓ ↓ ↓  diameter
↓ ↓ aspect ratio
↓ construction type
section width
```

and, before radials were introduced, just a single beginning letter. In this scheme, often referred to as the alphanumeric in contrast to its predecessor, the two numeric elements were as defined earlier, and the load range compatible with the above, but the width was coded alphabetically (with nearest equivalents in modern 75% aspect ratio instead of the then common '78 Series') as follows:

A	6.75 in	171 mm	175/75
B	7 in	178 mm	185/75
C	7.25 in	184 mm	185/75
D	7.5 in	190 mm	195/75
E	7.75 in	197 mm	195/75
F	8 in	203 mm	205/75
G	8.25 in	210 mm	215/75
H	8.5 in	216 mm	225/75
J	8.75 in	222 mm	225/75
L	9 in	229 mm	235/75

Yet earlier, the tyre cross-section was represented as though it were circular, by its diameter in inches to two decimal places, followed by rim diameter in inches, usually followed by a ply rating e.g. 7.75 – 15 4PR else 7.75 – 15 B, the ply rating being explicit, as 4PR for 4-ply, or coded alphabetically, with A for two, B for four (the typical car tyre) and so on (excluding the letter I) in steps of two plies. (While those examples would be pronounced as 'seven seventy-five … ', a value of no fractional part would be pronounced as though hundreds, e.g. 5.00 as 'five hundred … ')

The load designation as a letter was usually called load rating. The higher ratings applied, of course, to trucks and similar heavy-use vehicles. Their denotations varied somewhat, but were similar to those illustrated. The load range may be expressed with the extra initials, as LRH, or shown in the combined style of plies and code, e.g. 14 (G). The service descriptor, for instance, can also be augmented to show load indices for single and, in dual arrangement, e.g. $\frac{143}{140}$, actual permitted loads depending on inflation as well.

U

u *mass SI* In a molecular/atomic context, **u**, *see* unified atomic mass unit.

   other As a graphic appearance, **u** is sometimes used for μ, i.e. micro-.

U factor, U value *engineering UK* For building materials, a measure of thermal conductivity per unit thickness, being B.t.u. per square foot per hour per degree Fahrenheit; a U value of $1 = 5.678\,3\sim W \cdot m^{-2} \cdot K^{-1}$. A value less than 1 has traditionally been seen as appropriate for energy efficiency; traditional brick cavity walls have values about 0.3, but many roofs exceed 1. For the greater efficiency wanted in the modern world, a value of 1 in metric wattage terms is more the target. It is the reciprocal, for a given time and area, of †R value.[169]

UHF *physics* The 'Ultra High Frequency' waves of the †electromagnetic spectrum, being those of about 1GHz frequency, used, *inter alia*, for UHF television and for satellite transmissions.
North America   470 to 512 MHz for TV channels 14 to 20,
                      512 to 806 MHz for TV channels 21 to 69.

ultraviolet Symbols UV, u.v. The waves of the †electromagnetic spectrum just outside the †visible light spectrum, of about 750 to 1 500 THz in frequency, just above the frequency range of (visible) violet light. The limit of UV light reaching Earth's surface from the Sun (the dangerous carcinogenic factor in sunlight) is about 1 015 THz. Divided into three sub-bands:
    u.v.A of wavelength 400 to 320 nm, i.e. frequency 749 to 937 THz,
    u.v.B of wavelength 320 to 280 nm, i.e. frequency 937 to 1 071 THz,
    u.v.C of wavelength 280 nm down, i.e. frequency 1 071 THz up.

unified atomic mass unit *physics*. Symbols u, m_u. $\frac{1}{12}$ the mass of an atom of the commonest carbon isotope ($^{12}_6C$) $= 1.660\,538\,73(13) \times 10^{-27}$ kg with †relative standard uncertainty 7.9×10^{-8}.[4] Its frequent use in the burgeoning world of molecular biology has prompted the use of the handier name †dalton.

History

Imprecisely the mass of a proton else neutron, the unit began as the
atomic mass unit or **amu**, defined as the mass of the hydrogen atom,
then, in 1885, as $\frac{1}{16}$ the mass of an atom of oxygen. However, chemists
came to see this as the naturally occurring mixture of oxygen isotopes,
and physicists the pure commonest isotope, the latter giving values
0.027~% below the former. The unifying carbon-based definition and the
current name were adopted in 1961 (with symbol u assigned in the SI).
See mole for elaboration.

unit *catalytic activity* μmol/min, but *see* katal.

unit distance *astronomy See* astronomical unit (of length).

unit magnetic pole, unit pole *electromagnetics* A conceptual
parallelism in magnetism to the unit electric charge, the unit pole $= 4\pi$
times the unit of magnetic flux:
Metric-c.g.s.-e.m.u. 4π maxwell (125.663 7~ nWb);
Metric-c.g.s.-e.s.u. 4π statmaxwell (3 767.303~ Wb).

unit of action *sub-atomic physics See* Planck constant.

unit of activity *chemistry See* international biological standards.

universal conversion factor *See* Boltzmann universal conversion
factor.

universal gas constant, universal molar gas constant *See* gas
constant.

Universal Time (UT, also UTC and distinctly UT_0, UT_1, UT_2) *time* The
name adopted in 1972 for the long-familiar †Greenwich Mean Time
(GMT), i.e. the time scheme set to have 12:00 noon when the Sun is
nominally overhead on the Greenwich meridian, but precisely
synchronized, at the level of the second, with the atomic-based †TAI. (As
discussed under day, the elapsed time from solar noon to noon is not
consistent throughout the year; being overhead relates to an average
position.)

Atomic clocks were first brought into practical use in the 1950s. The
'atomic' second as the essential unit of time was defined in 1967, and an
international time scheme based thereon, labelled TAI (Temps Atomique
International), was adopted in 1970. TAI builds an 'atomic' day of 24
'atomic' hours of 60 'atomic' minutes of 60 such seconds, i.e. a 'day' is
precisely 86 400 'atomic' seconds. However, just as a year that is a regular

multiple of days does not stay in tune with the seasons, the regular day of TAI does not stay in tune with the diurnal pattern. The day varies considerably in length compared with the year, rhythmically over the year by more than 30 s (*see* equation of time), but its overall mean value also varies, irregularly and with an overall progressive slowing down. To avert the Sun's zenith creeping progressively from its familiar place, UT incorporates '†leap seconds' as appropriate. All adjustments are made at the close of a month, preferably of December else of June, but possibly others, as necessary to keep the discrepancy between TAI and reality (in the form of UT_1 shown below) less than 0.9 s. To date, because of the progressive slowing down (the mean solar day being now about 86 400.0028 s long), it has always meant an insertion, but it could require removal of a second, and either way it could be multiple. Starting in 1972, one leap second was added in 22 of the 28 years up to the end of 1998, putting UT 32 s behind TAI (hence 64.184 s behind TDT at the start of 1999). The following three years were devoid of leap seconds.[170]

Universal Time exists in a superfine gradation of forms. The above is distinguished as Co-ordinated Universal Time (**UTC**). **UT_0** is the rotational time scale based on observation and standard longitude, **UT_1** is UT_0 adjusted for the effect on functional longitude of the polar wobble caused by crust/mantle movements within Earth, and which results in the angle of tilt varying from its nominal value (currently nearly $23\frac{1}{2}$ degrees, but also subject to periodic variation, nutation, over 41 000 years of about 1 degree either side of this value). **UT_2** is UT_1 adjusted for the seasonal variations in Earth's rotation. The differences between the three versions are utterly trivial for everyday living.

UT is routinely used by satellites and related equipment that need to be mutually synchronized. However, to avoid the complication of leap seconds, each separate system is set permanently to the UTC at the epoch of its initialization. Thus the OMEGA system, its epoch being 01 January 1972, is only 10 s behind TAI, while the Global Positioning System, its epoch being 06 January 1980, is 19 s behind TAI for the computation of global position (the displayed time being full UTC).

u_r *physics See* relative standard uncertainty.

US-C, US-C-ap, US-C-av, US-C-troy *See* US Customary system.

US Customary system (US-C) The system for weights and other measures in general use in the USA. Founded on the measures in use in Britain before the introduction of †BI in 1825, the US-C is essentially the

BI system except for volumetric measures, where the old British wine †gallon of 231 cubic inches became the standard for liquids and the Winchester †bushel of England the standard for dry goods. The last, specified as dimensions of a round container, bore no simple relationship to the wine gallon.

The first formal action to establish a national system of 'Weights and Measures' in the USA did not occur until 1830, i.e. five years after the introduction of BI. The foot, etc., for length, and the pound, etc., for 'weight', were intended to equal the British values, though various related terms were used differently, notably the hundredweight, reset to its original 100 lb, and the ton. However, minor variation of sizing did occur, particularly through the statute of 1866 that led to the †Mendenhall Order. Like BI, the US-C has three distinct (though closely interrelated) schemes for mass: †avoirdupois scale (**US-C-av**), †troy scale (**US-C-troy**) and †apothecaries' scale (**US-C-ap**). *See* grain.

US number *photography* A measure of aperture size, hence of light-gathering capacity. Unlike the more familiar †f number, US number is inversely proportional to area rather than to diameter. The scale starts with 1 equalling f4 (technically f/4), then proceeds with whole numbers for the familiar British f values. *See also* film speed.

UT, UT$_0$, UT$_1$, UT$_2$, UTC *See* Universal Time.

UV, u.v., UVA, u.v.A, UVB, u.v.B, UVC, u.v.C *See* ultraviolet; SPF.

V, v [representing one hand, of five fingers] The †Roman numeral for 5.
electromagnetics Metric As **V**, *see* volt; *see also* SI alphabet for prefixes.

VAC, V.a.c. *electromagnetics* Volts alternating current: *see* volt; r.m.s.

variance *statistics* See standard deviation.

verber *electrics* An early name for a unit of charge similar to the
coulomb.

VHF *physics* The 'Very High Frequency' waves of the †electromagnetic
spectrum, being those of about 30 to 300 MHz frequency, used, *inter alia*,
for VHF television and for FM radio. Assignments vary between
countries; the following are representative.

North America	70 to	50 MHz for police, public safety, railroads,
	50 to	54 MHz for '6-metre' amateur band,
	54 to	72 MHz for TV channels 2 to 4,
		75 MHz for aeronautical markers,
	76 to	88 MHz for TV channels 5 and 6,
	88 to	108 MHz for FM radio,
	108 to	135.95 MHz for aeronautical communications,
	144 to	148 MHz for '2-metre' amateur band,
	174 to	216 MHz for TV channels 7 to 13.

The frequencies originally intended for TV channel 1 were redirected to
other use, leaving the phenomenon that there is no TV channel 1.

vibration The repeated movement to and fro of an object, of air, etc.
Normally, any count of vibrations (often †French vibrations) is the count
of movements away from the central position in either direction, hence
equalling twice the number of full †cycles and twice the frequency.

Vickers hardness number, Vickers pyramid number See
hardness numbers.

Victorian period *time* Re calendar patterns, *see* Dionysian period.

vigesimal [a composite adjective from Greek and Latin for 20] With divisor/multiplier steps of 20, in contrast with the steps of 2 for †binary, of 10 for †decimal, of 60 for †sexagesimal, etc.

violle [J. L. G. Violle; 1841–1923] *photics* Early name for candela.

Virgo *astronomy* See zodiac; right ascension.

viscosity grade (VG) *engineering* A †geometric scale for the viscosity of an oil, with 1 representing 2.2 centistokes and increments of 1 corresponding to a multiplication by 1.5, up to value 18, ≈ 1 500 cSt.

viscosity index *engineering* An index showing the stability of the viscosity of an oil under temperature change, with values towards +100 for highly stable, and towards −100 for highly variable. *See also* Engler degree; MacMichael degree.

visible light The waves of the †electromagnetic spectrum of about 430 to 750 THz frequency are visible to the human eye (other species vary in their sensitivities to such waves). The prism has long demonstrated that such light is a mixture of colours, a continuously changing range; Table 58 gives the nominal range of frequencies and wavelengths for the major colours, plus the nominal wavelength used for each (but see distinct values for cardinal illuminants in the †CIE system).

Table 58

Frequency THz	Wavelength nm	Nomenclature
		(ultraviolet)
749.5~	400.0	
731.2~	410.0	violet
707.1~	424.0	
637.9~	470.0	blue
610.3~	491.2	
576.5~	520.0	green
521.4~	575.0	
516.9~	580.0	yellow
512.5~	585.0	
499.7~	600.0	orange
463.4~	647.0	
461.2~	650.0	red
428.3~	700.0	
		(infrared)

VLF *physics* The 'Very Low Frequency' waves of the †electromagnetic spectrum, being those of about 10 kHz frequency, used notably for underwater radio transmissions.

volt [A. G. A. Anastassio, Count Volta; Italy 1745–1827] *voltage, electromotive force, potential difference*. Symbol V. The electromotive force or potential difference in volts between two points of a †steady current equals the ratio of power produced in watts to the current strength in amperes; equally the product of current strength in amperes and resistance in ohms. $V = W \cdot A^{-1} = A \cdot \Omega$.

SI, Metric-m.k.s.A. ($= m^2 \cdot kg \cdot s^{-3} \cdot A^{-1}$ in base terms). The following are among the coherent derived units:

- $V \cdot m^{-1}$ for electric field strength;
- $V \cdot s =$ weber for magnetic flux;
- $V \cdot A$ for apparent power;
- $V \cdot A^{-1} =$ ohm for electric resistance.

Metric-c.g.s. See abvolt (a base unit for the e.m.u. system); statvolt. *See also* practical unit.

UK 1 MV = 1 crocodile.

History
The volt as used today was originally defined in 1881 at the first International Electrical Conference,[33] among the †practical units derived from the absolute units of the †e.m.u. system. It was set at 10^8 times the volt of that system i.e. 1 **practical volt** $= 10^8$ abvolt. Like all the e.m. units, the abvolt was itself defined in terms of purely mechanical units; it had been initiated in the 1860s as the **ohma**.

Initially an explicit laboratory specification of the volt was established by the International Electrical Conference, making it a base unit instead of a derived unit. Expressed in terms of the potential of the common voltaic cell of the time, the specification was subsequently shown to have made the volt slightly larger than intended and not precisely consistent with the independently defined ampere and ohm.[5] The IEC of 1908 took away the inconsistency by making the volt a derived unit from those others, but this still had a derived discrepancy from intent, so the Conference adopted the distinct name **international volt**, lacking reference to being absolute or practical (though it was the latter). Because of experimental vagaries, the value for conversions is normally referred to as the mean international volt, $V_M = 1.000\,34\,V$.[162] There is also the **US international volt**, $= 1.000\,330\sim V$, defined by Congress in an Act of 1894.

With the implementation of the †m.k.s.A. system in 1948, and its basing of electrical units on an ampere compatible with the original absolute units, the modern volt became essentially the old practical volt; this became identically the volt of the SI.

The calibration of reference electrical instruments from the fundamental definition presents obvious practical problems of accuracy. Until the 1980s the method involved weighing on a balance the magnetic force between two coils of carefully measured copper wire; this gave an accuracy of barely 1 in 10^5. For maximum accuracy, the volt is now realized using the Josephson effect that applies at very low temperatures with superconductors, via the †Josephson constant, which relates volts to frequency.[6] Together with subsequent development of the moving-coil balance, this allows accuracy better than 1 in 10^8.

Revised conventional values for the constants involved in laboratory realizations using these effects, adopted internationally beginning in 1990, resulted in reducing the resistance value of many laboratory standards by about 0.000 8%.[130]

1946 CIPM 'Volt (unit of potential difference and of electromotive force) The volt
      is the potential difference between two points of a conducting wire
      carrying a constant current of 1 ampere, when the power dissipated
      between these points is 1 watt.'[8]

volume unit *telecommunications.* Symbol vu. *USA 1938* A measure of the magnitude of complex signals, such as analogue representations of speech and music, which, even for pure notes, have a multitude of different frequencies, each with its own amplitude at any moment. It is quantitatively defined as the number of decibels relative to a standard signal that is normally taken as a sinusoidal wave developing 1 milliwatt of power against an impedance of 600 ohms.[171]

von Klitzing constant [K. von Klitzing; Germany 1943–] *electromagnetics.* Symbol R_K. The conventional value, agreed in 1988 for implementation in 1990, is $R_{K-90} = 25\,812.807\,\Omega$. The true value, denoted R_K, is evaluated as $25\,812.807\,572(95)\,\Omega$ with †relative standard uncertainty 3.7×10^{-9}.[4]

vu *telecommunications See* volume unit.

vulgar fraction *See* number.

W *electromagnetics Metric See* watt; *see also* SI alphabet for prefixes.

watt [J. Watt; UK 1736–1819] *power, radiant flux.* Symbol W. The power that in 1 second gives rise to energy of 1 joule, identically the power dissipated for each joule of energy expended per second. $W = J \cdot s^{-1}$.

SI, Metric-m.k.s.A. ($= m^2 \cdot kg \cdot s^{-3}$ in base terms). The following are among the coherent derived units:

- $W \cdot m^{-2}$ for heat-flux density, irradiance;
- $W \cdot m^{-2} \cdot sr^{-1}$ for radiance;
- $W \cdot sr^{-1}$ for radiant intensity;
- $W \cdot s =$ joule for energy, work, quantity of heat;
- $W \cdot m^{-1} \cdot K^{-1}$ for thermal conductivity;
- $W \cdot A^{-1} =$ volt for voltage, electromotive force, potential difference.

See also practical unit.

History

The watt was recognized internationally in 1889, at the second International Electrical Conference, as an addition to the †practical units of the †c.g.s. system, hence the 'practical watt'. Discrepancies, for the underlying ampere and ohm, between measured absolute values (in centimetre-gram-second terms) and their laboratory specifications, led the IEC in 1908 to rename units based on the latter as unadorned international units, hence the **international watt**, 0.02~% larger than the practical. In 1948 recognition was transferred, along with that for the ampere, etc., to the **absolute watt**, with a decrease of just 0.019~% in size.

With the implementation of the †m.k.s.A. system in 1948, and its basing of electrical units on an ampere compatible with the original absolute units, the modern watt became essentially the old practical watt. Sometimes called the absolute watt, this became identically the watt of the SI.

Discovery of the Josephson effect, then of the quantum Hall effect, applying at very low temperatures with superconductors, together with the subsequent development of the moving-coil balance and related work with the volt, improved accuracies about a thousand-fold for the volt and other electrical units, including the watt, which is an intermediary in realizing the ampere.[7]

1946 CIPM 'Watt (unit of power) The watt is the power which in one second gives rise to energy of 1 joule.'[8]

wave number *physics* The reciprocal of wavelength, expressed in a known context of unit of length as a simple number. Introduced by the British Association in the 1870s as waves per millimetre, the term came to be equated with the kayser, i.e. waves per centimetre. Visible light has 1.4 to 2.5 $\times 10^3$ wave·cm^{-1}. It is equally appropriate to have waves per metre, but most radiations involved have millions of waves per metre.

Wb *electromagnetics* See weber. Also prefixed, as in kWb = kiloweber; *see* SI alphabet.

We *rheology* See Weber number.

weber [W. E. Weber; Germany 1804–91] *electric current strength*. Symbol Wb. *Metric* An 1880s name for the ampere.
   magnetic flux SI, *Metric-m.k.s.* Identically V·s, i.e. the amount of magnetic flux that in 1 second produces 1 volt per turn of a linked circuit, identically A·H (= m^2·kg·s^{-2}·A^{-1} in base terms). The following are among the coherent derived units:

- W·m^{-1} for magnetic vector potential;
- W·m^{-2} = tesla for magnetic flux density;
- W·A^{-1} = henry for electromagnetic inductance;
- W·H^{-1} = ampere for electric current strength.

History
The name 'weber' was in use in Britain prior to the first International Electrical Conference in 1881 when the ampere, the coulomb, and the farad joined the volt and the ohm as internationally recognized elements of the †c.g.s. system. However, that use was for a unit only one-tenth the size †coherent within c.g.s., while Weber himself was reported as using, and his colleagues calling 'weber', a unit a tenth smaller again.[33] So adoption was deferred. The International Electrotechnical Commission in 1933 accepted the current definition, which re-named the pramaxwell.[116]

1946 CIPM 'Weber (unit of magnetic flux) The weber is the magnetic flux which, linking a circuit of one turn, would produce in it an electromotive force of 1 volt if it reduced to zero at a uniform rate in 1 second.'[8]

magnetic pole strength Metric-c.g.s.-e.m.u. and -*Gaussian* An 1890s name for the magnetic pole strength that produces, in air, 1 gauss (oersted) at 1 centimetre.

Weber number *rheology*. Symbol We. Relating to momentum transport, in a fluid with an open surface, the dimensionless ratio of the product of the volumic mass, a representative length and the square of a representative speed to the surface tension,[61] hence proportional to the ratio of inertia force to surface-tension force. Depending on the units incorporated, this equals the ratio of the product of the volumic mass of the fluid and the square of the speed of the waves to the surface tension.

weigh, weighing, weight These familiar words have for many centuries been ubiquitous in everyday speech and in the formal laws dealing with measure. They have, to some extent, been set apart from the word 'measure'; the expression 'weights and measures' occurs in the title of every pertinent Act of the British parliament, and of the international office for metric measures (*see* BIPM), but the reason for that seems to have been the difference in the method of measurement for weight.

Weighing is the procedure routinely used to ascertain the (amount of) mass of something, yet it is repeatedly stressed that the gram, for example, is a unit of mass, not of weight. Technically the two are most certainly different; weight is a force, the product of mass and acceleration, the unit for which in metric is the newton. An object of a given (non-zero) mass at Earth's surface achieves its weight through being subject to gravitational acceleration. The actual acceleration causing weight is the true gravitational acceleration minus the amount of it consumed in providing the centripetal acceleration that keeps the object from flying off the rotating Earth. The latter is relatively trivial near the surface, even at the Equator.

Gravitational acceleration is inversely proportional to the square of the distance from the centre (approximately) of Earth, while centripetal acceleration is proportional to the distance from the axis of rotation. Hence, moving the object down a mine shaft increases its weight and moving it up a mountain decreases its weight (though such changes are minor). Elevating it further will reduce the gravitational acceleration; the concomitant impact on demand for centripetal acceleration will depend on circumstances. As centripetal acceleration is also proportional to the angular speed of rotation, flying fast above the surface can increase the demand considerably. A satellite at 600 km above the surface completing its orbit in two hours has a centripetal demand equal to the gravitational acceleration at that height, and hence it stays aloft. Likewise a satellite at 37 500 km orbiting in 24 hours stays aloft; if positioned above the Equator and travelling eastwards, like the points on Earth, such a satellite stays above the same point continuously (given a few small nudges from time to time to make it exactly so), so is said to be geostationary. The objects in the satellites, as well as the satellite itself, are then said to be 'weightless'. In such a context, 'weight' is unequivocally a force.

The pound and its kin of BI and US-C, including the pennyweight and the hundredweight, were also seen as weight in the sense of force. However, physicists have long regarded them as units of mass, so now they have a wider public. The pound is routinely equated with 454.~ g, and the kilogram with 2.2~ lb, at any elevation and speed.

For centuries any measurement of weight employed a balance (scales, steelyard, auncel) of some form that actually compared an object of unknown weight with a selection of objects of supposedly known weight; such measurement actually compared masses and was independent of the local net gravitational acceleration. Usage of the phrase 'weights and measures' reflected this measurement by comparison rather than by reading a graduated scale. Only with the evolution of modern spring-loaded devices (ironically occurring after the introduction of metric with its emphasis on mass) did measurement of true weight become generally practicable – and these had adjustments to correct for altitude, subverting this. Now weight and mass can be measured just like electric current, etc., by using a meter, though balances continue to be used, from the finest laboratories to fitness centres.

Mass is identically inertia, the resistance of a body to acceleration, yet also the source of gravitational acceleration. Although it can be measured accordingly, the only generally practicable means for measuring it is via weighing. The spring-loaded weigh-scale measures the weight against the force of the spring. The balance, although its known weights are fixed only in terms of mass and its operation independent of the exact gravitational acceleration, cannot function without such acceleration; one truly balances the weights, but never knows their common weight.

See further under †gravitational system.

It should be noted that the buoyancy of the surrounding medium affects the weight of an object. This was the essence of Archimedes' famous discovery, with water as the medium, but even the miniscule buoyancy of air has become significant in modern science. The 1963 definition of the BI †gallon specified the volumic mass of the weights and of the air.

Weston [E. Weston; UK, USA 1850–1936, via name of company] *photography* See film speed.

whetstone [a place in England] *informatics* A specific computer program package representative of scientific calculation of common proportions in the 1970s, used as a benchmark to measure the power of computers.

whole space *solid angle* See sphere.

Wien displacement law constant [W. C. W. O. F. F. Wien; Germany 1864–1928] *physics*. Symbol b. In Wien's displacement law, the product of the wavelength of maximum intensity from a black body and the thermodynamic temperature, $= 2.897\,768\,6(51) \times 10^{-3}$m·K with †relative standard uncertainty 1.7×10^{-6}.[4]

windchill (factor) *Canada* A composite of temperature and wind speed, constructed to reflect the greater freezing effect, particularly on hairless human skin and flesh, of low temperatures when accompanied by significant wind. The initial form consisted merely of subtracting the Fahrenheit temperature from the wind speed measured in miles per hour, e.g. a wind of 20 m.p.h. and a temperature of $-30°$F gave a windchill of 50, equatable with $-50°$F in calm conditions. The real effect is far from linearly additive in such a simple (and convenient) manner. Careful study has produced more complex formulations, convertible to a rate of heat loss and to rapidity of the freezing of the flesh.

Windisch See relative volumic mass.

wire gage, wire gauge See gauge.

Wobbe index *engineering* An indicator of the quality of a fuel gas, measured from the heat produced by burning through a defined orifice in standard temperature and pressure conditions, quantitatively defined as the ratio of Btu's per cubic foot to the square root of the specific gravity of the gas. Traditional 'town gas', which is mostly carbon monoxide (CO), typically has an index in the range 701 to 760 and is given the classification label G4; inferior forms can be much lower, with **G number** G5 for 641 to 700, G6 for 591 to 640, G7 for 531 to 590, and G8 below that. Natural gas, which is mostly methane (CH_4) at the well and almost entirely methane after refining for public use, typically has an index of 1 300 or more. Most bills for gas involve a heat-value factor to correct for variations in quality; measured centrally to represent average quality fed into the distribution network, this is applied to the measured volume consumed by each customer to establish the energy charge. The factor could be the Wobbe index, but may be in common energy units or the ratio of current heat–energy content to the reference value used in setting the tariff.

Wolf number [R. Wolf; Switzerland] *astrophysics See* sunspot number.

word *informatics* In telegraphy, etc., also in typing, 5 visible characters + 1 space character. In computers a word represents the amount of data stored, transferred, or such as one package, = 8 to 66 or even more bits, commonly between 16 and 48, and now usually taken to be 32 bits, even in a computer that moves data in other groupings.

X, x [representing two hands, ten fingers] The †Roman numeral for 10.

X stimulus *colorimetry* See CIE system.

X-unit, x-unit (**siegbahn unit**) *radiation physics*. Symbol xu. An SI-deprecated unit of length for the expression of wavelengths in X-ray spectroscopy, $= 0.100\,205\,6(\pm5)$ pm.[172] It is defined by the spacing of planes in a calcite crystal. As with the initial angstrom, the unit is very close to a straight decimal fraction of the metre; however, unlike for the angstrom, this definition is meant to be the fundamental definition rather than a practical interpretation.

Y, y *Metric* As an upper-case prefix, **Y-**, *see* yotta-, e.g. Yg = yottagram. As a lower-case prefix, y-, *see* yocto-, e.g. yg = yoctogram.

Y-, y- *Metric* As symbol, *see* Y, y.

yard ['stick'] *length* Internat, BI, US-C 3 ft = 36 in = 914.4~ mm (precisely that since 1959); *see* inch for greater details, including reference to **coast** or **survey yard** of $\frac{36}{39.37}$ m = 0.914 401 829~ m.

> *area* British For tiling, etc., sq. yd; for carpet and other rolled materials, 1 yd × standard width (usually 12 ft, making the square measure = 4 yd^2 = 3.344 5~m^2.
>
> *volume* British For gravel, etc., usually the cubic yard (0.764 6~ m^3).

yarn units *textiles* The measurement of yarn size is essentially on the basis of mass to length, this being termed a 'yarn count'. Traditional units have mostly been of length to unit mass; hence the larger the number the thinner the yarn (so measures of thinness); while thinner yarns are usually seen as superior, measurement on this basis was effectively indirect, taking the reciprocal from a weighing of a fixed length. Such an approach has been largely discarded in favour of one directly expressing mass to unit length (measures of thickness), the familiar denier being such a unit, the less familiar **tex** being now the standard.[173] Many different units have been used, for different materials and just for the products of different regions. Examples, showing the measurement base for each and the relationship to *t* as the equivalent number of tex, include those shown in Table 59.

yd. *See* yard.

year *time*. Symbol a, annum. Any of the various periods equated with one passage of Earth about the Sun, and hence of roughly 365 days. The familiar †calendar has a mixture of 365- and 366-day years, reflecting the fact that the time for one complete passage takes about 365$\frac{1}{4}$ days; the precise value for this figure depends on the manner of defining the year.

Table 59(a)

| Direct approach, measures of thickness: | | index = units of mass per set length | |
Symbol		Unit of mass	Set length
Tt	tex	1 g	$1\,000\,m = 1\,t$
Td	denier	1 g	$9\,000\,m = 9\,t$
	drex	1 g	$10\,000\,m = 10\ t$
	poumar	1 lb	$1\,000\,000\,yd = 2.015{\sim}\,t$
Tj	jute (also hemp, dry spun linen)	1 lb	spindle of $14\,400\,yd = 0.029\,03{\sim}\,t$
Ta	Aberdeen	1 lb	spindle of $14\,400\,yd = 0.029\,03{\sim}\,t$

Table 59(b)

| Indirect approach, measures of thinness: | | index = units of length per set mass | |
Symbol		Unit of length	Set mass
	American cut	cuts of 300 yd	lb	$= 1654{\sim}/t$
	American run	lengths of 100 yd	oz	$= 310.0{\sim}/t$
Ne_C	cotton	hanks of 840 yd	lb	$= 590.5{\sim}/t$
Ne_L	linen	leas of 300 yd	lb	$= 1654.{\sim}/t$
Nm	metric	lengths of 1000 m	kg	$= 1000/t$
Nt	typp	lengths of 1000 yd	lb	$= 496.1{\sim}/t$
New	worsted	lengths of 560 yd	lb	$= 885.8{\sim}/t$

As discussed under †time, there are various lengths of year (and of day).
The obvious year is the interval between successive mid-summers or
other seasonal mark. Because of particulars of astronomical motions,
this interval varies from one passage to the next, and progressively over
time. Throughout historical times and continuingly, the underlying
fluctuations are most pronounced around the time of the solstices, so the
astronomers bury most of these variations within the year by regarding a
year as beginning at an equinox. Occurring when the conceptual line of
intersection between Earth's equatorial plane and the plane of the
ecliptic (the plane in which Earth travels) passes through the Sun,
equinoxes are also inherently easier to recognize for a given level of
accuracy. The northern spring (or vernal) equinox is the chosen one, with
the marker being the noon position of the Sun as it crosses the Equator
(termed the †First Point of Aries), and the year beginning at that point is
called the **equinoctial year** or **tropical year** (a_{trop}); the mean value is
365.242 19~ mean solar days. This is the true year of the seasons; any
discrepancy from it in the average year of any calendar results in

progressive creep of the seasons, the factor that produced the drastic change of the European calendars from Julian to Gregorian, a few centuries ago. (But the †Julian year of precisely 365.25 days remains in use in astronomy.)

Because Earth's axis itself is turning relative to the stars ('precessing'), in a manner contrary to Earth's movement about the Sun, the successive vernal equinoxes occur about 20 minutes earlier each orbital passage, so the tropical years understate the time for a complete orbit by that much. The effect of precession on the registration of time can be avoided by demarcating the year at a point on the elliptical orbit regardless of seasons and axial tilt. The point of closest approach to the Sun is the chosen point, a unique one because the Sun lies at a focus of the ellipse, not the centre. This point is called perihelion, which marks the **anomalistic year** (a_{anom}). Currently perihelion occurs in early January, (and its opposite, aphelion, early in July) bringing in some of the variations around the solstices, but the average is a truer representation of the time for one orbit, and obviously much more appropriate than the tropical year for orbital calculations. However, this year too is discrepant in some sense, for the ellipse itself is slowly changing position relative to the wider context; the axes of the ellipse are rotating relative to the stars, in the same direction as Earth about the Sun, making a_{anom} about five minutes greater than the true time for one revolution.

The truest measure of one revolution is the **sidereal year** (a_{sid}), the time for one passage measured relative to the stars; this is the most stable year, but this too, because of overall motion of the solar system within the Galaxy, is not absolutely fixed. The **Gaussian year** (a_{gauss}) is the theoretical year derived from Kepler's law.

The values of these various years (in mean solar days) at the opening of the year 2002 were[10]

a_{trop} = 365.242 189 7∼ days = 365 days 5 h 48 min 45.19∼ s
a_{sid} = 365.256 363∼ days   = 365 days 6 h 9 min 9.8∼ s
a_{anom} = 365.259 635∼ days   = 365 days 6 h 13 min 52.5∼ s
a_{gauss} = 365.256 90∼ days   = 365 days 6 h 9 min 56.∼ s

with the a_{trop} shortening by a second every 380 years, a_{anom} lengthening at half that rate, and a_{sid} adding nearly a tenth of a second per millennium, but the a_{gauss} changing much less.

The original †Besselian year, being a period over which the right ascension of the Sun increases by 360°, was essentially the same as the tropical year; the name **astronomical year** seems to be applied to either. The term Besselian year now refers to the Julian year.

The **calendar year**, in contrast to the above years, is necessarily a whole number of complete days. *See* calendar for further discussion.

By natural extension, the term year is used, appropriately qualified (and often parenthesized), for the equivalent periods for other planets, e.g. the Martian year. *See also* eclipse year.

yoct- *SI* Contracted form of †yocto-.

yocto- [Lat: octo, 'eight'] Symbol y-. *SI* The $10^{-24} = 1\,000^{-8}$ multiplier, e.g. 1 yoctogram = $1\,\text{yg} = 10^{-24}$ g; contractable to yoct- before a vowel.

yott- *SI* contracted form of †yotta.

yotta- [Lat: octo, 'eight'] Symbol Y-. *SI* The $10^{24} = 1\,000^{8}$ multiplier, e.g. 1 yottagram = $1\,\text{Yg} = 10^{24}$ g; contractable to yott- before a vowel.

Y stimulus *colorimetry*. Symbol Y. *See* CIE system.

Z

Z, z *Metric* As an upper-case prefix, **Z-**, *see* zetta-, e.g. Zg = zettagram. As a lower-case prefix, **z-**, *see* zepto-, e.g. zg = zeptogram.

Z-, z- *Metric* As symbol, *see* Z, z.

Z_0 *physics* Characteristic impedance of a vacuum.

Zeeman splitting constant [P. Zeeman; Netherlands 1865–1943] *photics* The ratio of the †elementary charge to 4π times the product of the mass of the †electron and the †speed of light in a vacuum, $= 46.686\,5\sim \times 10^6 \mathrm{m}^{-1} \cdot \mathrm{T}^{-1}$. The number equals the †wave number for the factor induced by the Zeeman effect (the effect of a magnetic field on light whereby a spectral frequency of f results in light of frequencies $f \pm \Delta f$ appearing along the field and those two plus the original transversely). Also called, in a c.g.s. context (as $46.686\,5\sim \mathrm{cm}^{-1}\mathrm{gauss}^{-1}$), the **Lorentz unit**.[174]

zept- *SI* Contracted form of †zepto-.

zepto- [Lat: septo, 'seven'] Symbol z-. *SI* The $10^{-21} = 1\,000^{-7}$ multiplier, e.g. 1 zeptosecond = 1 zs = 10^{-21} s; contractable to zept- before a vowel.
  The prefix is used most notably as the **zeptomole** of molecular chemistry, being the amount of substance that is equal to $10^{-21} \times$ †Avogadro's number $\approx 602.\sim$ molecules.

zett- *SI* Contracted form of †zetta-.

zetta- [Lat: septo, 'seven'] Symbol Z-. *SI* The $10^{21} = 1\,000^7$ multiplier, e.g. 1 zettagram = 1 Zg = 10^{21} g; contractable to zett- before a vowel.

Zhubov scale [N. N. Zhubov; USSR 1895–1960] *oceanography Russia* For ice coverage of ocean surfaces, effectively the multiplier of 10 closest to the percentage of ice coverage.[175]

zodiac *astronomy* The band of our sky within which the Sun and its other planets appear to travel, divided traditionally into twelve equal-sized segments named sequentially Aries, Taurus, Gemini, Cancer, Leo,

Virgo, Libra, Scorpio, Sagittarius, Capricorn, Aquarius, and Pisces. However, while each of these is classically the name of a specific star pattern, still used, along with 76 outside this band, as one of the variously sized polygons ('constellations') into which the heavens are fixedly divided, they have a distinct meaning for the zodiac.

The apparent annular path of the Sun derives from the orbit of Earth in its plane, the ecliptic, hence is a line circling Earth. The apparent paths of the other planets are composites of their individual orbits, each of which is tilted somewhat relative to the ecliptic, and Earth's orbit; hence they range through a band of sky straddling the ecliptic. This band, the zodiac, has an angular width to an observer on Earth of only ±9°. Convention uses these twelve names distinctively such that the Aries segment of the zodiac begins at the point in the sky wherein sits the Sun at the moment of the northern spring equinox; this is the †First Point of Aries. Since the precessional gyration of Earth's axis causes the year of the seasons, the tropical year, to arrive about 20 minutes before completion of one full orbit relative to the stars, the zodiac fails by 50.2~ arcsecs to complete a full circuit of stars. Thus, while the zodiac originated with the Sun in the constellation Aries, it is now in, and well through, the constellation Pisces, creeping one constellation back in a little over two millennia, back through the whole twelve segments in about 25 800 years. It should enter Aquarius soon after 2500.

With Aries starting on the northern spring equinox, Cancer starts with the northern summer solstice and Capricorn with the southern summer solstice, i.e. the points at which the Sun is directly overhead on the Tropics of Cancer and Capricorn respectively.

See also right ascension.

Z stimulus *colorimetry See* CIE system.

Symbols

! *mathematics* The factorial expression, i.e. 2! equals the product 2×1, 3! $= 3 \times 2 \times 1$, etc.

!! *mathematics* The double-step factorial expression, i.e. 3!! equals the product 3×1, 5! $= 5 \times 3 \times 1$, etc., used in radiative transition probabilities.[176]

" *See* second.

The 'numero' symbol, used widely in North America as a prefix for both an identifying number and a numeric count; because it replaces the £ sign on many keyboards, it is often called 'pound'.

% *See* percentage.

' *See* prime.

(g)² *See* square grade.

(°)² *See* square degree.

* A multiply symbol, e.g. $a*b$ means a multiplied by b.

** An exponentiation symbol, e.g. $a**b$ means a to the power b.

+ The 'plus' symbol, applicable to the addition operation and simple signage (where it is usually omitted, implicit).

, Used in British tradition to punctuate integers, at every third position left-wards from the decimal point, but used in European practice as the decimal point. Consequently abjured within this text in favour of the space character (which is used likewise to punctuate the fractional part).

− The 'minus' symbol, applicable to the subtraction operation and simple signage. *See* negative number for form with logarithms.

. Identically the 'full stop' of British tradition and the 'period' of North America, now universally also the 'dot', but also the 'decimal point' used in English-speaking practice for separating integer and fractional parts of a number, even of a non-decimal number. Used likewise in this text.

: The ratio symbol.

^ Used in emails and some other typographically constrained situations as an exponentiation symbol, e.g. $a \wedge b$ means a to the power b.

/ Called 'solidus' properly but 'slash' vernacularly, it is used as a ratio or divider symbol, including within the formalities of the SI, e.g. a/b means a divided by b.

~ The 'tilde' of Spanish orthography; used within this text, in a lowered position, to represent further unstated digits in a number.

‰ *See* permille, i.e. parts per thousand, akin to †percentage.

° *See* degree; for $(°)^2$ *see* square degree.

• Historically the decimal point of British practice, now a multiply symbol, e.g. $a \cdot b$ means a multiplied by b.

∞ *See* infinity.

|z| *See* modulus.

□° *See* square degree.

α *sub-atomic physics See* fine-structure constant.

γ [Gk letter 'g', its name Anglicized as gamma] *See* Newtonian gravitational constant.
    mass Metric Old symbol for microgram, now properly µg.

γ_e *sub-atomic physics* Electron gyromagnetic ratio. *See* electron.

γ_n *sub-atomic physics* Neutron gyromagnetic ratio. *See* neutron.

γ_p *electromagnetics* Proton gyromagnetic ratio. *See* proton.

ϵ_0 *sub-atomic physics See* electric constant.

φ [Gk letter 'f', its name Anglicized as phi] *mathematics See* golden ratio.
    geology Relates to phi scale; *see* particle size.

Φ_0 *electromagnetics See* magnetic flux quantum.

λ [Gk letter 'l', its name Anglicized as lambda] *length See* lambda.
    volume Metric Old symbol for microlitre, now properly µL.
    sub-atomic physics Bohr magneton; *see* magneton.

λ_C *sub-atomic physics See* Compton wavelength.

$\lambdabar$ *sub-atomic physics See* Compton wavelength.

μ [Gk letter 'm', its name Anglicized as mu, pronounced 'mew'] *Metric* Official symbol for the prefix †micro-, i.e. for 10^{-6}. Originally adopted to mean the micron, i.e. the micrometre (10^{-6} m). Sometimes represented by the μ untouched – a deprecated practice. Often substituted by mc- with microgram in North America; *see* mcg.

μ₀ *electromagnetics See* magnetic constant.

μ_B *sub-atomic physics* Bohr magneton. *See* magneton.

μ_b *pressure* An improper representation of microbar.

μ_d *sub-atomic physics* Deuteron magnetic moment. *See* deuteron.

μEq Microequivalent. *See* equivalent weight.

μ_e *sub-atomic physics* Electron magnetic moment. *See* electron.

μ_N *sub-atomic physics* Nuclear magneton. *See* magneton.

μ_n *sub-atomic physics* Neutron magnetic moment. *See* neutron.

μ_p *sub-atomic physics* Proton magnetic moment. *See* proton.

μμ *Metric* Old symbol for †millimicron.

π (pi) [Gk letter 'p', its name Anglicized as pi, pronounced 'pie'] *See* pi.

σ [Gk letter 's', its name Anglicized as sigma] *See* sigma.
   statistics See standard deviation.
   fundamental physical constants See Stefan–Boltzmann constant.
   See also stigma; symmetry number.

σ_e *sub-atomic physics See* Thomson cross-section.

Ω [Gk letter 'O', the final letter of that alphabet, its name Anglicized as omega] *electromagnetics Metric* Official symbol for ohm. Also prefixed, as in mΩ = milliohm; *see* SI alphabet.

א [The first letter of the Hebrew alphabet, its name Anglicized as aleph.] *See* infinity.

Endnotes

1 Bullock M. L. *Amer. J. Phys.* **22**, 293–9 (1954)

2 Frederick H. A. *J. Acoust. Soc. Amer.* **9**, 60–71 (1937)

3 Kennelly A. G., Cook J. H. *J. Acoust. Soc. Amer.* **9**, 336–40 (1938)

4 Cohen E. R., Taylor B. N. *Phys. Today* **49**:8, 9–13 (1996)
Mohr P. J., Taylor B. N. *J. Phys. Chem. Ref. Data* **28**:6 1713–1822 (1999)
Mohr P. J., Taylor B. N. *Rev. Mod. Phys.* **72**:351–495 (2000)
Mohr P. *Phys. Today* **53**:7, 11–16 (2000)
For latest recommended values, see http://physics.nist.gov/cgi-bin/cuu/

5 *Nature* **78**, 678–81 (1908)

6 Hartland A. *Contemp. Phys.* **29**, 477 (1988) http://www.npl.co.uk/npl/publications/electricity/

7 Taylor B. N. *Metrologia* **21**, 37–9 (1985)

8 *Le Système International d'Unités* (Sèvres, France: Bureau International de Poids et Mesures, 1985)

9 Barrell H. *Proc. Roy. Soc. London Ser. A* **186**, 164–70 (1946)

10 *The Astronomical Almanac for the Year 2002* (Washington: US Government Printing Office and London: HMSO, 2001), also *Explanatory Supplement to the Astronomical Almanac* (Mill Valley, CA: University Science Books, 1994).

11 Glazebrook R. T. (ed.) *Dictionary of Applied Physics*, Vol. 1: *Mechanics, Engineering, Heat* (London: Macmillan, 1922)

12 Bjerknes V. *Dynamic Meteorology and Hydrography* (Washington: 1911)

13 Denne E. *Nature* **156**, 146–7 (1945)

14 Husseke R. E. (ed.) *Glossary of Meteorology* (Boston: Amer. Met. Soc., 1959)

15 McNish A. G. *Phys. Today* **10**:4, 19–25 (1957)

16 Kinitsky V. A. *Amer. J. Phys.* **30**, 89–93 (1962)

17 Tuninsky V. S. *Metrologia* **36**, 1–7 (1999)

18 McWeeny R. *Nature* **243**, 196–8 (1973)

19 Ludovici B. F. *Amer. J. Phys.* **24**, 400–7 (1956)

20 Dirac P. A. M. *Proc. Roy. Soc. London Ser. A* **165**, 199–208 (1938)

21 Allissy A. *Metrologia* **31**, 267–79 (1995)

22 *McGraw-Hill Dictionary of Scientific and Technical Terms*, 5th edn (New York, McGraw-Hill, 1994)

23 Fry G. A. *Illum. Eng.* **48**, 406–10 (1953)

24 Filon L. P. G. *Proc. Roy. Soc. London Ser. A* **83**, 572–9 (1910)

25 Moore J. B. *Elect. Engng* **73**, 959–60 (1954)

26 Powell R. W. *Nature* **149**, 525–6 (1942)

27 Darwin C. *Nature* **164**, 262–4 (1949)

28 The UK Weights and Measures Act 1985 explicitly excluded from use for trade the bushell, cental, chain, drachm, dram, fluid drachm, furlong, grain, hundredweight, ounce apoth., peck, pennyweight, quarter, quintal, rood, scruple, stone, ton, the square mile, cubic inch, cubic foot, cubic yard, and the term 'metric ton'. However, the legal status of the bushell, fluid drachm, and peck had been repealed, along with all apothecaries' units and troy units other than ounce, by Order in 1970. Besides the remaining BI units and the simple SI units, the Act included the kilometre, decimetre, centimetre and millimetre, the square metre, square decimetre, square centimetre and square millimetre, the hectare and decare along with the are, the cubic metre, cubic decimetre and cubic centimetre, the hectolitre decilitre, centilitre and millilitre, the tonne (or 'metric tonne'), kilogram, hectogram, milligram and carat (metric). All had been included in the similar Act of 1963, but with some variation of name: -gram was -gramme, decare was dekare, the tonne appeared only as metric ton.

29 Joly J. *Nature* **52**, 4 (1895)

30 *Proc. Roy. Soc. London Ser. A* **186**, 204–7 (1946)

31 Lewis G. N. *J. Amer. Chem. Soc.* **35**, 1–30 (1914)

32 *Nature* **39**, 18–19 (1873)

33 *Nature* **24**, 512 (1881)

34 ISO/CIE 10526:1991 *CIE Standard Illuminants*

35 Hardy A. C. *Handbook of Colorimetry* (Cambridge, MA: Technology Press, 1936)

36 Gagge A. P., Burton A. C., Bazett M. C. *Science* **94**, 428 (1941)

37 Jones L. A. *J. Opt. Soc. Amer.* **27**, 207–13 (1937)

38 Billmayer F. W., Saltzman M. *Principles of Color Technology* (New York: Interscience, 1966)

39 Ridgway R. *Color Standards and Color Nomenclature* (Washington, DC; author, 1912)

40 BS 381C *Specification for Colours for Identification and Special Purposes* (1988)

41 Glazebrook R. T. *Nature* **128**, 17–28 (1931)

42 Curtiss L. F., Evans R. D., Johnson W., Seaborg G. T. *Rev. Sci. Instrum.* **21**, 94 (1950)

43 Rutherford E. *Nature* **84**, 430–1 (1910)

44 Glasser O. *Physical Foundations of Radiology*, 2nd edn (New York: Harper, 1952)

45 Wyckoff R. D., Botset H. G., Muskat M., Reed D. W. *Rev. Sci. Instrum.* **4**, 395 (1933)

46 Fairbrother F. *Nature* **134**, 458 (1934)

47 Hartley R. V. L. *Elect. Comm.* **3**, 34–42 (1924)

48 Martin W. H. *Bell Sys. Tech. J.* **8**, 1 (1929)

49 http://www.npl.co.uk/npl/publications/acoustics/

50 Rose F. C. *Nature* **156**, 268 (1945)

51 Baldwin C. C., Tonks L. *Nature* **203**, 633–4 (1964)

52 Horton J. W. *Elect. Engng* **73**, 550–5 (1954)

53 Rao V. V. L., Lakshminarayanan S. *J. Acoust. Soc. Amer.* **27**, 376–8 (1955)

54 Green E. I. *Elect. Engng* **73**, 597–9 (1954)

55 Parker H. C., Parker E. W. *J. Amer. Chem. Soc.* **46**, 312–35 (1924)

56 Witmer E. E. *Phys. Rev. Ser. 2* **71**, 126 (1947)

57 Newton H. A. *Nature* **9**, 312 (1874)

58 Dirac P. A. M. *Nature* **139**, 323 (1937)

59 Petley B. W. *The Fundamental Constants and the Frontier of Measurement* (Boston, MA: Adam Hilger 1985)

60 Noyes W. A., Leighton P. A. *The Photochemistry of Gases* (New York: Reinhold, 1941; republished with minor corrections by Dover, 1966)

61 International Standards Association
        ISO 31-12:1992 *Quantities and Units: Characteristic Numbers*
        Mills I., Cvitas T., Homan K., Kuchitsu K. *Quantities, Units and Symbols in Physical Chemistry*, 2nd edn (Oxford: Blackwell, 1993)

62 Daintith J., Nelson R. D. (eds) *The Penguin Dictionary of Mathematics* (London: Penguin Books, 1989)

63 Guggenheim E. A. *Nature* **148**, 751 (1941)

64 Kingslake R. (ed.) *Applied Optics and Optical Engineering* (1969)

65 Bayliss N. S. *Nature* **167**, 367–8 (1951)

66 Burn J. H., Finney D. J., Goodwin L. C. *Biological Standardization*, 2nd edn (Oxford: Oxford University Press, 1950)

67 Ipsen D. C. *Units, Dimensions and Dimensionless Numbers* (New York: McGraw-Hill, 1960)

68 Hoggett P., Chorley R. J., Stoddart D. R. *Nature* **205**, 844–7 (1965)

69 Chree C. *Nature* **69**, 6 (1903)

70 ANSI/ASME B32.3M 1984

71 *Nature* **126**, 252 (1930)

72 *Nature* **62**, 414 (1900)

73 Connor R. D. *The Weights and Measures of England* (London: HMSO, 1987)

74 Koch W., Kaplan D. *Nature* **159**, 273 (1947)
Koch W., Kaplan D. *Nature* **161**, 247 (1948)

75 Tabor D. *The Hardness of Metals* (Oxford: Clarendon Press, 1951)

76 See ASTM E10-66 for specifications

77 See latest *Machinery's Handbook* (New York: Industrial Press, 2000 or later) for more

78 Hartree D. R. *Proc. Camb. Phil. Soc.* **24**, 89 (1927)

79 Gould F. A. *Proc. Roy. Soc. London Ser. A* **186**, 195–200 (1946)

80 Patterson J. B., Prowse D. B. *Metrologia* **21**, 107–13 (1985)
Ambrose D. *Metrologia* **27**, 233–47 (1990)

81 Gamow G. *Nature* **219**, 765 (1968)

82 Canada: An Act respecting Weights and Measures assented to 20 June 1951

83 Anderson H. L., Fermi E., Wattenberg A., Weil G. L., Zinn W. H. *Phys. Rev. Ser. 2* **72**, 16–23 (1947)

84 http://www.nibsc.ac.uk

85 Preston-Thomas H., Quinn T. J., Hudson R. P. *Metrologia* **21**, 75–9 (1985)

86 Hall J. A. *Proc. Roy. Soc. London Ser. A* **186**, 179–84 (1946)

87 Barber C. R. *Metrologia* **5**, 35–44 (1968)

88 Preston-Thomas H. *Metrologia* **27**, 3–10 and 27 (1990)

89 Jones T. P., Topping J. *Metrologia* **8**, 4–11 (1972)

90 Harris W. S. *Phil. Mag. (GB)* **4**, 436 (1834)

91 Dybkær R., Jørgensen K. *Quantities and Units in Clinical Chemistry* (Baltimore: Williams and Wilkins, 1967)

92 http://www.bipm.org/enus/2_Committees/cgpm21/resolutions.html

93 Meggers W. F. *J. Opt. Soc. Amer.* **41**, 1064 (1951)
Meggers W. F. (as reporter) *J. Opt. Soc. Amer.* **43**, 410–13 (1953)

94 Stott V. *Nature* **124**, 622–3 (1929)

95 UK Weights and Measures Act 1963

96 Firestone F. A. *J. Acoust. Soc. Amer.* **4**, 249–56 (1933)

97 Aldrich L. B. *et al. Nature* **160**, 327 (1947)

98 *Nature* **65**, 538 (1902)

99 Stott V. *Proc. Roy. Soc. London Ser. A* **186**, 200–4 (1946)

100 Fletcher H. *J. Acoust. Soc. Amer.* **9**, 275–93 (1938)
  J. Acoust. Soc. Amer. **14**, 105 (1942)

101 Moon P. *J. Opt. Soc. Amer.* **32**, 348–62 (1942)

102 Shercliff J. A. *A Textbook of Magnetohydrodynamics* (London: Pergamon, 1965)

103 Madelung E. *Physik. Zeits.* **19**, 524 (1919)

104 Haxel O., Jensen J. H. D., Suess H. D. *Phys. Rev.* **75**, 1766 (1949)

105 Richards T. W., Glucker T. F., *J. Amer. Chem. Soc.* **47**, 1876–93 (1925)

106 Stevens S. S., Volkmann J., Newman E. B. *J. Acoust. Soc. Amer.* **8**, 185–90 (1937)

107 Mercalli G. *Boll. Soc. Sismologica Italiana* **8**, 184–91 (1902)

108 Wood H. O., Neumann F. *Bull. Seis. Soc. Amer.* **21**, 277–83 (1931)

109 Richter C. F. *Elementary Seismology* (San Francisco: W. H. Freeman, 1958)

110 Jerrard H. G., McNeil D. B. *Dictionary of Scientific Units*, 6th edn (London: Chapman and Hall, 1992)

111 Barrell H. *Nature* **189**, 195–6 (1961)

112 Hartshorn L. *Proc. Roy. Soc. London Ser. A* **186**, 185–91 (1946)

113 Harkins W. D., Roberts L. E. *J. Amer. Chem. Soc.* **44**, 653–70 (1922)

114 Burington R. S. *Amer. Math. Monthly* **48**, 188–9 (1941)

115 Farnwell H. W. *Amer. J. Phys.* **13**, 349 (1939)

116 Kennelly A. E. *Elect. Engng* **53**, 402–5 (1934)

117 Glazebrook R. T. (ed.) *Dictionary of Applied Physics* Vol. 3: *Meteorology, Metrology and Measuring Apparatus* (London: Macmillan, 1923)

118 McGlashan M. L. *Metrologia* **31**, 247–55 (1995)

119 Wichers E. *Nature* **194**, 621–4 (1962)

120 See www.skypub.com/sights/moonplanets/9905bluemoon.html

121 Munsell A. H. *A Color Notation* (Baltimore MD: Munsell Color Co., 1936 onwards)

122 Nickerson D. *J. Opt. Soc. Amer.* **30**, 575–86 (1940)

123 Newhall S. M., Nickerson D., Judd D. B. *J. Opt. Soc. Amer.* **33**, 385–418 (1943)

124 Pollard E. C., Davidson W. L. *Applied Nuclear Physics* (New York: Wiley, 1945)

125 Hartshorn L., Vigoureux P. *Nature* **136**, 397 (1935)

126 Gillies G. T. *Metrologia* **24** (suppl), 1–56 (1986)

127 Gundlach J. H., Merkowitz S. M. *Phys. Rev. Letters* **85**, 2869–72 (2000)

128 Cavendish H. *Phil. Trans. Roy. Soc.* **83**, 385 (1798)

129 Harris C. M. *Handbook of Noise Control* (New York: McGraw-Hill, 1957)

130 Kibble B., Hartland A. *New Scientist* **1715**, 48–51 (1990)

131 Ferguson W. B. *Photographic Researches of Hurter and Driffield* (London: Roy. Photographic Soc., 1920)

132 Jacobson E. *Basic Color: An Interpretation of the Ostwald Color System* (Chicago: P. Theobald, 1948)

133 ISO 216:1975 *Writing Paper and Certain Classes of Printed Matter in Trimmed Sizes – A Series and B Series*

134 Quinn T. J., Mills I. M. *Metrologia* **35**, 799–906 (1998)
Giacomo P. *Metrologia* **37**, 93–4 (2000)

135 Bates R. G. *Determination of pH: Theory and Practice* (New York: John Wiley, 1st edn 1964, 2nd edn 1973)

136 *Nature* **140**, 370 (1937)

137 ISO 131:1979 *Acoustics – Expression of Physical and Subjective Magnitudes of Sound and Noise in Air*

138 Potnis V. R., Chai A.-T. *Amer. J. Phys.* **40**, 767 (1972)

139 Wilczek F. *Phys. Today* **52**:8, 11–16 (1999)

140 Gross D. J. *Phys. Today* **42**:6, 9–10 (1989)

141 Bigg P. H. *Nature* **194**, 719–21 (1962)

142 *Nature* **183**, 80–1 (1959)

143 International Standards Organisation
ISO 3:1973 *Preferred Numbers – Series of Preferred Numbers*
ISO 17:1973 *Guide to the Use of Preferred Numbers and of Series of Preferred Numbers*
ISO 497:1973 *Guide to the Choice of Preferred Numbers and of Series of Preferred Numbers*

144 Van Dyck A. *Proc. Inst. Radio Engrs* **24**, 159–79 (1936)

145 American National Standards Institute
ANSI/EIA RS-385 *Preferred Values*, 1971
ANSI Z17.1-1973 *American National Standard for Preferred Numbers*

146 Gould F. A. *Proc. Roy. Soc. London Ser. A* **186**, 171–9 (1946)

147 Bayly B. de F. *Proc. Inst. Radio Engrs* **19**, 873–9 (1931)

148 Eder W. E. *Metrologia* **18**, 1–12 (1982)
Eder W. E. *Metrologia* **19**, 1–8 (1983)
Torrens A. B. *Metrologia* **22**, 1–7 (1986)
Torrens A. B. *Metrologia* **23**, 57–8 (1986)

149 Giacomo P. *Metrologia* **17**, 69–74 (1981)

150 ISO 649-1 and -2:1981 *Laboratory Glassware – Density Hydrometers for General Purposes*

151 Richter C. F. *Bull. Seis. Soc. Amer.* **25**, 1–32 (1935)

152 Barker R. E. jnr. *Nature* **203**, 513 (1964)

153 Melaragno M. G. *Severe Storm Engineering for Structural Design* (Australia and Luxembourg: Gordon and Breach, 1996)

154 Hunt F. V. *J. Acoust. Soc. Amer.* **11**, 38–40 (1939)

155 Bergquist J. C., Jefferts S. R., Wineland D. J. *Phys. Today* **54**:3, 37–44 (2000) on-line at http://physicstoday.org/pt/vol-54/iss-3/p37.html

156 http://www.bipm.org/pdf/si-supplement2000.pdf

157 Fletcher H., Steinberg J. C. *Phys. Rev. Ser. 2* **24**, 306–17 (1924)

158 Steinberg J. C. *Phys. Rev. Ser. 2* **26**, 507–23 (1925)

159 ISO 3310–1 and 2:1990 *Test Sieves – Technical Requirements and Testing*

160 Kryter K. D. *J. Acoust. Soc. Amer.* **31**, 1415–29 (1959)

161 Clark J. S. *Proc. Roy. Soc. London Ser. A* **186**, 192–5 (1946)

162 *Nature* **163**, 427–8 (1949)

163 See http://www.gfy.ku.dk/~iag/

164 Judd D. B. *J. Opt. Soc. Amer.* **23**, 359–74 (1933)

165 Stoney G. J. *Phil. Mag. (GB)* **36**, 138 (1868)

166 Sears F. W. *Amer. J. Phys.* **28**, 167 (1960)

167 Pierce F. T., Rees W. M. *J. Textile Ind.* **7**, 181 (1946)

168 Huxley L. G. H., Crompton R. W., Elfod M. T. *Brit. J. App. Phys.* **17**, 1237 (1966)

169 Von Straaten J. F. *Thermal Performance of Buildings* (London: Elsevier, 1967)

170 For latest information see ftp://tycho.usno.navy.mil/pub/series/ or http://tycho.usno.navy.mil/leap.html

171 Chinn H. A., Gannett D. K., Morris R. M. *Proc. Inst. Radio Engrs* **28**, 1 (1940) *Acoustical Terminology* **1**, 405 American Standards Assoc. Z 24.1 (1951)

172 Bearden J. A. *Phys. Rev. Ser. 2* **137B**, 455–61 (1965)

173 ISO 2947:1973 *Textiles – Integrated Conversion Tables for Replacing Traditional Yarn Numbers by Rounded Values in the Tex System*

174 White H. E. *Introduction to Atomic Spectra* (New York: McGraw-Hill, 1934)

175 Fairhall D. *Russia Looks to the Sea* (London: Deutsch, 1971)

176 Weisskopf V. F. *Phys. Rev. Ser. 2* **83**, 1073 (1951)

177 For further values see any edition of *C. R. C. Handbook of Chemistry and Physics* (Cleveland, OH: C. R. C. Press)